Magnetoelectric Sensor Systems and Applications

Magnetoelectric Sensor Systems and Applications

Editors

Gerhard Schmidt
Eckhard Quandt
Nian X. Sun
Andreas Bahr

MDPI • Basel • Beijing • Wuhan • Barcelona • Belgrade • Manchester • Tokyo • Cluj • Tianjin

Editors
Gerhard Schmidt
Kiel University
Germany

Eckhard Quandt
Kiel University
Germany

Nian X. Sun
Northeastern University
Germany

Andreas Bahr
Kiel University
Germany

Editorial Office
MDPI
St. Alban-Anlage 66
4052 Basel, Switzerland

This is a reprint of articles from the Special Issue published online in the open access journal *Sensors* (ISSN 1424-8220) (available at: https://www.mdpi.com/journal/sensors/special_issues/Magnetoelectric_Sensors).

For citation purposes, cite each article independently as indicated on the article page online and as indicated below:

LastName, A.A.; LastName, B.B.; LastName, C.C. Article Title. *Journal Name* **Year**, *Volume Number*, Page Range.

ISBN 978-3-0365-3553-1 (Hbk)
ISBN 978-3-0365-3554-8 (PDF)

© 2022 by the authors. Articles in this book are Open Access and distributed under the Creative Commons Attribution (CC BY) license, which allows users to download, copy and build upon published articles, as long as the author and publisher are properly credited, which ensures maximum dissemination and a wider impact of our publications.

The book as a whole is distributed by MDPI under the terms and conditions of the Creative Commons license CC BY-NC-ND.

Contents

About the Editors vii

Preface to "Magnetoelectric Sensor Systems and Applications" ix

Eric Elzenheimer, Christin Bald, Erik Engelhardt, Johannes Hoffmann, Patrick Hayes, Johan Arbustini, Andreas Bahr, Eckhard Quandt, Michael Höft and Gerhard Schmidt
Quantitative Evaluation for Magnetoelectric Sensor Systems in Biomagnetic Diagnostics
Reprinted from: Sensors **2022**, 22, 1018, doi:10.3390/s22031018 1

Ron-Marco Friedrich and Franz Faupel
Adaptive Model for Magnetic Particle Mapping Using Magnetoelectric Sensors
Reprinted from: Sensors **2022**, 22, 894, doi:10.3390/s22030894 29

Christin Bald and Gerhard Schmidt
Processing Chain for Localization of Magnetoelectric Sensors in Real Time
Reprinted from: Sensors **2021**, 21, 5675, doi:10.3390/s21165675 51

Johannes Hoffmann, Eric Elzenheimer, Christin Bald, Clint Hansen, Walter Maetzler and Gerhard Schmidt
Active Magnetoelectric Motion Sensing: Examining Performance Metrics with an Experimental Setup
Reprinted from: Sensors **2021**, 21, 8000, doi:10.3390/s21238000 69

Phillip Durdaut, Cai Müller, Anne Kittmann, Viktor Schell, Andreas Bahr, Eckhard Quandt, Reinhard Knöchel, Michael Höft and Jeffrey McCord
Phase Noise of SAW Delay Line Magnetic Field Sensors
Reprinted from: Sensors **2021**, 21, 5631, doi:10.3390/s21165631 85

Lars Thormählen, Dennis Seidler, Viktor Schell, Frans Munnik, Jeffrey McCord and Dirk Meyners
Sputter Deposited Magnetostrictive Layers for SAW Magnetic Field Sensors
Reprinted from: Sensors **2021**, 21, 8386, doi:10.3390/s21248386 113

Jana Marie Meyer, Viktor Schell, Jingxiang Su, Simon Fichtner, Erdem Yarar, Florian Niekiel, Thorsten Giese, Anne Kittmann, Lars Thormählen, Vadim Lebedev, Stefan Moench, Agnė Žukauskaitė, Eckhard Quandt and Fabian Lofink
Thin-Film-Based SAW Magnetic Field Sensors
Reprinted from: Sensors **2021**, 21, 8166, doi:10.3390/s21248166 131

Benjamin Spetzler, Patrick Wiegand, Phillip Durdaut, Michael Höft, Andreas Bahr, Robert Rieger and Franz Faupel
Modeling and Parallel Operation of Exchange-Biased Delta-E Effect Magnetometers for Sensor Arrays
Reprinted from: Sensors **2021**, 21, 7594, doi:10.3390/s21227594 139

Benjamin Spetzler, Elizaveta V. Golubeva, Ron-Marco Friedrich, Sebastian Zabel, Christine Kirchhof, Dirk Meyners, Jeffrey McCord and Franz Faupel
Magnetoelastic Coupling and Delta-E Effect in Magnetoelectric Torsion Mode Resonators
Reprinted from: Sensors **2021**, 21, 2022, doi:10.3390/s21062022 157

Cara Broß, Carolin Enzingmüller, Ilka Parchmann and Gerhard Schmidt
Teaching Magnetoelectric Sensing to Secondary School Students—Considerations for
Educational STEM Outreach
Reprinted from: *Sensors* **2021**, *21*, 7354, doi:10.3390/s21217354 . . . . **175**

About the Editors

Gerhard Schmidt received his Dipl.-Ing. degree in 1996 and his Dr.-Ing. degree in 2001, both from Darmstadt, University of Technology, Germany. After his Ph.D., he worked in the research groups of the acoustic signal processing departments at Harman/Becker Automotive Systems and at SVOX, both in Ulm, Germany. Parallel to his time at SVOX, he was a part-time professor at Darmstadt, University of Technology. Since 2010, he has been a full professor at Kiel University, Germany. His main research interests include adaptive methods for audio, SONAR, and medical signal processing.

Eckhard Quandt began studying at Kiel University in 1979 and, in 1980, switched to a physics degree at Technische Universität Berlin, where he completed his doctoral degree in 1990. In 2000, he completed his habilitation in materials science at Universität Karlsruhe. After his doctoral degree, he worked in materials science at the Kernforschungszentrum Karlsruhe (KfK, Karlsruhe Nuclear Research Center) until 1999 and at the Center of Advanced European Studies and Research (caesar) in Bonn until 2007. In December 2006, he was offered a W3 professorship for inorganic functional materials at the Faculty of Engineering at Kiel University.

At the CAU, from 2009 to 2012, he was a spokesperson of the priority research area "Nano Science and Surface Research"; from 2010 to 2014, he was spokesperson of the Collaborative Research Centre 855 "Magnetoelectric Composites"; and from 2010 to 2012, he was a member of the University Senate at Kiel University. From 2012 to 2014, he was firstly Vice Dean and then, from 2014 to 2016, Dean of the Faculty of Engineering. Since 2016, he has been a spokesperson of the Collaborative Research Centre 1261 "Magnetoelectric Sensors".

In 2016, he received personal funding from the German Research Foundation (DFG) for a five-year Reinhart Koselleck project to research ceramic shape memory materials. In 2018, together with the start-up Acquandas GmbH, he was awarded the first-place prize for the Petersen Innovation Transfer Award for an "innovative technology platform for new bioelectronic and microtechnological medical technology products". He was a member of the Science review board of the German Research Foundation (DFG) for two terms and is a member of the German Academy of Science and Engineering (acatech).

Nian X. Sun is a professor of Electrical and Computer Engineering and of Bioengineering; Director of the Advanced Materials and Microsystems Laboratory; Director of the W.M. Keck Laboratory for Integrated Ferroics, Northeastern University; and founder and chief technical advisor of Winchester Technologies, LLC. He received his Ph.D. degree from Stanford University. Prior to joining Northeastern University, he was a Scientist at IBM and Hitachi Global Storage Technologies. Dr. Sun was the recipient of the Humboldt Research Award, the NSF CAREER Award, the ONR Young Investigator Award, the SBuus Outstanding Research Award, the Outstanding Translational Research Award, and more. His research interests include novel magnetic, ferroelectric and multiferroic materials, devices and microsystems, novel gas sensors and systems, etc. He has over 280 publications and over 20 patents and patent applications. One of his papers was selected as the "ten most outstanding full papers in the past decade (2001–2010) in Advanced Functional Materials". Dr. Sun has given over 180 plenary/keynote/invited presentations and seminars. He is an editor of

Sensors and *IEEE Transactions on Magnetics* and a fellow of the IEEE, Institute of Physics, and of the Institution of Engineering and Technology.

Andreas Bahr has been Professor of Sensor System Electronics at Kiel University since November 2017. Previously, he was a research associate at the Institute of Nano- and Medical Electronics at the Technical University of Hamburg. In 2017, he received his doctorate at the TU Hamburg. His research interests include analog and mixed-signal integrated circuit design, low-noise and low-power design, sensor system electronics, and biomedical signal processing.

Preface to "Magnetoelectric Sensor Systems and Applications"

In the field of magnetic sensing, a wide variety of different magnetometer and gradiometer sensor types, as well as corresponding read-out concepts, are available. Well-established sensor concepts such as Hall sensors and magnetoresistive sensors based on giant magnetoresistances (and many more) have been researched for decades. The development of these types of sensors has reached maturity in many aspects (e.g., performance metrics, reliability, and physical understanding), and these types of sensors are established in a large variety of industrial applications.

Magnetic sensors based on the magnetoelectric effect are a relatively new type of magnetic sensor. The potential of magnetoelectric sensors has not yet been fully investigated. Especially in biomedical applications, magnetoelectric sensors show several advantages compared to other concepts for their ability, for example, to operate in magnetically unshielded environments and the absence of required cooling or heating systems.

In recent years, research has focused on understanding the different aspects influencing the performance of magnetoelectric sensors. At Kiel University, Germany, the Collaborative Research Center 1261 "Magnetoelectric Sensors: From Composite Materials to Biomagnetic Diagnostics", funded by the German Research Foundation, has dedicated its work to establishing a fundamental understanding of magnetoelectric sensors and their performance parameters, pushing the performance of magnetoelectric sensors to the limits and establishing full magnetoelectric sensor systems in biological and clinical practice. The research questions range from fundamental material modelling aiming to understand the underlying principles and physical limits, to the development of innovative sensor concepts and the establishment of thin-film processes technology, and to the usage of entire sensor systems in biomedical applications.

In many applications, magnetic sensors have several advantages if they are used either in addition or even instead of electric measurements. The advantages have been proven in science and research using magnetic sensors such as superconducting quantum interference devices (SQUIDs) or optically pumped magnetometers (OPMs). Application examples include spatially and temporally high-resolution medical analyses such as magnetocardiography (MCG) and combined electro- and magnetoencephalography (EEG/MEG). The drawbacks of these sensor technologies are mainly their high cost and their limited robustness against environmental influences. External magnetic fields, such as the magnetic field of the Earth or the fields created by power supplies, saturate SQUID and OPM sensors, which requires expensive and difficult-to-install magnetic shielding. Furthermore, SQUID sensor technology absolutely needs expensive liquid He cooling.

The magnetoelectric sensor principle—as a relatively new principle—has the potential to overcome these limitations at a very low cost. This would facilitate the transfer of medical research results into clinical practice. Recent advances, in terms of magnetic layer optimization, low-noise readout and dedicated signal processing for new read-out principles can potentially enhance the sensitivity of magnetoelectric sensor principles and bring them very close to that of OPMs or SQUIDs without robustness problems. Additional advantages are the large dynamic range—the requirement being insensitive to large external fields—and the very high bandwidth of certain magnetoelectric sensor approaches.

This book reports the latest research on magnetoelectric sensor systems and corresponding applications. The bandwidth of contributions ranges from biomedical application examples, specially tailored readout schemes for ME sensors, low-noise amplification circuits, and advances in the material science and improved understanding of the magnetic processes that are involved in magnetoelectric layers.

Gerhard Schmidt, Eckhard Quandt, Nian X. Sun, Andreas Bahr
Editors

Article

Quantitative Evaluation for Magnetoelectric Sensor Systems in Biomagnetic Diagnostics

Eric Elzenheimer [1], Christin Bald [1], Erik Engelhardt [1], Johannes Hoffmann [1], Patrick Hayes [2], Johan Arbustini [3], Andreas Bahr [3], Eckhard Quandt [2], Michael Höft [4] and Gerhard Schmidt [1,*]

1. Digital Signal Processing and System Theory, Institute of Electrical Engineering and Information Technology, Faculty of Engineering, Kiel University, Kaiserstr. 2, 24143 Kiel, Germany; ee@tf.uni-kiel.de (E.E.); cbal@tf.uni-kiel.de (C.B.); eren@tf.uni-kiel.de (E.E.); jph@tf.uni-kiel.de (J.H.)
2. Inorganic Functional Materials, Institute for Materials Science, Faculty of Engineering, Kiel University, Kaiserstr. 2, 24143 Kiel, Germany; pah@tf.uni-kiel.de (P.H.); eq@tf.uni-kiel.de (E.Q.)
3. Sensor System Electronics, Institute of Electrical Engineering and Information Technology, Faculty of Engineering, Kiel University, Kaiserstr. 2, 24143 Kiel, Germany; jrsa@tf.uni-kiel.de (J.A.); ab@tf.uni-kiel.de (A.B.)
4. Microwave Engineering, Institute of Electrical Engineering and Information Technology, Faculty of Engineering, Kiel University, Kaiserstr. 2, 24143 Kiel, Germany; mh@tf.uni-kiel.de
* Correspondence: gus@tf.uni-kiel.de; Tel.: +49-431-880-6125

Abstract: Dedicated research is currently being conducted on novel thin film magnetoelectric (ME) sensor concepts for medical applications. These concepts enable a contactless magnetic signal acquisition in the presence of large interference fields such as the magnetic field of the Earth and are operational at room temperature. As more and more different ME sensor concepts are accessible to medical applications, the need for comparative quality metrics significantly arises. For a medical application, both the specification of the sensor itself and the specification of the readout scheme must be considered. Therefore, from a medical user's perspective, a system consideration is better suited to specific quantitative measures that consider the sensor readout scheme as well. The corresponding sensor system evaluation should be performed in reproducible measurement conditions (e.g., magnetically, electrically and acoustically shielded environment). Within this contribution, an ME sensor system evaluation scheme will be described and discussed. The quantitative measures will be determined exemplarily for two ME sensors: a resonant ME sensor and an electrically modulated ME sensor. In addition, an application-related signal evaluation scheme will be introduced and exemplified for cardiovascular application. The utilized prototype signal is based on a magnetocardiogram (MCG), which was recorded with a superconducting quantum-interference device. As a potential figure of merit for a quantitative signal assessment, an application specific capacity (ASC) is introduced. In conclusion, this contribution highlights metrics for the quantitative characterization of ME sensor systems and their resulting output signals in biomagnetism. Finally, different ASC values and signal-to-noise ratios (SNRs) could be clearly presented for the resonant ME sensor (SNR: −90 dB, ASC: 9.8×10^{-7} dB Hz) and also the electrically modulated ME sensor (SNR: −11 dB, ASC: 23 dB Hz), showing that the electrically modulated ME sensor is better suited for a possible MCG application under ideal conditions. The presented approach is transferable to other magnetic sensors and applications.

Keywords: application specific signal evaluation; magnetoelectric sensors; quantitative sensor system characterization; sensor system performance

Citation: Elzenheimer, E.; Bald, C.; Engelhardt, E.; Hoffmann, J.; Hayes, P.; Arbustini, J.; Bahr, A.; Quandt, E.; Höft, M.; Schmidt, G. Quantitative Evaluation for Magnetoelectric Sensor Systems in Biomagnetic Diagnostics. *Sensors* **2022**, *22*, 1018. https://doi.org/10.3390/s22031018

Academic Editor: Nicolò Marconato

Received: 10 December 2021
Accepted: 25 January 2022
Published: 28 January 2022

Publisher's Note: MDPI stays neutral with regard to jurisdictional claims in published maps and institutional affiliations.

Copyright: © 2022 by the authors. Licensee MDPI, Basel, Switzerland. This article is an open access article distributed under the terms and conditions of the Creative Commons Attribution (CC BY) license (https://creativecommons.org/licenses/by/4.0/).

1. Introduction

Medical diagnostics based on electrical signal acquisition methods such as electrocardiography (ECG) or electroencephalography (EEG) are an established routine in clinical practice. These methods have been researched for decades [1,2]. Nowadays, room-temperature magnetic field sensors are being investigated, such as optically pumped magnetometers [3,4], xMR sensors [5], orthogonal fluxgates [6], and many more. These sensor concepts promise several advantages and enable contactless signal acquisition by detecting the magnetic field strength or the magnetic flux density. Obtaining biomagnetic signals is beneficial compared to the standard electrical methods for several reasons. Magnetic sensing promises increased spatial resolution [7], it enables better positioning with less exogenous signal artifacts and the nearly constant relative permeability [8], which prevents physiologic signals from being changed by the elements of the body (tissue, bones, etc). In particular, the ongoing research of thin-film magnetoelectric (ME) sensors enables new areas of signal acquisition in medicine since they do not require cryogenic cooling or thermal heating for sensor operation [9–11]. These sensors are easy to use, provide unprecedented flexibility and are operational in the presence of interference fields such as the Earth's magnetic field [12,13]. Magnetic recording techniques have the potential to support and replace traditional electrode-based (electrical) methods by default [14]. The performance of a magnetic field sensor is usually described by its sensor-specific properties, e.g., operation temperature, inherent noise, dynamic range (in the sense of amplitude range of operation), bandwidth and sensitivity [15], as exemplified for two current biomagnetic ME sensor types in Table 1.

Table 1. Two researched ME sensors with their individual metrics given by publications.

Metrics	Exchange Bias ME Sensor [13]	Electrically Modulated ME Sensor [10,16]
Operation Temperature	Room temperature	Room temperature
Inherent Noise	$\approx 4\,\text{pT}/\sqrt{\text{Hz}}$ at 7.684 kHz	$\approx 70\,\text{pT}/\sqrt{\text{Hz}}$ at 10 Hz
Bandwidth	$\approx 12.5\,\text{Hz}\,(-6\,\text{dB})$	unknown
Sensitivity	$\approx 98\,\text{kV/T}$	$\approx 40\,\text{kV/T}$
Availability	under development	under development

For medical applications, it is not sufficient to consider only the sensing element specification because the overall performance of a sensor system is a superposition of all its subsystems and their individual performances. This includes especially the sensor readout electronics. Since the application of magnetic sensors is a relatively new field of research, often only the sensing element's specifications are provided. The specification of the entire sensor system must be taken into consideration for determining if a sensor is appropriate for a specific application. To exemplify, in a medical applications the question could be asked, whether a signal of interest, for example, the heartbeat, could be measured for diagnostic purposes. From the viewpoint of a medical application, it does not matter where potential disturbances originate. Therefore, a sensor system in a biomagnetic application can be considered a black box. This black box is evaluated with its corresponding system metrics. A simplified representation of such an approach is shown in Figure 1.

In this contribution, magnetic field signals created by physiological means are considered the input signals of interest $b_\text{d}(t)$ (desired input signals). This is exemplified by the signal generated from the human heart. The system input $b_\text{in}(t)$ consists of an additive undesired magnetic signal $b_\text{u}(t)$ from environmental disturbances (coexisting magnetic fields). The available field at the system input can be converted with a magnetic field sensor into a proportional measurand, typically a time-dependent voltage. The sensor signal is read out in analog form within the sensor system, digitally processed, and provided as a signal at the output. The output can also be taken in form of a sample dependent field strength $b_\text{out}(n)$ after unit conversion (voltage → magnetic flux density) or analog as time-dependent voltage $u_\text{out}(t)$. In the overall system, each process step has an individual transfer or conversion function and noise characteristic. At the digital output $b_\text{out}(n)$, the signal can be considered as the sum of the input signal $b_\text{in}(n)$ and a noise superposition of all involved noise components $v_0, \ldots, v_3$. The noise at the system output is a superposition of different uncorrelated random processes [17]. For an application, it is not decisive from where noise contributions originate. As a consequence, the noise power spectral densities or, respectively, the noise amplitude densities superimpose [18]. Finally, this view allows a quantitative description of a sensor system from a user perspective and permits comparing sensor systems for a specific medical discipline or new biomagnetic applications. Since diagnostic information depends mainly on signal characteristics, an application-specific signal evaluation scheme will be presented. This enables an improved quantitative description of the system's suitability. In summary, this contribution highlights metrics for magnetic sensor systems and offers an application-oriented signal evaluation scheme for biomagnetic applications. The remainder of this contribution is organized as follows: In Section 2, different metrics for sensor system evaluation will be introduced. Since diagnostic information depends mainly on signal characteristics, figures of merit for signal evaluation will be supplementary defined in Section 3. Then, in Section 4, an exemplary evaluation will be executed for two different ME sensor systems: a exchange bias magnetoelectric sensor and an electrically modulated ME sensor. Based on these findings, a signal evaluation will be performed, exemplified by a cardiovascular application in Section 4.3. Finally, in Section 5 the individual results will be discussed.

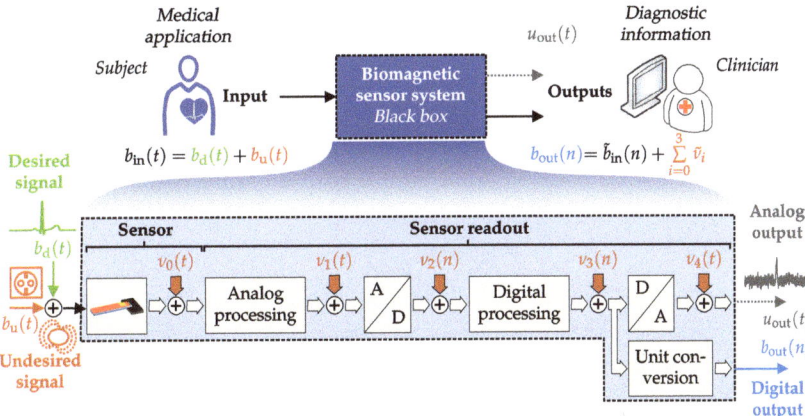

Figure 1. Schematic representation of a typical biomagnetic sensor system.

2. Evaluation Metrics for Magnetoelectric Sensor Systems

Quantitative evaluation metrics are of importance for the characterization and comparison of biomagnetic sensor systems. Key metrics are the *Input–Output–Amplitude–Relation* and *frequency response* [19]. First, the Input–Output–Amplitude–Relation will be discussed, since no explicit system knowledge and presumptions are required.

2.1. Input–Output–Amplitude–Relation

A typical *Input–Output–Amplitude–Relation* of a sensor system is illustrated in Figure 2. It can be divided into three different regions. In the first region (I; gray shaded area), the magnetic signal is so small that the system noise dominates at the output. If the magnetic signal is large enough and exceeds the system noise, the system output increases linearly with the input amplitude (II; green shaded area) until it is limited by compression and saturation effects (III; red shaded area). Limiting factors can be the sensor's dynamic range or the readout electronics characteristics and limitations, e.g., operating voltages, sensor's dynamic range (DR). In addition, transition areas can be identified (cyan shaded area) which cannot be unambiguously assigned to one of the areas mentioned above. For a quantification of the Input–Output–Amplitude–Relation at a particular excitation frequency (typically 10 Hz, 1 kHz, or resonance frequency), $P \in \mathbb{N}$ pairs of root-mean-square (RMS) input and output values are required.

$$\boldsymbol{b}_{\text{in}}^{\text{rms}} = [b_{\text{in}}^{\text{rms}}(0),\ b_{\text{in}}^{\text{rms}}(1),\ \ldots\ ,b_{\text{in}}^{\text{rms}}(P-1)]^{\text{T}}, \tag{1}$$

$$\boldsymbol{b}_{\text{out}}^{\text{rms}} = [b_{\text{out}}^{\text{rms}}(0),\ b_{\text{out}}^{\text{rms}}(1),\ \ldots\ ,b_{\text{out}}^{\text{rms}}(P-1)]^{\text{T}}, \tag{2}$$

where $\boldsymbol{b}_{\text{in}}^{\text{rms}}$ are the input RMS values of the sensor system and $\boldsymbol{b}_{\text{out}}^{\text{rms}}$ are the acquired RMS output values. Since an additional DC offset has, in general, a significant influence on the curve progression including the derived quantities, it should be identified and minimized for functional determination for all system output values within $\boldsymbol{b}_{\text{out}}^{\text{rms}}$. Characteristic quantities such as *Limit-of-Detection*, *Limit-of-Quantification*, *maximum value*, and *1-dB-compression-point* can be defined for quantitative description of the Input–Output–Amplitude–Relation.

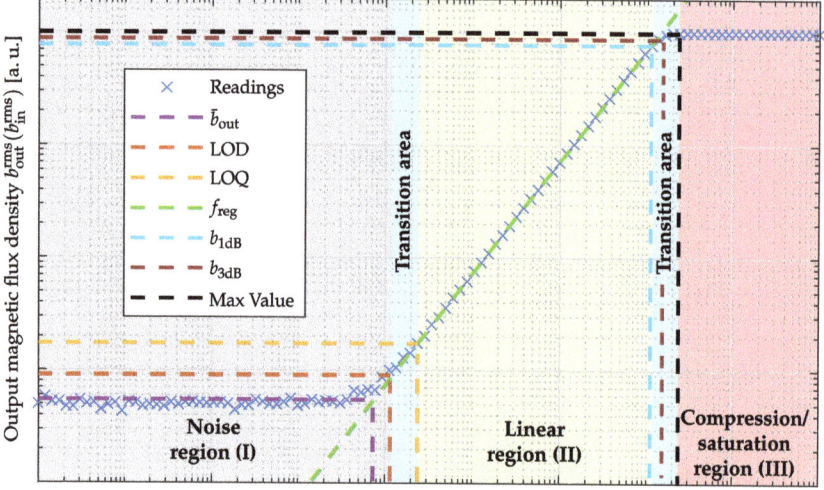

Figure 2. Input–Output–Amplitude–Relation with labeling of the typical regions (noise region (I), linear region (II) and compression/saturation region (III)). Transition areas are marked in cyan. Furthermore, characteristic quantities, mean value within noise region, *Limit-of-Detection* (LOD), *Limit-of-Quantification* (LOQ), *1-dB-compression-point*, *3-dB-compression-point* and *maximum value* are marked in different colors.

2.1.1. Limit-of-Detection

The *Limit-of-Detection* (LOD) of a biomagnetic sensor system describes the smallest measurable magnetic flux density where a magnetic field can be reliably detected [20]. For LOD estimation, the values $\boldsymbol{b}_{\text{out}}^{\text{rms}}$, where no signal can be reliably detected and the system noise dominates, are of interest. This condition is in general fulfilled if the desired signal is less than the effective magnetic noise amplitude $b_{\text{n}}^{\text{rms}}$ corresponding to

$$b_{\text{out}}^{\text{rms}}(i) < b_{\text{n}}^{\text{rms}}. \tag{3}$$

The LOD can be determined from $K \in \mathbb{N}$ measurement points of interest, where the system noise dominates [21,22]. The LOD can be estimated by the mean value μ_n plus three times the standard deviation σ_n of the predefined measurement points with:

$$LOD = \underbrace{\frac{1}{K} \cdot \sum_{i=0}^{K-1} b_{out}^{rms}(i)}_{\bar{b}_{out} = \mu_n} + 3 \cdot \underbrace{\sqrt{\frac{\sum_{i=0}^{K-1} \left[b_{out}^{rms}(i) - \bar{b}_{out}\right]^2}{K-1}}}_{\sigma_n}. \quad (4)$$

The LOD serves as a criterion of reliable evidence and is provided in magnetic sensor systems as an RMS value with the unit T for a particular excitation frequency. It defines the lowest magnetic field that the sensor system can reliably detect [19]. A spectral LOD consideration in $T/\sqrt{Hz}$ is occasionally used instead, especially for modulated magnetic sensors. The amplitude density with unit $T/\sqrt{Hz}$ is related to the RMS value with unit T by Parseval's theorem [23,24]. In general, the LOD value relies on the applied measurement procedure, as illustrated in Figure 3 and has to be defined in detail. Implicit filtering of the output signal by averaging methods [11] will result in very optimistic LOD determinations, which are only realizable in applications with equal bandwidth requirements. Consequently, the LOD will only be reproducible in an experimental setup with identically chosen parameters. Therefore, additional applied filters (lowpass, highpass, bandpass filters) should be specified by their cutoff frequencies. As an assessing bandwidth, the supported frequency range of the sensing element should be chosen. In addition, the window length for RMS amplitude calculation must be stated, whereby one period of the magnetic excitation signal should be used. The resulting RMS value corresponds to the standard deviation of a noise process with zero mean. Since the calculation can also be performed in the frequency domain, the RMS amplitude can be estimated by determining the square root out of the sum from power spectral density (PSD) values. The obtainable results are equivalent [23,24].

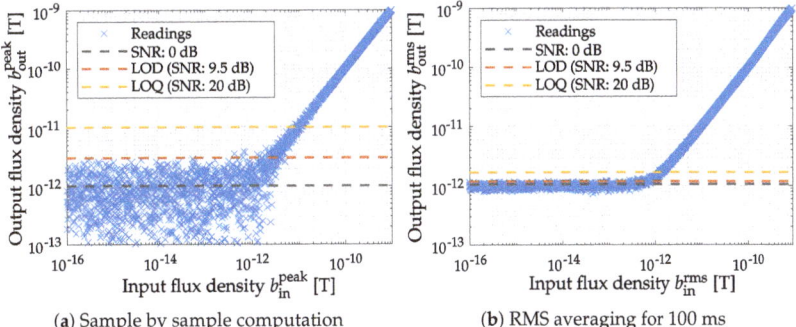

(a) Sample by sample computation (b) RMS averaging for 100 ms

Figure 3. Different methods for LOD computation for the same signal. The simulated sample by sample computation (**a**) of the standard deviation and the mean value yields the same results as the stochastic parameters of the applied random process ($\mu_n = 0$, $\sigma_n = 1\,\text{pT}$). The averaging window applied in (**b**) results in a reduced spread and therefore an SNR gain at the cost of a reduction in bandwidth ($\mu_n = 1\,\text{pT}$, $\sigma_n = 70\,\text{fT}$).

2.1.2. Limit-of-Quantification

Another quantity in connection with the detection limit is the *Limit-of-Quantification* (LOQ), which defines the boundary from which a measured value can be reliably quantified [21,22]. The LOQ can be expressed as follows:

$$LOQ = \bar{b}_{out} + 10 \cdot \sqrt{\frac{\sum_{i=0}^{K-1} \left[b_{out}^{rms}(i) - \bar{b}_{out}\right]^2}{K-1}}. \tag{5}$$

Compared to the previously defined LOD (see Equation (4)) the required standard deviation is set by a factor of 10/3 higher [21,22]. This corresponds to a signal-to-noise ratio (SNR) of 20 dB, ensuring that signal amplitudes above LOQ can be detected at the system output in the time domain.

2.1.3. Linear Region

The linear region given by the Input–Output–Amplitude–Relation (Figure 2) can be described using a linear function approximation [19]. For this purpose, an affine linear regression $f_{reg}(\cdot)$ can be performed using the measurement values b_{out}^{rms} as a function of the excitation signal amplitudes b_{in}^{rms} given by

$$b_{out}^{rms}(i) = f_{reg}\left(b_{in}^{rms}(i)\right) = \alpha + \beta \cdot b_{in}^{rms}(i) \; \forall \; b_{in}^{rms}(i), LOQ < b_{out}^{rms}(i) < b_{1dB}. \tag{6}$$

In the case of curve compression for large input amplitudes originating from system limitations based on saturation effects (nonlinearities), corresponding data points should be excluded. Therefore, signal amplitudes b_{in}^{rms} should be bigger than LOQ and smaller than the *1-dB-Compression-Point* b_{1dB} (cf. Section 2.1.4). Unique solutions for α and β can be found by minimizing the sum of squared deviations as follows for the remaining W data points:

$$\beta = \frac{\sum_{i=0}^{W-1} b_{in}^{rms}(i) \cdot b_{out}^{rms}(i) - W \cdot \mu_{in} \cdot \mu_{out}}{\sum_{i=0}^{W-1} \left[b_{in}^{rms}(i)\right]^2 - W \cdot [\mu_{in}]^2} \quad \text{and} \quad \alpha = \mu_{out} - \beta \cdot \mu_{in} \tag{7}$$

$$\text{with} \; \mu_{in} = \frac{1}{W} \cdot \sum_{i=0}^{W-1} b_{in}^{rms}(i) \; \text{and} \; \mu_{out} = \frac{1}{W} \cdot \sum_{i=0}^{W-1} b_{out}^{rms}(i).$$

Finally, it is required that not all values in b_{in}^{rms} are equal, which ensures that the denominator of β is different from zero [25]. This primary requirement is fulfilled due to the performed amplitude variation at the system input.

2.1.4. 1-dB-Compression-Point and 3-dB-Compression-Point

In general, sensor systems must have high linearity in their operating range to prevent unwanted signal components at the output [26]. Therefore, if a limitation of the curve progression is perceived and compression exists, the maximum system output b_{max} could be determined. For this purpose $L \in \mathbb{Z}$ data points are used, which lay in this specific region (compression/saturation region, see Figure 2). For identification of b_{max} the mean value of those data points can be calculated by:

$$b_{max} = \frac{1}{L} \cdot \sum_{i=0}^{L-1} b_{out}^{rms}(i). \tag{8}$$

For most standard magnetic field sensors, b_{max} is limited by the operating voltage of the readout electronics. In the case of ultra-sensitive magnetic field sensors, the system limitation results from saturation effects. Both effects have the consequence that, above a certain input level, the amplitude of the system output is limited with co-occurring non-linearities. A quantitative measure of linearity can be obtained using the *1-dB-Compression-Point* ($b_{1\text{dB}}$) and the *3-dB-Compression-Point* ($b_{3\text{dB}}$), which specify the input level at which the real transfer characteristic deviates from the regression function with ideal characteristic (see Equation (6)) by 1 dB or 3 dB, respectively. The $b_{1\text{dB}}$ and $b_{3\text{dB}}$ point can be determined as follows:

$$b_{1\text{dB}} = b_{\text{in}}^{\text{rms}}(i) \quad \text{with} \quad 20 \cdot \log_{10}\left(\frac{b_{\text{out}}^{\text{rms}}(i)}{f_{\text{reg}}(b_{\text{in}}^{\text{rms}}(i))}\right) \stackrel{!}{=} -1\,\text{dB}, \tag{9}$$

$$b_{3\text{dB}} = b_{\text{in}}^{\text{rms}}(i) \quad \text{with} \quad 20 \cdot \log_{10}\left(\frac{b_{\text{out}}^{\text{rms}}(i)}{f_{\text{reg}}(b_{\text{in}}^{\text{rms}}(i))}\right) \stackrel{!}{=} -3\,\text{dB}. \tag{10}$$

2.1.5. Dynamic Range

The supported *dynamic range* (*DR*) of the system can be specified using *LOQ* as the lower limit and $b_{1\text{dB}}$ as the upper limit. The dynamic range is given by:

$$DR = 20 \cdot \log_{10}\left(\frac{b_{1\text{dB}}}{LOQ}\right)\,\text{dB}, \tag{11}$$

and is provided in dB units [15].

2.1.6. Determination of the Input–Output–Amplitude–Relation

For the determination of the *Input–Output–Amplitude–Relation*, precise amplitude knowledge of the excitation signal $b_{\text{d}}(t)$ and a measurement of the output signal (u_{out}; b_{out}) are necessary. Therefore, a high precision A/D converter combined with a known magnetic reference field with frequency f_{exc} is used. The magnetic field is generated with a calibrated cylindrical coil within a magnetically shielded environment (permalloy cylinder). The calibrated coil is excited with an alternating current $i_{\text{ac}}(i,t)$ at the frequency f_{exc} generated by an ultra-low-noise current source. The parameters i_{ac} and f_{exc} have to be chosen such that the following relation is valid:

$$b_{\text{in}}^{\text{rms}}(i) \propto \frac{\hat{i}_{\text{ac}}(i)}{\sqrt{2}} \quad \text{with} \quad i_{\text{ac}}(i,t) = \hat{i}_{\text{ac}}(i) \cdot \sin(2\,\pi\,f_{\text{exc}}\,t). \tag{12}$$

The current source serves as the generator for the coil and as the reference signal. The resulting magnetic flux density should be varied linearly from zero to the maximum assessable flux density of the system. The saturation region may not be reachable for all sensor types. The sensor's sensing area should be placed in the center of the coil. This approach enables the identification of the detection limit, the system behavior, and saturation effects through operating voltage or sensor dynamic limits. Finally, the measured RMS magnetic flux density at system output $b_{\text{out}}^{\text{rms}}(i)$ is plotted against the applied AC magnetic field amplitude $b_{\text{in}}^{\text{rms}}(i)$.

2.2. Frequency Response (Magnitude and Phase Response)

For the following considerations, it has to be assumed that the sensor system is a linear time-invariant (LTI) system that is analyzed in the discrete time-domain. Even though most sensor systems, which in some way rely on ferromagnetic material, do not have strictly linear behavior, it is convenient to assume that the sensor systems are at least approximately linear in their operation regime for small input signals (*small signal consideration*). Furthermore, an existing DC offset in the *Input–Output–Amplitude–Relation*, especially for $b_{\text{in}}^{\text{rms}}(i) = 0$, must be identified and minimized. The precondition of time-invariance is not fulfilled by default because the sensor system performance varies in time due to changes in environmental conditions (e.g., Earth's magnetic field) and changes in their internal environment (e.g., operating temperature). That being said, time-invariance can be assumed for a short period of system evaluation. In conclusion, the LTI conditions are achievable under the given assumptions, and consequently, the sensor system can be uniquely characterized by its causal real-valued impulse response h with $N \in \mathbb{N}$ sample values:

$$h = [h(0),\ h(1),\ \ldots,h(N-1)]^\mathrm{T}. \tag{13}$$

The output $b_{\mathrm{out}}(n)$ of a sensor system (see Figure 4) can be generally determined by applying a convolution with the impulse response $h(n)$ to any input signal $b_{\mathrm{in}}(n)$ corresponding to:

$$b_{\mathrm{out}}(n) = h(n) * b_{\mathrm{in}}(n) = \sum_{i=0}^{N-1} h(i) \cdot b_{\mathrm{in}}(n-i). \tag{14}$$

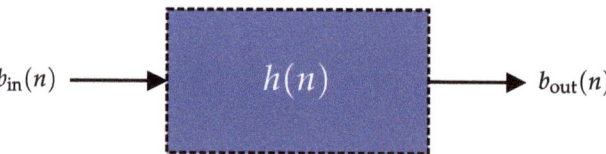

Figure 4. Biomagnetic LTI sensor system with impulse response $h(n)$.

Consequently, the system signal output will be a sum of time-shifted versions of the input signal each weighted by an impulse response coefficient. The complex-valued frequency response $H(e^{j\Omega})$ is the frequency domain representation of the impulse response given by:

$$b_{\mathrm{out}}(n) = h(n) * b_{\mathrm{in}}(n) \circ\!\!-\!\!\bullet B_{\mathrm{out}}(e^{j\Omega}) = H(e^{j\Omega}) \cdot B_{\mathrm{in}}(e^{j\Omega})$$
$$\Rightarrow H(e^{j\Omega}) = \frac{B_{\mathrm{out}}(e^{j\Omega})}{B_{\mathrm{in}}(e^{j\Omega})}, \tag{15}$$

whereby symbol $\circ\!\!-\!\!\bullet$ abbreviates a Fourier transform for discrete signals in the one direction and its inverse counterpart in the other. The frequency response $H(e^{j\Omega})$ of the system significantly influences the signal characteristics. Therefore, $H(e^{j\Omega})$ is of particular interest for the determination of the transfer characteristic of a sensor system, because a system impact on the magnitude and phase exists and must be considered for any application [27]. A commonly used approach for frequency response estimation can be performed by exciting the sensor system in the steady-state (transient effects are no longer present in the system) with a sinusoidal alternating magnetic field. A successive excitation with $M \in \mathbb{N}$ different discrete angular frequencies Ω_μ with $\mu = 0, \ldots, M-1$ in the frequency range of interest enables estimation of the absolute magnitude of $H(e^{j\Omega})$ (amplitude response) represented by

$$|\hat{H}(e^{j\Omega_\mu})| = \frac{|\hat{B}_{\mathrm{out}}(e^{j\Omega_\mu})|}{|\hat{B}_{\mathrm{in}}(e^{j\Omega_\mu})|}, \tag{16}$$

and the corresponding phase estimation (phase response) given by:

$$\hat{\Phi}(e^{j\Omega_\mu}) = \arg\{\hat{H}(e^{j\Omega_\mu})\} = \arctan\left(\frac{\Im\{\hat{H}(e^{j\Omega_\mu})\}}{\Re\{\hat{H}(e^{j\Omega_\mu})\}}\right). \tag{17}$$

The amplitude response $|\hat{H}(e^{j\Omega_\mu})|$ is usually presented in dB units and plotted in a double logarithmic scale [26]. The phase angle $\hat{\Phi}(e^{j\Omega_\mu})$ is provided in degree units and presented in a semi-logarithmic scale. The phase information is essential since it indicates the phase change introduced by the sensor system, which is required for phase-sensitive applications, special readout schemes, and medical signal evaluations.

Subsequently, the result is influenced by choice of supporting points Ω_μ, which limits the accuracy for amplitude and phase response. Furthermore, an exact knowledge of the excitation signal $b_{in}(n)$ and a phase-synchronous signal evaluation are essential prerequisites. This fact necessitates the use of a lock-in-amplifier. The accuracy of the system identification process can be improved if the complete frequency range of interest is excited simultaneously by a broadband signal, for example, white noise or a maximum length sequence. Based on the recorded output signal and the input signal, the transfer characteristic can then be estimated in the frequency domain [24]. Another system identification approach could be realized by determining the impulse response with a gradient-based method based on the input signal and the corresponding output signal [28]. Due to the necessity of a detailed phase response evaluation, the frequency response identification process is commonly performed with lock-in amplifiers by a successive mono-frequent excitation. Therefore, this standard method is established in current analyzers systems and has been applied successfully for years [29].

In general, it is helpful to describe the amplitude response of a sensor system with quantitative metrics because the magnitude behavior can be predominantly assigned to a bandpass or lowpass characteristic. A typical amplitude response of a sensor system with bandpass characteristic is shown in Figure 5. For this, the following metrics can be defined [24]:

- Mean Passband Amplitude

$$\bar{a} = \frac{1}{M} \sum_{\mu=0}^{M-1} |\hat{H}(e^{j\Omega_\mu})| \quad \text{for } \Omega_{p1} \leq \Omega_\mu \leq \Omega_{p2} \tag{18}$$

- Passband Ripple

$$\delta_p = 20 \cdot \log_{10}\left(\frac{\bar{a} + \delta_{p,\max}}{\bar{a} - \delta_{p,\min}}\right) \text{ with } \delta_{p,\max} = \max\left\{|\hat{H}(e^{j\Omega_\mu})|\right\}$$
$$\text{and } \delta_{p,\min} = \min\left\{|\hat{H}(e^{j\Omega_\mu})|\right\} \text{ for } \Omega_{p1} \leq \Omega_\mu \leq \Omega_{p2} \tag{19}$$

- Passband Edge Frequencies

$$\Omega_{p1} = \arg\left\{|\hat{H}(e^{j\Omega_\mu})| \stackrel{!}{=} \bar{a} - \delta_{p,\min}\right\} \text{ for } \Omega_\mu < \Omega_z \tag{20}$$

$$\Omega_{p2} = \arg\left\{|\hat{H}(e^{j\Omega_\mu})| \stackrel{!}{=} \bar{a} - \delta_{p,\min}\right\} \text{ for } \Omega_\mu > \Omega_z \tag{21}$$

- Stopband Edge Frequencies

$$\Omega_{s1} = \arg\left\{|H(e^{j\Omega_\mu})| \stackrel{!}{=} \max\left(|\hat{H}(e^{j\Omega_\mu})| \text{ for } \Omega_\mu < \Omega_{s1}\right)\right\} \text{ for } \Omega_\mu \ll \Omega_{p1} \tag{22}$$

$$\Omega_{s2} = \arg\left\{|H(e^{j\Omega_\mu})| \stackrel{!}{=} \max\left(|\hat{H}(e^{j\Omega_\mu})| \text{ for } \Omega_\mu > \Omega_{s2}\right)\right\} \text{ for } \Omega_\mu \gg \Omega_{p2} \tag{23}$$

- Transition Bands

$$\Delta\Omega_1 = \Omega_{p1} - \Omega_{s1} \tag{24}$$

$$\Delta\Omega_2 = \Omega_{s2} - \Omega_{p2} \tag{25}$$

- $-3\,\text{dB}$ Angular Frequencies, Bandwidth

$$\Omega_{-3\text{dB},1} = \arg\left\{|\hat{H}(e^{j\Omega_\mu})| \stackrel{!}{=} \frac{1}{\sqrt{2}}\bar{a}\right\} \text{ for } \Omega_{s1} \leq \Omega_\mu \leq \Omega_{p1} \tag{26}$$

$$\Omega_{-3\text{dB},2} = \arg\left\{|\hat{H}(e^{j\Omega_\mu})| \stackrel{!}{=} \frac{1}{\sqrt{2}}\bar{a}\right\} \text{ for } \Omega_{p2} \leq \Omega_\mu \leq \Omega_{s2} \qquad (27)$$

$$w = \Omega_{-3\text{dB},2} - \Omega_{-3\text{dB},1}. \qquad (28)$$

Sensor systems with predominant resonator behavior in magnitude can be better described by resonance angular frequency (Ω_{res}) and -3-dB-bandwidth ($\Omega_{-3\text{dB},1}$; $\Omega_{-3\text{dB},2}$). These angular frequencies are related to their time-continuous counterparts f_{res}, $f_{-3\text{dB},1}$ and $f_{-3\text{dB},2}$. In this case, the -3-dB-bandwidth is related to the magnitude maximum in resonance according to the condition:

$$|\hat{H}(e^{j\Omega_{res}})| = 1. \qquad (29)$$

Finally, also the quality (Q) factor can be determined [2] corresponding to

$$Q = \frac{\Omega_{res}}{\Omega_{-3\text{dB},2} - \Omega_{-3\text{dB},1}}. \qquad (30)$$

Other metrics are not required for an adequate resonator description. For sensor systems with predominant lowpass behavior, the provided bandpass metrics ($\bar{a}$, δ_p, Ω_{p2}, Ω_{s2}, $\Delta\Omega_2$, $\Omega_{-3\text{dB},2}$) can be modified, because only the right half of the magnitude response according to Figure 5 with $\Omega_{s1} = \Omega_{p1} = \Omega_z = 0$ has to be considered.

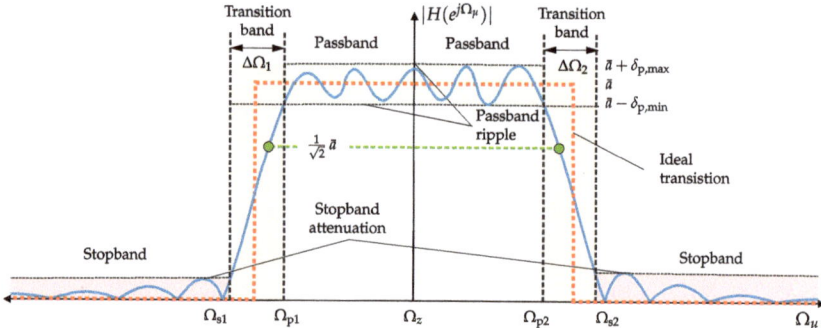

Figure 5. Amplitude response (asymmetrical passband) with predominant bandpass characteristic including metrics labeling. For a response with lowpass characteristic, only the right abscissa axis is required.

Frequency Response Determination

Following Figure 6 the *Frequency Response* can be determined if a monofrequent signal (sinusoid) is applied in a shielded environment (permalloy cylinder) via a calibrated coil to the magnetic sensor. The current source serves as a sweep generator [30] and also as the reference signal for the lock-in-amplifier. The normalized discrete angular frequency Ω_μ of the excitation signal is varied linearly in a predefined frequency range $f_{start} \leq f_\mu \leq f_{end}$. A common frequency range of biomagnetic signals extends from 0.01 Hz to 10 kHz [31]. In most biomedical applications, a supported bandwidth of approximately 1 kHz (0.01 Hz to 1 kHz) is sufficient to record fast time-dependent field variations [31]. For special applications, such as nerve activity detection and muscle spontaneous activity detection, the required bandwidth is even higher [4,32]. The excitation signal amplitude must be chosen such that the resulting magnetic flux density lies in the typical linear region (cf. Figure 2). The sensor's sensing volume should be placed in the center of the calibrated coil. Finally, the signal at the system output is analyzed and compared to the applied AC magnetic field with regard to amplitude and phase change.

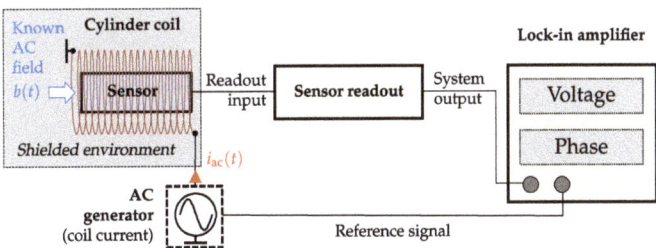

Figure 6. Frequency response measurement of a sensor system in a magnetically shielded environment by lock-in-amplifier, current source and cylinder coil.

3. Figures of Merit for Sensor Signal Evaluation

The quantities introduced in Section 2 are targeted at comparing different sensors to each other. It is theoretically possible to evaluate the suitability of a sensor for a specific application according to these metrics. However, doing so requires experience and expertise in dealing with sensor characteristics and the application in question. In this section, we will move away from considering sensors systems on their own and start introducing metrics that can be used to evaluate them for specific applications. Therefore, figures of merit for sensor signal evaluation will be introduced, primarily influenced by the desired biomedical signal itself and the noise present in the overall system. Furthermore, an application-specific capacity is presented, which ensures a quantitative evaluation in the frequency domain. This approach is essential since diagnostic information depends mainly on signal characteristics. Therefore, a biomedical signal should remain as unaffected as possible by the sensor system; otherwise, a signal feature change will occur purely due to a technical limitation and has no pathophysiological or physiological origin. As an exemplary desired magnetic signal within the following sections, a prototype heart signal is applied. Compared to other biomedical sources like nerves or the brain, the magnetic field of the heart is by far the strongest [33]. The prototype signal is based on magnetocardiography (MCG) measurements with super conducting quantum interference device (SQUIDs) (cf. Figure 7a,c, Appendix A) recorded at the Physikalisch Technische Bundesanstalt (PTB) in Berlin. For signal generation, characteristic data points (cf. Table A1) of the SQUID-MCG recording and a cubic Hermite spline are applied. Using a sampling frequency of $f_s = 2000\,\text{Hz}$ results in the signal $s(t)$ (cf. Figure 7b,d, which is used in the following experiments. At low frequencies, the PSDs of the prototype and SQUID signals are very similar. The deviations at higher frequencies might be explained by the absence of additive noise in the prototype signal. For the estimation of the PSD, Welch's method is used in this section with a signal length of 5 s, a Hanning window of 256 samples width, an overlap of 50 percent, and an FFT length of 4096. In the next section, the system noise is introduced, which is fundamental for all upcoming metrics.

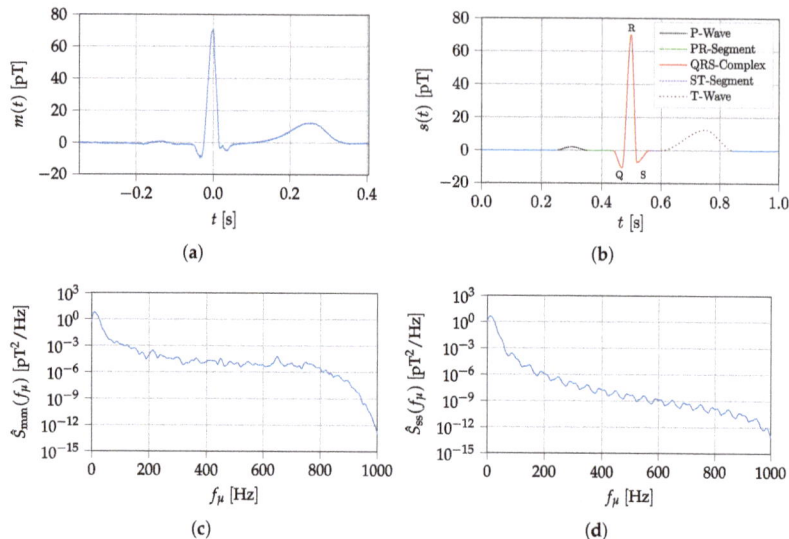

Figure 7. MCG prototype signal based on SQUID-MCG-Data. (**a**) SQUID Measurement—time domain. (**b**) Prototype MCG—time domain. (**c**) SQUID Measurement—power spectral density. (**d**) Prototype MCG—power spectral density.

3.1. System Noise

The desired biomagnetic signal is usually superimposed by undesired noise. These stochastic signal components can be characterized in the frequency domain with the frequency-dependent PSD. As a consequence, PSD measurements are performed after the required post-processing steps, e.g., *filtering, demodulation*, and *A/D conversion*, in order to evaluate the entire noise characteristics (cf. Figure 1). A noise-free system could theoretically acquire arbitrarily small measurement signals and optimally adapt them to the dynamic range of digitization. The detection limit of sensor systems is constrained by noise processes present at the system output, whereby a distinction of the noise sources is not considered in this analysis. This aggregated noise describes unwanted signals and processes of all components within the signal chain, which results in decisive limitations [15]. The PSD for a stationary random process can be determined from the Fourier Transform of the autocorrelation function (ACF) by applying the Wiener–Khintchine theorem [18]. In practice, for PSD estimation $\hat{S}_{xx}(\Omega_\mu)$ of a digitized sequence $x(n)$, the well-known Welch's algorithm [34] is mainly used, where Ω_μ describes the normalized frequency bins (Furthermore, using the sampling frequency f_s and the relation $f_\mu = \Omega_\mu \cdot f_s/(2\pi)$, the estimated power spectral density can also be related to the discrete frequency bins f_μ). Welch's algorithm guarantees a reduction of the variance in the frequency domain based on multiple windowed and squared Discrete Fourier Transform (DFT, *periodograms*) averages. In order to ensure traceability of the results, acquisition time and the total amount of averages should be provided. The resulting power density spectrum estimated from the noise is called noise power spectral density and allows the additive superposition of uncorrelated noise sources. Typically for sensor system specification, the amplitude spectral density (ASD) is provided instead, which is the square root of the power density spectrum $\sqrt{\hat{S}_{xx}(\Omega_\mu)}$. It represents the RMS value as a physical unit of the measurand with respect to a frequency bandwidth of 1 Hz [24]. Moreover, the density spectra (PSD/ASD) are related to the common power spectra and amplitude spectra (PS/AS) via the equivalent noise bandwidth (ENBW) [20]. At the output of a sensor system, voltages are directly acquired so that the noise power density spectrum in V^2/Hz can be represented as amplitude spectral density,

$$\sqrt{\hat{S}_{uu}(\Omega_\mu)}, \tag{31}$$

in the unit V/$\sqrt{\text{Hz}}$. On the other hand, magnetic noise,

$$\sqrt{\hat{S}_{bb}(\Omega_\mu)}, \tag{32}$$

is given in the unit T/$\sqrt{\text{Hz}}$ [15]. Finally, to guarantee a unit conversion from the electric output quantity to the magnetic input quantity, a description of the overall system *sensitivity* ϵ_{sys} is necessary in the unit V/T [19]. The frequency-dependent sensitivity $\epsilon_{sys}(\Omega_\mu)$ is the ratio of the output voltage to the change of a known predefined magnetic flux density ($B_{ext} \neq 0$) so that the following equation holds:

$$\epsilon_{sys}(\Omega_\mu) = \frac{U_{out}(\Omega_\mu)}{B_{ext}(\Omega_\mu)}, \tag{33}$$

whereby $U_{out}(\Omega_\mu)$ denotes the RMS sensor output voltage and $B_{ext}(\Omega_\mu)$ the RMS magnetic flux density as input quantity at a particular normalized angular frequency Ω_μ. It should be mentioned that only a sensitivity determination is performed to get the also required physical unit conversation factor, while the frequency response (cf. Equation (15)) is usually dimensionless. After all, the magnetic field noise can be determined in the unit T/$\sqrt{\text{Hz}}$. Therefore, the noise voltage spectral density is divided by the frequency-depended sensor sensitivity according to

$$\sqrt{\hat{S}_{bb}(\Omega_\mu)} = \frac{\sqrt{\hat{S}_{uu}(\Omega_\mu)}}{\epsilon_{sys}(\Omega_\mu)}. \tag{34}$$

Thus for the achievement of an overall low magnetic field noise density, high sensitivity and low noise are required. The specification of the noise as a parameter must always be related to the bandwidth w. The *effective magnetic noise amplitude* $b_n^{rms}(\Omega_{-3dB,1}, \Omega_{-3dB,2})$, which is available within a given bandwidth (-3 dB or -6 dB sensor bandwidth are commonly used, cf. Equations (26) and (27)), can be determined from the estimated frequency-dependent power spectral density $\hat{S}_{bb}(\Omega_\mu)$ by:

$$b_n^{rms}(\Omega_{-3dB,1}, \Omega_{-3dB,2}) = 2 \cdot \sqrt{\lim_{\Delta\Omega_\mu \to 0} \sum_{\Omega_\mu = \Omega_{-3dB,1}}^{\Omega_{-3dB,2}} \hat{S}_{bb}(\Omega_\mu) \cdot \Delta\Omega_\mu} \tag{35}$$

$$\forall \ 0 \leq \Omega_{-3dB,1} \leq \Omega_\mu \leq \Omega_{-3dB,2}.$$

The lower normalized angular cutoff-frequency $\Omega_{-3dB,1}$ and the upper normalized angular cutoff-frequency $\Omega_{-3dB,2}$ are quite crucial for effective noise amplitude determination. Therefore, a bandwidth reduction usually results in a decrease in noise amplitude. Figure 8 illustrates the ambiguity of this metric without a given bandwidth specification.

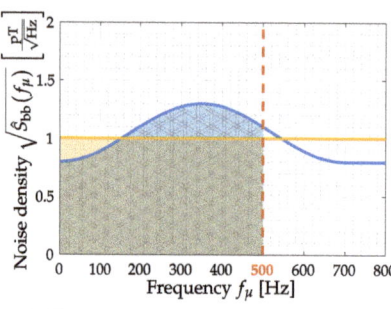 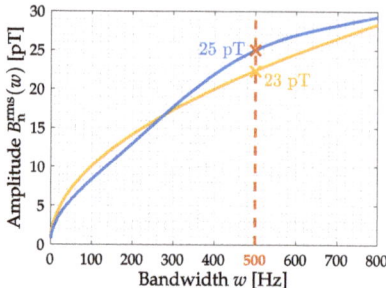

(a) Exemplary noise amplitude densities.   (b) Bandwidth-dependent RMS amplitude.

Figure 8. Summation of noise amplitude densities. Two exemplary noise amplitude densities (constant and arbitrary shape) are provided (a). Summation is performed from DC up to an increasing upper cutoff frequency to obtain the corresponding RMS value (b) for both densities. Assuming a sensor −3 dB cutoff frequency of 500 Hz, the colored areas under curve (a) yield RMS amplitudes of 25 pT and 23 pT, which will vary if a different upper frequency limit is applied (b).

Consequentially, the considered frequency range/bandwidth is another characteristic value provided for effective noise amplitude considerations. Please note that the noise consideration within 3-dB-bandwidth is meant as a general sensor system performance metric. Any practical (biomedical) application might produce varying noise characteristics due to its respective bandwidth requirements and application-specific prefilters.

Measurement

The overall system noise $n(t)$ of a sensor system can be determined with a minor change of the experimental setup shown in Figure 6. For this purpose, the external magnetic excitation, including the coil, is no longer required and should be removed entirely from the experimental setup to avoid unnecessary additional noise components. Finally, the sensor is operated in an almost zero field environment $b(t) \approx 0$, for example, in a permalloy cylinder, and the system output voltage is continuously analyzed.

3.2. Signal-to-Noise Ratio

A quantity commonly used to describe signal quality is the SNR. The SNR quantitatively describes the differences in power between signal and noise by the quotient:

$$SNR = \frac{P_s}{P_n} \approx \frac{\hat{\sigma}_s^2}{\hat{\sigma}_n^2} \circ\!\!-\!\!\bullet \frac{\int_0^\infty \hat{S}_{ss}(f)\,df}{\int_0^\infty \hat{S}_{nn}(f)\,df} \approx \frac{\sum_{\Omega_\mu=\Omega_l}^{\pi} \hat{S}_{ss}(\Omega_\mu)}{\sum_{\Omega_\mu=\Omega_l}^{\pi} \hat{S}_{nn}(\Omega_\mu)}, \tag{36}$$

where P_s is the average power of the signal and P_n is the average power of the noise. The SNR could also be estimated by the ratio between estimated signal variance $\hat{\sigma}_s^2$ and estimated noise variance $\hat{\sigma}_n^2$ of the time domain signals. Another approximation could be made in the spectral domain by using the application-specific PSD of the signal $\hat{S}_{ss}(\Omega_\mu)$ and the noise PSD of the sensor $\hat{S}_{nn}(\Omega_\mu)$ (cf. Equation (34)).

Since it is not possible to measure the pure signal component in the absence of noise, it can be more practical to calculate the signal-plus-noise to noise ratio [35] (SNNR) instead:

$$SNNR = \frac{P_s + P_n}{P_n} = \frac{P_m}{P_n} \approx \frac{\hat{\sigma}_m^2}{\hat{\sigma}_n^2} \circ\!\!-\!\!\bullet \frac{\int_0^\infty \hat{S}_{mm}(f)\,df}{\int_0^\infty \hat{S}_{nn}(f)\,df} \approx \frac{\sum_{\Omega_\mu=\Omega_l}^{\pi} \hat{S}_{mm}(\Omega_\mu)}{\sum_{\Omega_\mu=\Omega_l}^{\pi} \hat{S}_{nn}(\Omega_\mu)}, \tag{37}$$

where P_m is the average power of the measured signal (superimposed with the noise), σ_m^2 is the estimated variance of the measured signal and $\hat{S}_{mm}(\Omega_\mu)$ is its estimated power spectral density. The *SNNR* contains the same information as the *SNR*. In order to convert one into the other, the following relationship can be used:

$$ SNNR = \underbrace{\frac{P_s}{P_n}}_{SNR} + \frac{P_n}{P_n} = SNR + 1. \tag{38}$$

To calculate the power from the PSD it is theoretically necessary to integrate from $f = 0$ Hz to $f \to \infty$. Since this calculation takes place digitally in practice, some approximations and restrictions have to be made. First, the integration over the frequency becomes a numerical integration over the support points Ω_μ and the upper integration limit is confined to $\Omega_u = \pi$, due to the sampling theorem. In practice, the lower integration limit can be confined by metrological constraints to a value of Ω_l. In the following simulations, Romberg's method is used for the numeric integration and $\Omega_l = 0$. The main problem of the SNR metric is explainable with Figure 9a,b. Both signals look qualitatively the same, but in one case the introduced prototype MCG-signal (cf. Figure 7b) is superimposed with white noise $n_w(t)$ and in the other case with high-pass (HP) filtered noise $n_{hp}(t)$. A frequency-independent performance metric like SNR does not sufficiently consider the ability of a filter to improve the signal quality by separating the desired and undesired signal components.

For the electromagnetic field of the human heart, it is known that the signal contains no significant power above frequencies of 100 Hz (cf. Figure 9c,d) [36,37]. Therefore, applying a suitable band limitation by a low-pass filter (FIR filter using the Remez exchange algorithm [38] with $N = 516$; bands $= [0, 100, 110, 1000]$ Hz; normalized gain $= [1, 1, 1 \times 10^{-4}, 1 \times 10^{-4}]$.), reveals that the high frequency noise can be easily suppressed, while the white noise can only be partially suppressed. In this particular example this results in two different superimposed signals, which had the same SNR (0 dB) at the beginning, but look very different after filtering (cf. Figure 9c,d). Calculating the SNR after applying the low-pass filter yields an SNR of 23 dB in the case of white noise and 121 dB in the case of high frequency noise.

Furthermore, the applied sampling frequency also influences the SNR result, because the entire frequency interval between 0 and $f_s/2$ is considered by default (cf. Equation (36)). The desired signal only has relevant components within a specific bandwidth, that are necessary to preserve the signal characteristics for diagnostic proposes. Increasing the sampling frequency will worsen the SNR, while in practice, a filter can be applied to limit the signal to the appropriate bandwidth. For a consistent system evaluation, the influence of the signal bandwidth and the spectral characteristics must be considered. A figure of merit used to describe potential signal quality after processing needs to either explicitly consider post-processing steps (i.e., applying the same band limitation to signal and noise) or take the frequency dependence of the PSDs into account. Since the required processing steps depend highly on the system output and the specific biomedical application, a metric that focuses on the individual power spectral densities and their predefined frequency limits is preferable.

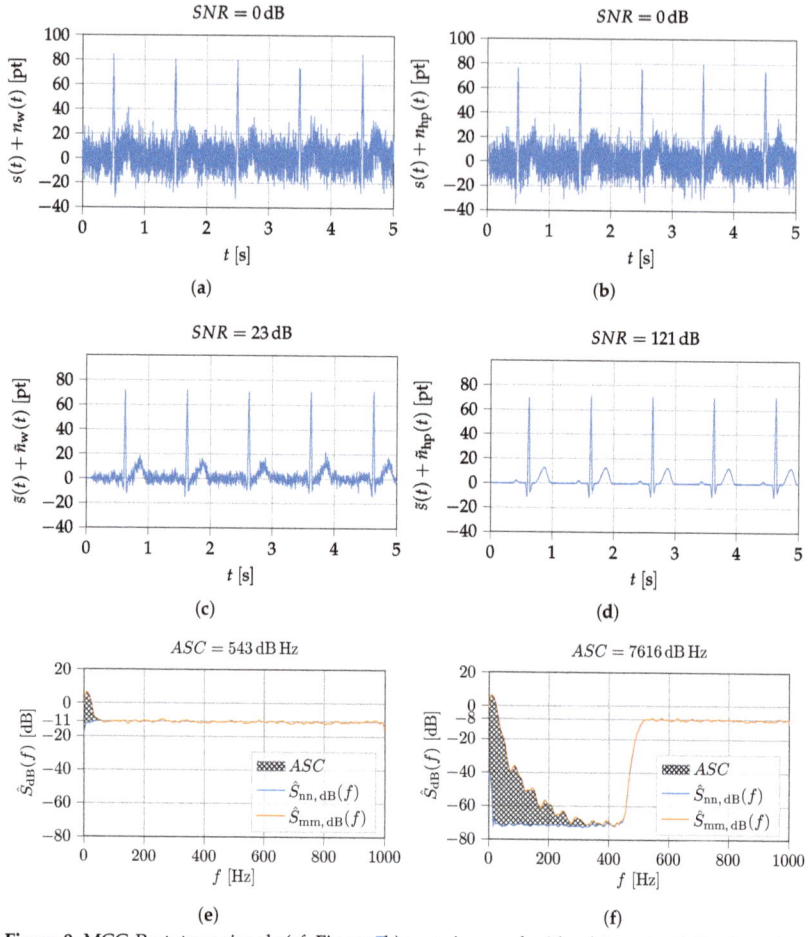

Figure 9. MCG-Prototype signals (cf. Figure 7b) superimposed with white noise (left column) and high-pass filtered noise (right column). The second row shows the low-pass filtered sum of the signal and noise, while the third row shows application specific capacity and the power spectral densities of the signal, noise and weighted noise (cf. Equation (39)). (**a**) MCG Signal plus white noise—time signal. (**b**) MCG Signal plus HP noise—time signal. (**c**) MCG Signal plus white noise—time signal, filtered. (**d**) MCG Signal plus HP noise—time signal, filtered. (**e**) MCG Signal and white noise—PSD. (**f**) MCG Signal and HP noise—PSD.

3.3. Application Specific Capacity

To take the frequency-dependent power into account, a suitable type of operation has to be applied on the PSDs before integration (cf. Equation (36)). The strength at which we consider the power of the noise at a certain frequency should be dependent on the power of the signal $\hat{S}_{ss}(\Omega_\mu)$ at that particular normalized frequency Ω_μ. If $\hat{S}_{ss}(\Omega_\mu)$ is low at a certain frequency, that frequency can be filtered out of the measurement without distorting the signal—resulting in a better signal quality. Therefore, high noise power at frequencies where the signal power is low should not negatively influence the quantity. High noise power at frequencies where the signal power is also significant, on the other hand, should reduce the quantity. The spectral contribution at those frequencies can not be removed from the measurement without disturbing the desired signal. This required constraint can be achieved by integrating over the logarithm of the ratio between the power spectral densities of the measured and the desired signal. Doing so results in an equation that is identical (besides a different basis of the logarithm and additional scaling) to that of the channel capacity C [39], which is why we introduce the term *Application Specific Capacity* and the symbol ASC for this quantity:

$$\begin{aligned} ASC &= \int_0^\infty 10 \cdot \log_{10}\left(\frac{\hat{S}_{ss}(f) + \hat{S}_{nn}(f)}{\hat{S}_{nn}(f)}\right) df \cdot \text{dB} \\ &\approx \frac{1}{\Omega_u - \Omega_l} \sum_{\Omega_\mu = \Omega_l}^{\Omega_u} 10 \cdot \log_{10}\left(\frac{\hat{S}_{ss}(\Omega_\mu) + \hat{S}_{nn}(\Omega_\mu)}{\hat{S}_{nn}(\Omega_\mu)}\right) \text{dB} \\ &\approx \frac{1}{\Omega_u - \Omega_l} \sum_{\Omega_\mu = \Omega_l}^{\Omega_u} 10 \cdot \log_{10}\left(\frac{\hat{S}_{mm}(\Omega_\mu)}{\hat{S}_{nn}(\Omega_\mu)}\right) \text{dB}, \end{aligned} \quad (39)$$

where $\hat{S}_{mm}(\Omega_\mu)$ is the PSD of the measured signal, Ω_u is the upper and Ω_l the lower normalized frequency limit of the numeric integral with $\Omega_{-3\text{dB}} < \Omega_u \leq \pi$. For the calculation of the ASC, the same considerations to PSD estimation and integration as mentioned in Section 3.2 apply. In the following simulations $\Omega_l = 0$ and $\Omega_u = \pi$ is used. To understand how this equation satisfies the abovementioned constraints, consider the following: In regions where $\hat{S}_{nn}(\Omega_\mu)$ is greater than $\hat{S}_{ss}(\Omega_\mu)$, the ratio between $\hat{S}_{mm}(\Omega_\mu)$ and $\hat{S}_{nn}(\Omega_\mu)$ will be close to one. Consequently, the logarithm of the ratio will be close to zero. These regions, therefore, do not contribute significantly to the ASC. If the signal power $\hat{S}_{ss}(\Omega_\mu)$ is high while the noise power $\hat{S}_{nn}(\Omega_\mu)$ is low, the dB power difference will be big, resulting in a significant contribution to the ASC.

Considering the ASC beyond the necessary bandwidth (determined by the bandwidth of the desired signal) will not noteably affect the ASC. This is a desired behavior since a filter can always be applied to the measured signal afterwards to reduce its bandwidth. In practical terms, this means that considering the PSD of the noise over a wider range of frequencies will not significantly change the ASC. Compared to the SNR this eliminates one potential cause for inconsistencies between different measurements.

Taking a look at the ASC values for the previous example (cf. Figure 9), it can be seen that the ASC exhibits the desired behavior. For the case of white noise the ASC equals 543 dB Hz and for the case of high-frequency noise, it is 7616 dB Hz. The SNR of the input signals is 0 dB in both cases. Consequently, after processing the signal superimposed with the high-frequency noise, it could have a better quality than the signal superimposed with the white noise (provided that the applied processing is sensible). This is in accordance with the results of the previous section (cf. Figure 9c,d).

4. Exemplary Evaluation of Magnetoelectric Sensor Systems

In this section, two different ME sensor systems will be assessed by applying the functional characteristics proposed in Sections 2 and 3. Both sensor concepts are investigated at Kiel University. The *exchange bias magnetoelectric sensor* is used to demonstrate a typical resonant ME sensor system. This sensor type is especially applicable for detecting narrowband signals, for example, coil signals utilized in novel ME localization [12] and ME movement detection applications [13]. In contrast to this, the *electrically modulated ME sensor* is potentially better suited for broadband biomedical signals due to a much higher bandwidth. Both sensors are shown in Figure 10 and their concepts will be separately introduced and evaluated in the following subsections. In addition, the SNR and the ASC, presented in Sections 3.2 and 3.3, are used as a figure of merit concerning a possible sensor usage for MCG. Therefore, the definitions are applied by using the noise measurements and the generated prototype MCG signal (cf. Figure 7b) with its spectral distribution. Both sensor systems will be compared and finally discussed in a results overview.

The measurements for evaluation have been performed in a magnetically, electrically, and acoustically shielded environment comprising a multilayer mu-metal cylinder (Model

ZG1 from Aaronia), further details are given in [11,40]. The noise measurements have been accomplished with the Dynamic Signal Analyzer SR785 from Stanford Research Systems [41]. All other measurements, where a magnetic signal is required, have been performed with a long solenoid driven by the low noise current source Keithley 6221 [30]. The coil was used to generate a mono-frequent signal with a magnetic field amplitude of $b_\text{in}^\text{rms} = 1\,\mu\text{T}$ (desired signal). The amplitude and phase responses, as well as the linearity curve of the sensors, have been measured with the lock-in amplifier SR830 from Stanford Research Systems [42]. For determining the linearity curve, the amplitude of the magnetic flux density within the solenoid was varied in the range from 0.1 pT to 100 µT. Consequently, the coil excitation signal has been used as the reference signal for the lock-in amplifier and the acquisition time was set to 100 ms.

In addition, it is essential to ensure a dedicated magnetic state of the ME sensor before the sensor system evaluation starts, especially considering hysteresis effects of the magnetostrictive layer. Magnetic saturation of the magnetostrictive layer can be achieved using a high constant field within the coil. Therefore, a DC current source (BOP 20-10ML from KEPCO) is used. The direction of magnetic saturation is sensor dependent and was be chosen such that the best sensor performance in terms of sensitivity and noise is reached. Finally, this dedicated magnetic state served as the starting point for the ME sensor system evaluation.

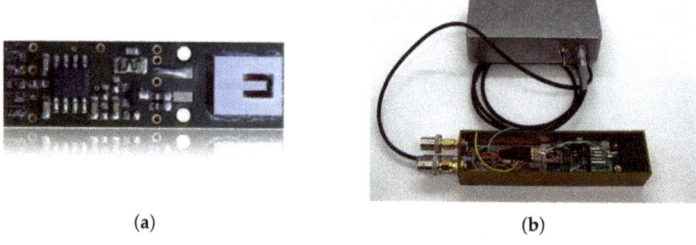

(a) (b)

Figure 10. Sensors systems used in this study: In (**a**) an exchange bias magnetoelectric sensor is shown with integrated readout electronics. In (**b**) an electrically modulated ME sensor is presented with integrated preamplifier and external shielded battery supply (gray box; ±9 V). (**a**) Exchange bias magnetoelectric sensor (cantilever) with integrated readout. (**b**) Electrically modulated ME sensor with integrated preamplifier and external battery.

4.1. Exchange Bias Magnetoelectric Sensor

ME thin film composite sensors are composed of mechanically coupled magnetostrictive and piezoelectric layers and utilize the mechanical resonance of a cantilever structure [11]. Hence, a resonator behavior (bandpass characteristic) is present when operating the sensor in direct detection, that is, without any modulation technique. Besides reading out the sensor directly in its mechanical resonance, various readout schemes can be applied to the sensor for measuring low-frequency signals. Recently investigated readout schemes for ME sensors are e.g., the ΔE-effect [43–45] or magnetic frequency conversion [46,47]. In this contribution, an exchange bias ME sensor in a so-called direct detection mode has been used as shown in Figure 10a. The cantilever sensing element has a size of 3 mm × 1 mm × 0.1 mm. The sensor is connected to a low-noise JFET (junction-gate-field-effect transistor) charge amplifier [48]. Due to the exchange biasing of the sensor, there is no need for an additional coil generating a magnetic bias field [49]. Further details about the sensor and the fabrication process can be found in [50]. The sensor was operated in direct detection and the output signal of the sensor system, including the charge amplifier, was taken into account. For comparability with the other ME sensor type (shielded printed-circuit-board (PCB) housing, cf. Figure 10b), this sensor is also operated with additional shielding. Therefore, the sensor has been provided with an extra electromagnetic compatible (EMC) braided cable and has been connected to the measurement ground. A sensor operation from negative saturation showed the best sensor performance at the resonance frequency in terms of sensitivity and noise. Therefore, the ME sensor was saturated before the ME system evaluation. Three representative measurements have been performed for the evaluation of this particular sensor system. The *amplitude and phase response*, the *noise spectrum*, and the *Input–Output–Amplitude–Relation* are shown in Figure 11. For determining the noise amplitude spectral density a frequency range of 800 Hz was observed around f_{res} with an FFT size (single-sided) of 800 points, resulting in a frequency resolution of 1 Hz and a total acquisition time of 1 s (60 RMS averages). In addition, especially for determining SNR and ASC, a noise amplitude spectral density from 4 to 800 Hz (the same FFT size) has been acquired with an identical acquisition time of 1 s (60 RMS averages) to cover the required MCG-Bandwidth (cf. Figure 7d). The Input–Output–Amplitude–Relation measurements have been performed in resonance of the sensor at $f_r = 7684$ Hz. As stated with the help of Figure 3 for a reproducible LOD determination an exact specification of the measurement routine is required. Here, a dedicated RC-lowpass filter with a slope of 24 dB/oct and a time constant of 100 ms have been chosen at the lock-in amplifier [42].

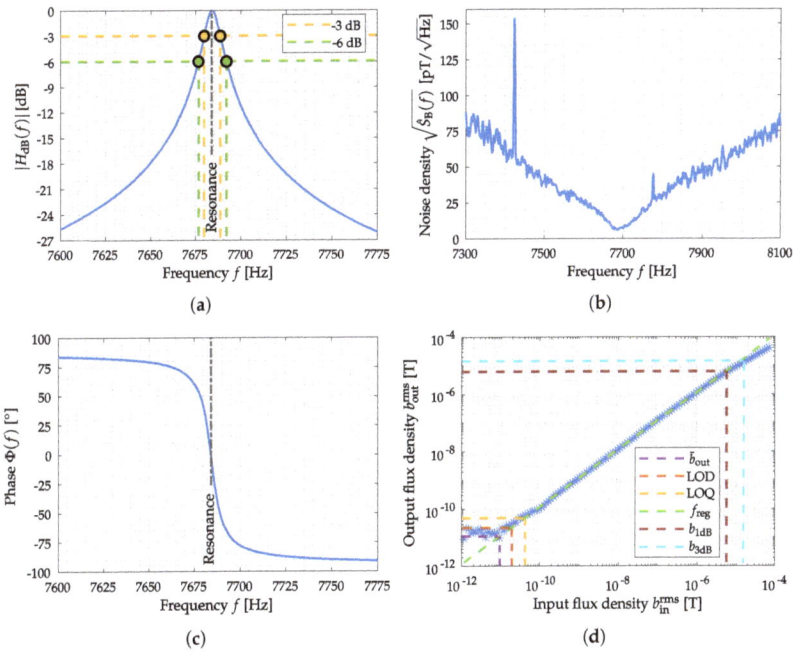

Figure 11. Measurements for the evaluation of the exchange bias magnetoelectric sensor system. In (**a**) the amplitude response and in (**c**) the phase response of the sensor system near the mechanical resonance are depicted. The noise measurement equalized with amplitude response is shown in (**b**). The Input–Output–Amplitude–Relation of the sensor, with an external magnetic field at $f = f_{res}$, is depicted in (**d**).

The expected resonator behavior of the ME sensor is visible in the amplitude spectrum in Figure 11a. The noise spectrum of the sensor is dominated by thermal-mechanical noise [51] as shown in Figure 11b. Looking at the Input–Output–Amplitude–Relation in Figure 11d, the quantities $\bar{b}_{out} = 11$ pT, $LOD = 22$ pT and $LOQ = 42$ pT can be determined. Furthermore, the linear region can be described by using a linear function approximation f_{reg}. Based on the magnetic hysteresis, nonlinearities occur bevor reaching the operating voltage (± 12 V), so it is helpful to determine the compression points from the Input–Output–

Amplitude–Relation. The *1-dB-compression-point* (b_{1dB}) is at $b_{in}^{rms} \approx 5.8\,\mu T$ and the *3-dB-compression-point* (b_{3dB}) is at $b_{in}^{rms} \approx 17.6\,\mu T$. Based on the noise spectral measurement covering the bandwidth from 4 to 800 Hz and the applied prototype MCG signal, an SNR of -90 dB and an ASC of 9.8×10^{-7} dB Hz could be determined quantitatively.

4.2. Electrically Modulated ME Sensor

Resonant magnetoelectric (ME) sensors combined with modulation techniques can be used to achieve high bandwidth at low frequencies. Consequently, it is favorable to use electric instead of magnetic modulation. Magnetic modulation has demonstrated its general high potential but suffers from high power consumption and possible crosstalk between sensors. For alternative ME sensor concept realization, the piezoelectric phase of thin film magnetoelectric composites is actively excited by an alternating voltage, thus exploiting the converse ME effect [10], remedying shortcomings of the direct ME effect. High frequency mechanical resonances between 500 kHz and 540 kHz are typically used for sensor operation. These resonances show high mechanical quality factors ($Q \approx 1000$) [10], which results in better SNRs at those frequencies. The resulting mechanical oscillation, being rigidly coupled into the magnetostrictive material phase leads to a voltage induced in a pickup coil surrounding the sensor composite. This converse ME voltage response with respect to small external fields shows high sensitivities in the order of kV/T. No external magnetic driving field is required, as is the case for the exchange bias ME sensor using magnetic frequency conversion techniques.

The ME sensor system (shielded housing) is shown in Figure 10b with the output preamplified by a low-noise operational amplifier (LT1128) in unity gain configuration to decouple the resonant circuit from the readout. This operational amplifier is connected to the additional shielding box that contains a ±9 V battery powered voltage supply and has been connected to the measurement ground. Further details about this particular ME sensor type and the fabrication process can be found in [9].

The electrically modulated ME sensor (cf. Figure 10b) is piezoelectrically excited at 514.249 kHz (2nd mechanical U-mode of oscillation). Therefore, a sinusoidal voltage with an amplitude of 500 mV has been used. Sensor excitation and the required synchronous demodulation of the coil signal are performed using a high frequency lock-in amplifier (HF2LI from Zurich Instruments). A 4th-order RC-lowpass filter (24 dB/oct or 80 dB/dec) with a cutoff frequency of 30 kHz has been chosen as the demodulation filter within the lock-in amplifier. The demodulated analog coil signal (lock-in amplifier output) is used with an additional output gain of one hundred for signal acquisition to optimally use the internal A/D converter dynamics. The results have been corrected for the applied amplification factor to show the correct (unamplified) values. The amplitude and phase response of the sensor have been measured between 1 Hz and 100 Hz with a resolution of 1 Hz and between 100 Hz and 30 kHz with a resolution of 100 Hz. The noise amplitude spectral density with a frequency range from 16 Hz to 12.656 kHz was determined with a FFT size of 800 points, resulting in a frequency resolution of 16 Hz and a total acquisition time of 1 min (960 RMS averages). For determining the SNR and ASC the noise amplitude spectral densities over the important frequency range from 4 to 800 Hz (same FFT size) has been acquired with an identical acquisition time of 1 min (60 RMS averages). The acquisition of the Input–Output–Amplitude–Relation has been performed changing the flux density of the external magnetic field at a frequency of 10 Hz. The measurements performed for evaluation are the three measurements shown in Figure 11, the *amplitude and phase response*, the *noise spectrum*, and the *Input–Output–Amplitude–Relation*.

By analyzing the amplitude response in Figure 12a, the lowpass-characteristic of the sensor with a larger supported signal bandwidth is visible. The phase response in Figure 12b is linear (even if being depicted with a logarithm scaled *x*-axis). Figure 11b shows the noise amplitude density spectrum of the sensor, where the excessive magnetization reorientation initiated by stress anisotropy and the emergence of eddy currents are the main known noise sources [10]. By using the Input–Output–Amplitude–Relation ($f = 10$ Hz) the defined metrics can be determined: $\bar{b}_{out}$ is 55 pT, *LOD* is 102 pT and *LOQ* is 210 pT. Saturation from external magnetic flux density after $b_{in}^{rms} \approx 26.70$ µT generate a decay in the output flux density. Therefore, the *1-dB-compression-point* (b_{1dB}) and *3-dB-compression-point* (b_{3dB}) could be determined to $b_{in}^{rms} \approx 18$ µT and $b_{in}^{rms} \approx 23$ µT. Based on the noise spectral measurement covering the bandwidth from 4 to 800 Hz and the applied prototype MCG signal, an SNR of -11 dB and an ASC of 23 dB Hz could be determined.

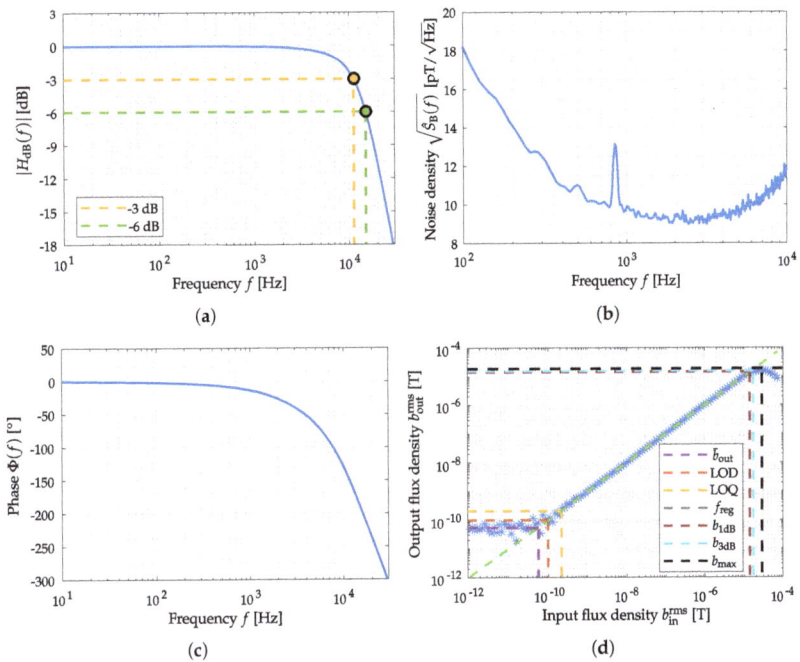

Figure 12. Measurements for the evaluation of the electrically modulated ME sensor. In (**a**) the amplitude response and in (**c**) the phase response of the sensor system are depicted. The noise measurement equalized with amplitude response is shown in (**b**). The Input–Output–Amplitude–Relation of the sensor, with an external magnetic field at $f = 10$ Hz, is depicted in (**d**).

4.3. Overview of the ME Evaluation Results

A comparison of the sensor systems evaluated within this contribution is provided in the following. Table 2 shows an overview of the essential sensor system metrics and also the application-specific values concerning an ME sensor system usage in cardiovascular medicine. By considering *SNR* and *ASC*, it can be determined if the sensor system can reliably detect a magnetic heart signal. As described in Section 4.3, the parameters used for estimating the PSDs and integrating them are of significant importance for the comparability of the results. Since the PSDs of the sensors are determined by measurement and those of the signals by simulation, the parameters of the simulation must be adapted to those of the measurement. The estimation of the PSD of the prototype signal is therefore done by Welch's method using a signal length of 5 s, a Flattop window of 1600 samples width, an overlap of 50 percent, and an FFT length of 1600. For the numerical integration Simpson's rule with the limits $f_l = 4$ Hz and $f_u = f_s/2 = 800$ Hz is used. The upper cutoff frequency does not have a significant effect on the *ASC* here, since the prototype signal has no significant components above this frequency. The lower cutoff frequency of 4 Hz, limited by the measurement setup, on the other hand, affects both *SNR* and *ASC*. Since this cutoff frequency was used identically for the measurements of both sensors, the results remain comparable here. In a comparison with other measurements or simulations; however, care would have to be taken to maintain the same integration limits.

Table 2. Comparison of magnetic field sensor systems evaluated within this contribution.

Parameters	Exchange Bias ME Sensor	Electrically Modulated ME Sensor
Amplitude Response		
f_{res}	7684 Hz	
f_{-3dB}	$f_{-3dB,1}$ = 7680 Hz (low) $f_{-3dB,2}$ = 7689 Hz (high)	11.3 kHz
f_{-6dB}	$f_{-6dB,1}$ = 7677 Hz (low) $f_{-6dB,2}$ = 7692 Hz (high)	15 kHz
Q	854	
$|Slope_{-3dB/-6dB}|$	0.94 dB/Hz (low) 0.92 dB/Hz (high)	0.805 dB/kHz
B_{3dB}	9 Hz (bandpass)	11.3 kHz (lowpass)
B_{6dB}	15 Hz (bandpass)	15 kHz (lowpass)
Sensitivity		
ϵ_{sys}	63 kV/T at f_{res}	5.76 kV/T at 10 Hz
Noise		
$\sqrt{\hat{S}_B(f)}$	6 pT/$\sqrt{Hz}$ at f_{res}	66 pT/$\sqrt{Hz}$ at 10 Hz
B_n^{rms}	20 pT ($f_{-3dB,low}$ to $f_{-3dB,high}$)	11.7 nT (1 Hz to f_{-3dB})
Input-Output-Relation		
$\bar{b}_{out}$	11 pT	55 pT
LOD	22 pT	102 pT
LOQ	42 pT	210 pT
b_{1dB}	6 µT	18 µT
b_{3dB}	18 µT	23 µT
b_{max}		27 µT
DR	103 dB	98 dB
Application (MCG) Specific Quantities		
SNR	−90 dB	−11 dB
ASC	9.8 ×10^{-7} dB Hz	23 dB Hz

Beginning with the exchange bias ME sensor, it is evident that the frequency range where the sensor is operating is too high and thus not compatible with the bandwidth of a magnetically detected heart signal. A frequency range from 0.01 to 100 Hz [31] is typically required for a signal-true MCG recording. This bandwidth specification could also be confirmed with an MCG performed with SQUIDs. In Figure 7, the time signal (a) and the resulting power spectral density (c) of a single MCG heartbeat recorded by a SQUID have been presented. Accordingly, this ME sensor system is not suitable for measuring heart signals when operating in direct detection. This can also be quantitatively confirmed with the given SNR and ASC. Nevertheless, this sensor type enables detecting narrowband signals with a center frequency of 7684 Hz, for example, modulated coil signals applied for ME localization and movement detection. Furthermore, using a modulation technique (e.g., ΔE-effect, magnetic frequency conversion), the sensors can measure low-frequency magnetic fields. In [33] magnetoelectric sensors have been evaluated concerning cardiologic applications. For example, the R-wave could be detected, averaging the time signal over 743 periods by using magnetic frequency conversion for ME sensor readout. Different signal enhancement stages for improving the quality of an MCG using uncooled magnetometers are additionally applicable as a post-processing step [52], but they have not to be considered primary for sensor evaluation.

Evaluating the electrically modulated ME sensor, it is evident that the bandwidth of the sensor is quite large in direct comparison to the ME sensor operating in direct detection. Therefore, more noise is picked up by the sensor, which results in an RMS noise amplitude of approximately 12 nT within bandwidth (1 Hz to f_{-3dB}). For the Input–Output–Relation, in general, only a mono-frequency consideration is performed at 10 Hz, and the extracted metrics were in good agreement with already published ones. When considering the available bandwidth and the LOD, the electrically modulated ME sensor is close to measuring a magnetic heart signal under ideal conditions [9]. This can be quantitatively confirmed by considering SNR (-11 dB) and ASC (23 dB Hz). The dominant noise source of this kind of sensor is the intense magnetization activity, practically limiting the LOD. Using sophisticated magnetic layer systems such as exchange bias have already shown to effectively lower the magnetically dominated noise [53], while maintaining the sensor performance. Finally, this also enables the possibility to bring magnetoelectric thin film sensors towards a cardiovascular application.

5. Discussion

Based on the proposed application-oriented comparison, two different ME sensor systems have been evaluated and compared here. First, the ME sensor systems were rated with common metrics, and additionally, a signal evaluation was performed for a cardiovascular application. While targeted at ME sensors here, the same methodologies are applicable to any magnetometer. By exemplary, applying the introduced evaluation methods, it can be concluded that especially the *electrically modulated ME sensor system* has the potential to measure a magnetic heart signal, at least by applying advanced averaging techniques [52]. The other presented ME sensor type is better suited for applications where only a small frequency range is necessary, such as active magnetoelectric motion sensing [13] or ME sensor localization [12]. Nevertheless, cardiovascular medicine requires an unaveraged MCG morphology (QRS complex, P-wave, and T-wave) in the time domain, especially as cardiac arrhythmias manifest from beat to beat [7,54]. For this reason, ME sensor systems (sensor and readout electronics) are currently being further developed interdisciplinarily to ensure their use in cardiology [9]. In Figure 13, an application-specific noise requirement for MCG is presented.

The novel introduced *application-specific capacity* is a new figure of merit that utilizes details about the dominant frequencies of the desired signal and sensor noise. One weakness of the ASC is for sure the unfamiliar unit of dB Hz. The SNR definition is quite common and one can easily imagine how a signal with an SNR of 0 dB will roughly look like, because signal power and noise power will be in the same order of magnitude. An equivalent concern cannot be said about the ASC in the current state of research, because there is no intuition value mapping of how high the ASC needs to be for a sensor system to be suitable for a specific application. In addition, no data are yet available regarding the expressiveness of the ASC. There is an awareness of this issue and is planned to be addressed it in future research, which means a definition of application-depended prototype signals, noise quantification for various sensor types and an ASC mapping to qualitative measures. Finally a detailed signal evaluation should be performed with medical specialists of the specific field in an iterative fashion. In further investigations, a clear presentation of the individual system and application-related metrics is desirable, which can be achieved, for example, with a special type of pie chart. This diagram is called *target performance profile* and a first prototype design is exemplary shown in Figure 14. Nevertheless, further studies are necessary for an adequate metric normalization and comparability of the different magnetometers.

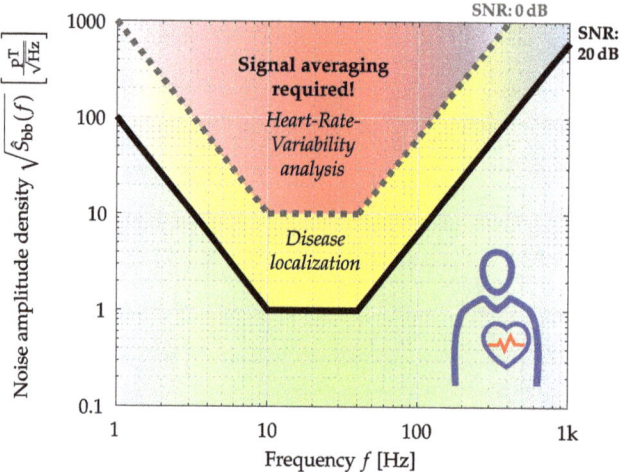

Figure 13. Application-specific noise requirements for MCG (Heart-Rate-Variability analyses by detection of the R-Peak; Disease localization by solving the inverse problem) specified by the amplitude density. Sensor positioned directly over the chest [55].

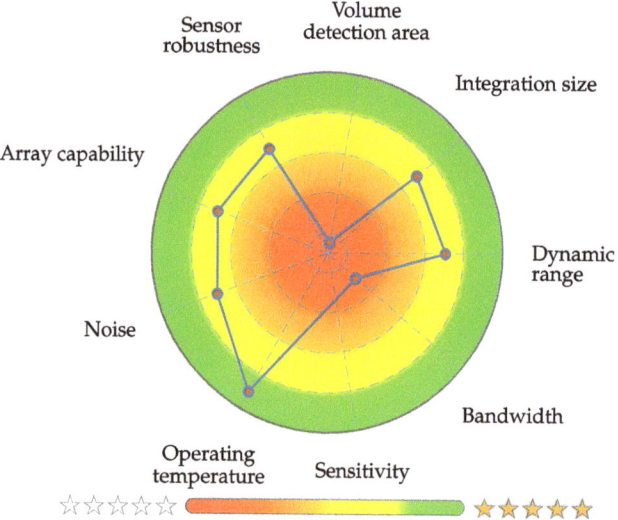

Figure 14. Exemplary prototype design for a clear presentation of application-related metrics [55].

6. Conclusions

In this work, essential metrics for sensor system assessment were initially defined and complemented by a novel application-specific signal evaluation. Currently, such an ME sensor system evaluation approach is not available, and the literature and standardization so far can be described as inadequate. Table 2 presents exemplary essential ME system evaluation metrics and values for a quantitative signal evaluation concerning a cardiological application. Certainly, practical considerations, for example, readout method, integration size, volume of the detection area, array capability and robustness, can also be included in a further step. These quantities are not considered in this contribution since it concerns purely qualitative quantities, where a fundamental definition of a uniform ordinal scale is indispensable. This paper is meant as a starting point for the assessment of biomagnetic sensor systems by supporting the accurate differentiation and use of sensor system metrics like *Limit-of-Detection, noise* and *bandwidth*. Additionally, it shows the possibility for end-users, for example, medical doctors, to select the appropriate sensor system. Beyond that, this novel approach could help to quantify and subsequently optimize the system performance of already available magnetometers (e.g., alkali OPMs [56] or Helium OPMs [57]) as well as new ones (e.g., ultra-sensitive xMR-Sensors [58]) for a particular biomagnetic application. Thus, it will enable the benchmarking of individual systems for both single and multi-sensor approaches. In conclusion, our presented system evaluation can lead to an application-oriented system comparison. In addition to its potential support in sensor research and development, the quantitative metrics might prove helpful for performance optimization. On this basis, the usability of available and upcoming sensor systems can be easily verified for new applications.

Author Contributions: Conceptualization, E.E. (Eric Elzenheimer) and G.S.; methodology, E.E. (Eric Elzenheimer) and G.S.; software, E.E. (Eric Elzenheimer), C.B. and E.E. (Erik Engelhardt); validation, C.B., J.H. and P.H.; formal analysis, E.E. (Eric Elzenheimer), C.B., J.H., E.E. (Erik Engelhardt); investigation, E.E. (Eric Elzenheimer), C.B. and E.E. (Erik Engelhardt); resources, E.E. (Eric Elzenheimer), C.B., E.E. (Erik Engelhardt), J.H., J.A. and P.H.; data acquisition, E.E. (Eric Elzenheimer), C.B., J.A. and P.H.; data curation, E.E. (Eric Elzenheimer); C.B. and J.H.; writing—original draft preparation, E.E. (Eric Elzenheimer), C.B. and E.E. (Erik Engelhardt); writing—review and editing, E.E. (Eric Elzenheimer), C.B., E.E. (Erik Engelhardt), J.H., J.A., P.H., G.S., M.H., A.B. and E.Q.; visualization, E.E. (Eric Elzenheimer) and E.E. (Erik Engelhardt); supervision, G.S.; project administration, M.H. and G.S.; funding acquisition, E.Q., M.H. and G.S. All authors have read and agreed to the published version of the manuscript.

Funding: This work was supported by the German Research Foundation (Deutsche Forschungsgemeinschaft, DFG) through the project Z2 of the Collaborative Research Centre CRC 1261 "Magnetoelectric Sensors: From Composite Materials to Biomagnetic Diagnostics".

Institutional Review Board Statement: Not applicable.

Informed Consent Statement: Not applicable.

Data Availability Statement: Not applicable.

Acknowledgments: The SQUID-MCG recording was made feasible by DFG funding of the core facility "Metrology of Ultra-Low Magnetic Fields" at Physikalisch-Technische Bundesanstalt. Furthermore, the authors would like to thank Phillip Durdaut for providing the low-noise amplifiers for the exchange bias ME sensor and Christine Kirchhof for creating the sensor.

Conflicts of Interest: The authors declare no conflict of interest. The founders had no role in the design of the study; in the collection, analyses, or interpretation of data; in the writing of the manuscript, or in the decision to publish the results.

Abbreviations

The following abbreviations are used in this manuscript:

A/D	Analog/Digital
ACF	Autocorrelation function
AS	Amplitude spectrum
ASC	Application specific capacity
ASD	Amplitude spectral density
DFT	Discrete Fourier transform
DR	Dynamic range
ECG	Electrocardiography
EEG	Electroencephalography

EMC	Electromagnetic compatibility
ENBW	Equivalent noise bandwidth
HP	Highpass
JFET	Junction-gate-field-effect transistor
LOD	Limit-of-Detection
LOQ	Limit-of-Quantification
LTI	Linear time-invariant
ME	Magnetoelectric
MCG	Magnetocardiography
OPM	Optically Pumped Magnetometer
PCB	Printed circuit board
PS	Power spectrum
PSD	Power spectral density
PTB	Physikalisch Technische Bundesanstalt
RMS	Root mean square
SNNR	Signal-plus-noise to noise ratio
SNR	Signal-to-noise ratio
SQUID	Super conducting quantum interference device

Appendix A. MCG Prototype Signal

Table A1. Support points for MCG prototype signal.

t [s]	0	0.25	0.3	0.35	0.44	0.47	0.5	0.52	0.56	0.6	0.75	0.85	1
$s(t)$ [pT]	0	0	2.1	0	0	-10.5	70	-7	0	0	12.6	0	0
$\frac{ds(t)}{dt}$ [$\frac{\text{pT}}{\text{s}}$]	0	0	0	0	0	0	0	0	0	0	0	0	0

References

1. Sanchez-Reyes, L.M.; Rodriguez-Resendiz, J.; Avecilla-Ramirez, G.N.; Garcia-Gomar, M.L.; Robles-Ocampo, J.B. Impact of EEG Parameters Detecting Dementia Diseases: A Systematic Review. *IEEE Access* **2021**, *9*, 78060–78074. https://doi.org/10.1109/ACCESS.2021.3083519.
2. Williamson, S.J. *Advances in Biomagnetism: Proceedings of the Seventh International Conference on Biomagnetism, New York, NY, USA, 13–18 August 1989*; Plenum Press: New York, NY, USA, 1989.
3. Boto, E.; Hill, R.M.; Rea, M.; Holmes, N.; Seedat, Z.A.; Leggett, J.; Shah, V.; Osborne, J.; Bowtell, R.; Brookes, M.J. Measuring functional connectivity with wearable MEG. *NeuroImage* **2021**, *230*, 117815. https://doi.org/10.1016/j.neuroimage.2021.117815.
4. Elzenheimer, E.; Laufs, H.; Schulte-Mattler, W.; Schmidt, G. Magnetic Measurement of Electrically Evoked Muscle Responses with Optically Pumped Magnetometers. *IEEE Trans. Neural Syst. Rehabil. Eng.* **2020**, *28*, 756–765. https://doi.org/10.1109/TNSRE.2020.2968148.
5. Shirai, Y.; Hirao, K.; Shibuya, T.; Okawa, S.; Hasegawa, Y.; Adachi, Y.; Sekihara, K.; Kawabata, S. Magnetocardiography Using a Magnetoresistive Sensor Array. *Int. Heart J.* **2019**, *60*, 50–54. https://doi.org/10.1536/ihj.18-002.
6. Janosek, M.; Butta, M.; Dressler, M.; Saunderson, E.; Novotny, D.; Fourie, C. 1-pT Noise Fluxgate Magnetometer for Geomagnetic Measurements and Unshielded Magnetocardiography. *IEEE Trans. Instrum. Meas.* **2020**, *69*, 2552–2560. https://doi.org/10.1109/TIM.2019.2949205.
7. Macfarlane, P.W.; van Oosterom, A.; Pahlm, O.; Kligfield, P.; Janse, M.; Camm, A.J. *Comprehensive Electrocardiology*, 2nd ed.; Springer: London, UK, 2010. https://doi.org/10.1007/978-1-84882-046-3.
8. Koch, H. Recent advances in magnetocardiography. *J. Electrocardiol.* **2004**, *37*, 117–122.
9. Hayes, P. Converse Magnetoelectric Resonators for Biomagnetic Field Sensing. Ph.D. Thesis, Kiel University, Kiel, Germany, 2020.
10. Hayes, P.; Jovičević Klug, M.; Toxværd, S.; Durdaut, P.; Schell, V.; Teplyuk, A.; Burdin, D.; Winkler, A.; Weser, R.; Fetisov, Y.; et al. Converse Magnetoelectric Composite Resonator for Sensing Small Magnetic Fields. *Sci. Rep.* **2019**, *9*, 16355. https://doi.org/10.1038/s41598-019-52657-w.
11. Jahns, R.; Knöchel, R.; Greve, H.; Woltermann, E.; Lage, E.; Quandt, E. Magnetoelectric sensors for biomagnetic measurements. In Proceedings of the 2011 IEEE International Symposium on Medical Measurements and Applications, Bari, Italy, 30–31 May 2011; pp. 107–110. https://doi.org/10.1109/MeMeA.2011.5966676.
12. Bald, C.; Schmidt, G. Processing Chain for Localization of Magnetoelectric Sensors in Real Time. *Sensors* **2021**, *21*, 5675. https://doi.org/10.3390/s21165675.
13. Hoffmann, J.; Elzenheimer, E.; Bald, C.; Hansen, C.; Maetzler, W.; Schmidt, G. Active Magnetoelectric Motion Sensing: Examining Performance Metrics with an Experimental Setup. *Sensors* **2021**, *21*, 8000. https://doi.org/10.3390/s21238000.

14. Sander, T.; Jodko-Władzińska, A.; Hartwig, S.; Brühl, R.; Middelmann, T. Optically pumped magnetometers enable a new level of biomagnetic measurements. *Adv. Opt. Technol.* **2020**, *9*, 247–251. https://doi.org/10.1515/aot-2020-0027.
15. Ripka, P. *Magnetic Sensors and Magnetometers*; Artech House: Boston, MA, USA, 2001.
16. Hayes, P.; Schell, V.; Salzer, S.; Burdin, D.; Yarar, E.; Piorra, A.; Knöchel, R.; Fetisov, Y.K.; Quandt, E. Electrically modulated magnetoelectric AlN/FeCoSiB film composites for DC magnetic field sensing. *J. Phys. D Appl. Phys.* **2018**, *51*, 354002. https://doi.org/10.1088/1361-6463/aad456.
17. Ohm, J.R.; Lüke, H.D. *Signalübertragung: Grundlagen der digitalen und analogen Nachrichtenübertragungssysteme*, 11th ed.; Springer-Lehrbuch; Springer: Berlin, Germany, 2010. https://doi.org/10.1007/978-3-642-10200-4.
18. Hänsler, E. *Statistische Signale: Grundlagen und Anwendungen*, 3rd ed.; Springer: Berlin/Heidelberg, Germany, 2001. https://doi.org/10.1007/978-3-642-56674-5.
19. Tumański, S. *Handbook of Magnetic Measurements*; Sensors Series; CRC Press: Boca Raton, FL, USA, 2011.
20. Brinkmann, B. *Internationales Wörterbuch der Metrologie: Grundlegende und Allgemeine Begriffe und Zugeordnete Benennungen (VIM) Deutsch-Englische Fassung ISO/IEC-Leitfaden 99:2007*, 4th ed.; Beuth Wissen; Beuth Verlag GmbH: Berlin, Germany, 2012.
21. DIN 32645:2008-11. Chemische Analytik: Nachweis-, Erfassungs- und Bestimmungsgrenze unter Wiederholbedingungen-Begriffe, Verfahren, Auswertung. Available online: https://www.beuth.de/de/norm/din-32645/110729574 (accessed on 21 January 2021). https://doi.org/10.31030/1465413.
22. Wenclawiak, B.W.; Koch, M.; Hadjicostas, E. *Quality Assurance in Analytical Chemistry*; Springer: Berlin/Heidelberg, Germany, 2010. https://doi.org/10.1007/978-3-642-13609-2.
23. Mitra, S.K. *Digital Signal Processing: A Computer-Based Approach*, 4th ed.; McGraw-Hill: New York, NY, USA, 2011.
24. Proakis, J.G.; Manolakis, D.G. *Digital signal processing*, 4th ed.; Pearson New International Edition ed.; Always Learning; Pearson: Harlow, UK, 2014.
25. Papula, L. *Mathematik für Ingenieure und Naturwissenschaftler: Band 3: Vektoranalysis, Wahrscheinlichkeitsrechnung, Mathematische Statistik, Fehler- und Ausgleichsrechnung*, 7th ed.; Springer Vieweg: Wiesbaden, Germany, 2016. https://doi.org/10.1007/978-3-658-11924-9.
26. Tietze, U.; Schenk, C.; Gamm, E. *Electronic Circuits: Handbook for Design and Application*, 2nd ed.; First Indian Reprint ed.; Springer: New Delhi, India, 2012. https://doi.org/10.1007/978-3-540-78655-9.
27. Norton, H.N. *Handbook of Transducers*; Prentice-Hall: Englewood Cliffs, NJ, USA, 1989.
28. Madisetti, V.K.; Williams, D.B. *The Digital Signal Processing Handbook*; The Electrical Engineering Handbook Series; CRC Press: Boca Raton, FL, USA, 1997.
29. Rohde & Schwarz GmbH & Co. KG. *R&S UPV Audio Analyzer Operating Manual*; Rohde & Schwarz GmbH & Co. KG: München, Germany, 2015.
30. Keithley Instruments Inc. *Model 6220 DC Current Source Model 6221 AC and DC Current Source Users Manual*; Keithley Instruments Inc.: Cleveland, OH, USA, 2008.
31. Sternickel, K.; Braginski, A.I. Biomagnetism using SQUIDs: Status and perspectives. *Supercond. Sci. Technol.* **2006**, *19*, S160–S171. https://doi.org/10.1088/0953-2048/19/3/024.
32. Elzenheimer, E.; Laufs, H.; Sander-Thömmes, T.; Schmidt, G. Magnetoneurograhy of an Electrically Stimulated Arm Nerve. *Curr. Dir. Biomed. Eng.* **2018**, *4*, 363–366. https://doi.org/10.1515/cdbme-2018-0087.
33. Reermann, J.; Durdaut, P.; Salzer, S.; Demming, T.; Piorra, A.; Quandt, E.; Frey, N.; Höft, M.; Schmidt, G. Evaluation of magnetoelectric sensor systems for cardiological applications. *Measurement* **2018**, *116*, 230–238. https://doi.org/10.1016/j.measurement.2017.09.047.
34. Welch, P. The use of fast Fourier transform for the estimation of power spectra: A method based on time averaging over short, modified periodograms. *IEEE Trans. Audio Electroacust.* **1967**, *15*, 70–73. https://doi.org/10.1109/TAU.1967.1161901.
35. Weik, M.H. Signal-plus-noise to noise ratio. In *Computer Science and Communications Dictionary*; Springer: Boston, MA, USA, 2001; p. 1583. https://doi.org/10.1007/1-4020-0613-6_17391.
36. Bald, C.; Elzenheimer, E.; Sander-Thömmes, T.; Schmidt, G. Amplitudenverlauf des Herzmagnetfeldes als Funktion des Abstandes. In Proceedings of the Workshop Biosignal Processing 2018—Innovative Processing of Bioelectric and Biomagnetic Signals, Erfurt, Germany, 21–23 March 2018.
37. Scher, A.M.; Young, A.C. Frequency Analysis of the Electrocardiogram. *Circ. Res.* **1960**, *8*, 344–346.
38. Schlichthärle, D. *Digital Filters: Basics and Design*; Springer eBook Collection; Springer: Berlin/Heidelberg, Germany, 2000. https://doi.org/10.1007/978-3-662-04170-3.
39. Shannon, C. Communication in the Presence of Noise. *Proc. IRE* **1949**, *37*, 10–21. https://doi.org/10.1109/JRPROC.1949.232969.
40. Salzer, S.D. Readout Methods for Magnetoelectric Sensors. Ph.D. Thesis, Kiel University, Kiel, Germany, 2018.
41. Stanford Research Systems. *Operating Manual and Programming Reference, Model SR785 Dynamic Signal Analyzer*; Stanford Research Systems Inc.: Sunnyvale, CA, USA, 2017.
42. Stanford Research Systems. *MODEL SR830 DSP Lock-In Amplifier*; Stanford Research Systems Inc.: Sunnyvale, CA, USA, 2011.
43. Zabel, S.; Reermann, J.; Fichtner, S.; Kirchhof, C.; Quandt, E.; Wagner, B.; Schmidt, G.; Faupel, F. Multimode delta-E effect magnetic field sensors with adapted electrodes. *Appl. Phys. Lett.* **2016**, *108*, 222401. https://doi.org/10.1063/1.4952735.
44. Reermann, J.; Zabel, S.; Kirchhof, C.; Quandt, E.; Faupel, F.; Schmidt, G. Adaptive Readout Schemes for Thin-Film Magnetoelectric Sensors Based on the delta-E Effect. *IEEE Sens. J.* **2016**, *16*, 4891–4900. https://doi.org/10.1109/JSEN.2016.2553962.

45. Ludwig, A.; Quandt, E. Optimization of the ΔE-effect in thin films and multilayers by magnetic field annealing. *IEEE Trans. Magn.* **2002**, *38*, 2829–2831. https://doi.org/10.1109/INTMAG.2002.1000626.
46. Durdaut, P.; Salzer, S.; Reermann, J.; Röbisch, V.; McCord, J.; Meyners, D.; Quandt, E.; Schmidt, G.; Knöchel, R.; Höft, M. Improved Magnetic Frequency Conversion Approach for Magnetoelectric Sensors. *IEEE Sens. Lett.* **2017**, *1*, 1–4. https://doi.org/10.1109/LSENS.2017.2699559.
47. Salzer, S.; Durdaut, P.; Röbisch, V.; Meyners, D.; Quandt, E.; Höft, M.; Knöchel, R. Generalized Magnetic Frequency Conversion for Thin-Film Laminate Magnetoelectric Sensors. *IEEE Sens. J.* **2017**, *17*, 1373–1383. https://doi.org/10.1109/JSEN.2016.2645707.
48. Durdaut, P.; Penner, V.; Kirchhof, C.; Quandt, E.; Knöchel, R.; Höft, M. Noise of a JFET Charge Amplifier for Piezoelectric Sensors. *IEEE Sens. J.* **2017**, *17*, 7364–7371. https://doi.org/10.1109/JSEN.2017.2759000.
49. Lage, E.; Kirchhof, C.; Hrkac, V.; Kienle, L.; Jahns, R.; Knöchel, R.; Quandt, E.; Meyners, D. Exchange biasing of magnetoelectric composites. *Nat. Mater* **2012**, *11*, 523–529. https://doi.org/https://doi.org/10.1038/nmat3306.
50. Spetzler, B.; Bald, C.; Durdaut, P.; Reermann, J.; Kirchhof, C.; Teplyuk, A.; Meyners, D.; Quandt, E.; Höft, M.; Schmidt, G.; et al. Exchange biased delta-E effect enables the detection of low frequency pT magnetic fields with simultaneous localization. *Sci. Rep.* **2021**, *11*, 5269. https://doi.org/10.1038/s41598-021-84415-2.
51. Durdaut, P.; Salzer, S.; Reermann, J.; Röbisch, V.; Hayes, P.; Piorra, A.; Meyners, D.; Quandt, E.; Schmidt, G.; Knöchel, R.; et al. Thermal-Mechanical Noise in Resonant Thin-Film Magnetoelectric Sensors. *IEEE Sens. J.* **2017**, *17*, 2338–2348. https://doi.org/10.1109/JSEN.2017.2671442.
52. Reermann, J.; Elzenheimer, E.; Schmidt, G. Real-Time Biomagnetic Signal Processing for Uncooled Magnetometers in Cardiology. *IEEE Sens. J.* **2019**, *19*, 4237–4249. https://doi.org/10.1109/JSEN.2019.2893236.
53. Urs, N.O.; Golubeva, E.; Röbisch, V.; Toxvaerd, S.; Deldar, S.; Knöchel, R.; Höft, M.; Quandt, E.; Meyners, D.; McCord, J. Direct Link between Specific Magnetic Domain Activities and Magnetic Noise in Modulated Magnetoelectric Sensors. *Phys. Rev. Appl.* **2020**, *13*. https://doi.org/10.1103/PhysRevApplied.13.024018.
54. Gussak, I.; Antzelevitch, C.; Wilde, A.A.M.; Friedman, P.A.; Ackerman, M.J. *Electrical Diseases of the Heart: Genetics, Mechanisms, Treatment, Prevention*, 1st ed.; Springer: London, UK, 2008.
55. Elzenheimer, E. Analyse Stimulationsevozierter Muskel- und Nervensignale Mithilfe Elektrischer und Magnetischer Sensorik. Ph.D. Thesis. Chair of Digital Signal Processing and System Theory, Kiel University, Kiel, Germany, 2022
56. QuSpin. Specification QZFM Gen-3 . Available online: https://quspin.com/products-qzfm/ (accessed on 9 December 2021).
57. Bertrand, F.; Jager, T.; Boness, A.; Fourcault, W.; Le Gal, G.; Palacios-Laloy, A.; Paulet, J.; Léger, J.M. A 4He vector zero-field optically pumped magnetometer operated in the Earth-field. *Rev. Sci. Instrum.* **2021**, *92*, 105005. https://doi.org/10.1063/5.0062791.
58. TDK Corporation. Ultrasensitive Magnetic Sensor-Nivio xMR. Available online: https://product.tdk.com/system/files/dam/doc/content/event/techfro2020/tech20_17.pdf (accessed on 9 December 2021).

Article

Adaptive Model for Magnetic Particle Mapping Using Magnetoelectric Sensors

Ron-Marco Friedrich and Franz Faupel *

Institute of Materials Science, Faculty of Engineering, Kiel University, 24143 Kiel, Germany; rmfr@tf.uni-kiel.de
* Correspondence: ff@tf.uni-kiel.de; Tel.: +49-431-880-6225

Abstract: Imaging of magnetic nanoparticles (MNPs) is of great interest in the medical sciences. By using resonant magnetoelectric sensors, higher harmonic excitations of MNPs can be measured and mapped in space. The proper reconstruction of particle distribution via solving the inverse problem is paramount for any imaging technique. For this, the forward model needs to be modeled accurately. However, depending on the state of the magnetoelectric sensors, the projection axis for the magnetic field may vary and may not be known accurately beforehand. As a result, the projection axis used in the model may be inaccurate, which can result in inaccurate reconstructions and artifact formation. Here, we show an approach for mapping MNPs that includes sources of uncertainty to both select the correct particle distribution and the correct model simultaneously.

Keywords: magnetoelectric; magnetic nanoparticle; imaging; inverse problem; blind deconvolution

1. Introduction

Imaging techniques often involve solving inverse problems in order to be able to image the entity of interest sufficiently [1,2]. The interplay between measured data and the model used to invert said data is often neglected, meaning that the model is assumed to reflect reality correctly. However, this poses a source of error for the inversion of the data—the formation of artifacts in the reconstruction due to the use of an incorrect model. To address this problem, techniques were invented that use additional information (a priori knowledge) on the models and source distribution of the inverse problem. One approach that describes the use of additional information on the inverse problem is called Blind Deconvolution, which is used when the impulse response is not exactly known [3–5].

Most often, an imaging system can be described via a convolution; hence, the impulse response (or Point-Spread Function) is the quantity that needs to be modeled correctly. Here, the impulse response can be shift-invariant or not, meaning that the shape of the impulse response depends on the position of the source in the underlying distribution. Given the measured data and assumptions on the underlying source distribution, the goal of Blind Deconvolution is to find the correct model that maps the source distribution to the measurement. For this, appropriate subspaces or constraint sets of the source distribution and models have to formulated [3].

Recently, an imaging system for magnetic nanoparticles (MNPs) using magnetoelectric sensors called Magnetic Particle Mapping (MPM) was developed. In MPM, MNPs are excited into the nonlinear magnetic regime using a homogeneous magnetic AC field. The frequency of the excitation field is chosen such that the higher harmonic excitations due to the magnetic nonlinearity coincides with a highly sensitive mechanical resonance of a cantilever-type magnetoelectric (ME) thin film sensor [6]. The magnetoelectric sensor's magnetic state determines a sensitive axis and is often not exactly known and varies from sensor to sensor [7]. Hence, prior calibration is performed to quantify the sensitive axis such that one can model the system appropriately. However, this axis may change if biasing is applied or the orientation of the sensor is not known exactly [7]. Thus, additional considerations should be taken when modeling the MPM imaging system.

Citation: Friedrich, R.-M.; Faupel, F. Adaptive Model for Magnetic Particle Mapping Using Magnetoelectric Sensors. *Sensors* **2022**, *22*, 894. https://doi.org/10.3390/s22030894

Academic Editor: Arcady Zhukov

Received: 10 December 2021
Accepted: 20 January 2022
Published: 24 January 2022

Publisher's Note: MDPI stays neutral with regard to jurisdictional claims in published maps and institutional affiliations.

Copyright: © 2022 by the authors. Licensee MDPI, Basel, Switzerland. This article is an open access article distributed under the terms and conditions of the Creative Commons Attribution (CC BY) license (https://creativecommons.org/licenses/by/4.0/).

The MPM imaging system can be described as a linear shift-invariant system, where the distribution of MNPs is nonnegative. The impulse response of the system is dependent on the orientation of the MNPs and the sensitive axis of the sensor. These two constraints can be used to create an adaptive scheme that finds the correct model and the underlying MNP distribution simultaneously by using a gradient descent procedure with alternating projections onto feasible sets. The presented approach is also applicable for other systems where projection axes are unknown and may, thus, be adapted for specific imaging systems.

2. Materials and Methods

2.1. Imaging MNPs

MNPs can be imaged in a variety of ways. MNP imaging techniques that use MNP as tracer material include Magnetic Particle Imaging (MPI) [8–10], Scanning Magnetic Particle Spectrometry (SMPS) [11–14], Magnetorelaxometry Imaging (MRXI) [15–18] and Magneto Acoustic Tomography (MAT) [19–21]. Another type of MNP imaging technique can be categorized into Magnetic Susceptibility Imaging (MSI) [22], one of which is called Susceptibility Magnitude Imaging (SMI) [23]. This technique was further developed using nonlinear MNP responses by spectroscopy AC susceptibility Imaging (sASI) [24] and nonlinear Susceptibility Magnitude Imaging (nSMI) [25].

Approaches for enhanced imaging based on figure of merit optimization for models were performed for several imaging techniques. For example, the authors in [18] performed optimized coil activation sequences in MRXI measurements to reduce the condition number of the model for easier inversion and robustness. Another example includes theoretical enhancements on the SMI setup in [23] as investigated in [26] based on geometric considerations on the figure of merits in inverse problems.

Currently, little investigations have been conducted explicitly with regards to Blind Deconvolution techniques in imaging MNP. In MPI, blind deconvolution techniques were proposed and investigated to address the issue of unknown impulse responses [27]. Most often, prior calibration of imaging systems are sufficient for imaging, yet the possibility for simultaneous characterization of the imaging system and accurate image reconstruction would be an attractive property for any imaging system.

Sensing MNPs with Magnetoelectric Sensors

Magnetoelectric sensors were recently used for detecting the magnetic response of MNPs. That this was possible was first shown by [28], where a laminate composite consisting of 500-micrometer-thick PZT with 18-micrometer-thick soft magnetic Ni-based Metglas ribbon. The research was performed in the context of clinical interventions. It was argued that ME sensors could be used for interventions such as sentinel lymph node biopsy (SLNB) for cancer detection, an application that is also argued for other imaging techniques, as mentioned earlier. With their setup, they performed a one dimensional measurement of the magnetic field, which can be thought of as a precursor for imaging MNP with ME sensors. To magnetize MNPs, a permanent magnet was used, and the sensor was aligned such that no in-plane magnetic field component affected the sensor. The smallest amount of MNP they were able to detect was 310 ng at a distance of 2 mm.

ME sensors were also used for the detection of tissue iron content for Biomagnetic Liver Susceptometry (BLS) [29]. In the study, permanent magnets were used to magnetize the sample and bias the sensor simultaneously. The sample is then moved periodically such that a low frequency magnetic signal is created that can be detected via the ME sensor. The source of the magnetic signal was the protein ferritin, for which its magnetic response is paramagnetic. This approach was reinvented in [30] for the imaging of MNP and is called Magnetic Susceptibility Particle Mapping (MSPM). To generate a low frequency signal of the MNP, a motion-modulated approach was taken, where the MNP distribution is mounted on a rotating disc, which is moved through a magnetic field of permanent magnets. The permanent magnets also bias the sensor for high sensitivity, while using the shape anisotropy of the sensor to adjust the extent of biasing.

In the context of this study, some of the just mentioned considerations were used in MPM as well [6]. The shape anisotropy of the sensors will be used for the separation of externally applied field and the MNP field. Imaging of MNP is performed in 2D using AC fields for magnetic excitation and linear stages for translation. Signal acquisition involves an AC field that magnetically excites the MNP to detect higher harmonic responses via the sensor, similar to MPI or sMPS.

The ME sensor only measures a projection of the magnetic field via its the sensitive axis. This axis, in turn, is dependent on the fabrication procedure and magnetic state of the sensor [7]. It is, thus, convenient for an imaging system, which uses ME sensors, to be able to address an unknown sensitive axis while operating in an imaging experiment.

2.2. Modeling the MPM Imaging System

In the following, the imaging system for MPM will be derived and important aspects will be highlighted. To model the MPM imaging system, we have to develop a mathematical relationship between the sources (MNPs) and the measurement positions (sensor positions). The MNPs have a magnetic (vector)field associated with them due to their magnetic dipole moment **m** and the sensor measures only a single projection of the magnetic field via the sensitivity axis **s**. The next section will deal with the role of the magnetic dipole field and the sensor in context of an imaging system for MNPs. Furthermore, we can expect the system to be linear as the MNPs' magnetic fields simply superimpose. In the following, lower case bold letters will denote vectors and upper case bold letters will denote matrices. A hat above a vector will denote a vector of unit length. Hence, **m** and **s** are the vectors describing the magnetic moment of the MNP and the sensitivity axis of the sensor, respectively. The vector **r** will denote a spatial position. Vectors $\hat{\mathbf{m}}$, $\hat{\mathbf{s}}$ and $\hat{\mathbf{r}}$ only describe the directions of said quantities. Matrix **I** is the identity matrix.

The magnetic field $\mathbf{B}_D$ is given by the following.

$$\mathbf{B}_D = \frac{\mu_0 \mathbf{m}^\mathsf{T}}{4\pi r^3}(3\hat{\mathbf{r}}\hat{\mathbf{r}}^\mathsf{T} - \mathbf{I}). \tag{1}$$

If we measure only a single projection of the magnetic field via the sensor's sensitive axis **s**, we have the following.

$$\frac{4\pi}{\mu_0}\mathbf{B}_D = \mathbf{m}^\mathsf{T}\frac{(3\hat{\mathbf{r}}\hat{\mathbf{r}}^\mathsf{T} - \mathbf{I})}{(\mathbf{r}^\mathsf{T}\mathbf{r})^{3/2}}\mathbf{s}$$

$$\frac{4\pi}{\mu_0\|\mathbf{m}\|\|\mathbf{s}\|}\mathbf{B}_D = \hat{\mathbf{m}}^\mathsf{T}\mathbf{H}\hat{\mathbf{s}}. \tag{2}$$

The projected magnetic field can, thus, be written in a bilinear form with magnetic moment direction $\hat{\mathbf{m}}$, sensitive direction of the sensor $\hat{\mathbf{s}}$ and a symmetric matrix **H**. The functions contained in **H** and their respective symmetries are important. Due to outer product $\hat{\mathbf{r}}\hat{\mathbf{r}}^\mathsf{T}$, the diagonal will yield symmetric functions while the off-diagonal elements exhibit antisymmetries with respect to the spatial coordinates. This fact will be important at a later stage, because they imply orthogonality of the measurable fields. Only a few correlated fields remain, making estimation of source distributions easier and allowing the possibility of adaptive models in the inverse problem, which will be shown further below.

We now introduce the relationship of the spatial distribution of MNPs and the resulting magnetic fields in space (regions where the field is measured). We denote the spatial magnetic particle distribution as ρ, and the region where particles are present is domain Ω over which they will be integrated. The measurement positions are denoted as $\mathbf{r}_m$. The resulting magnetic fields B_m at the measurement positions $\mathbf{r}_m$ are then given by the following.

$$\int_\Omega \rho(\mathbf{r}) B_D(\mathbf{r}_m, \mathbf{r}) \mathrm{d}^3\mathbf{r} = B_m(\mathbf{r}_m) \tag{3}$$

The equation above is a Fredholm integral equation of the first kind [31]. In fact, in this case, the integral equation is a convolution of the following.

$$\int_\Omega \rho(\mathbf{r}) B_D(\mathbf{r}_m - \mathbf{r}) d^3\mathbf{r} = B_m(\mathbf{r}_m) \quad (4)$$

The projected dipole field B_D is called the kernel of this equation and is, in this case, equivalent to an impulse response of the system or is also commonly known as the Point-Spread-Function (PSF) used in optics/imaging systems. The mapping from the spatial particle distribution to the measurable signal is the forward model. In this case, the system is linear and shift-invariant, assuming that the Point-Spread-Function does not change depending on the sample position. Linearity stems from the assumption that the particles do not interact with each other (thus not altering the PSF depending on, e.g., local concentrations) and that the magnetic fields linearly superimpose. This assumption can be assumed if homogeneous fields are used, which is to a sufficient degree the case. Therefore, one can describe this system, similar to linear time-invariant (LTI) systems, as a *linear space-invariant* system.

Discretization of Equation (4) yields a system of linear equations, i.e., in the following.

$$\mathbf{A}\mathbf{x} = \mathbf{b} \quad (5)$$

Here, $\mathbf{A}$ denotes the model matrix, which incorporates the orientation of magnetic dipoles and sensor sensitive axis (compare Equation (2)). Vector $\mathbf{x}$ is the spatial MNP distribution for which its entries are non-negative ($\mathbf{x} \geq 0$), and $\mathbf{b}$ denotes the superposition of magnetic fields from the MNPs for each measurement position. Here, it is beneficial to explicitly write out the dependence of the model matrix on the magnetic moment direction $\mathbf{m}$ and the sensor sensitive axis $\mathbf{s}$.

$$\begin{aligned}\mathbf{A} =& m_1(s_1\mathbf{A}_{11} + s_2\mathbf{A}_{12} + s_3\mathbf{A}_{13}) \\ &+ m_2(s_1\mathbf{A}_{21} + s_2\mathbf{A}_{22} + s_3\mathbf{A}_{23}) \\ &+ m_3(s_1\mathbf{A}_{31} + s_2\mathbf{A}_{32} + s_3\mathbf{A}_{33}).\end{aligned} \quad (6)$$

The model matrix can be more compactly represented by using the Kronecker matrix product (denoted by $\otimes$):

$$\mathbf{A} = \left(\mathbf{m}^T \otimes \mathbf{I}_{m\times m}\right) \mathbf{A}_B (\mathbf{s} \otimes \mathbf{I}_{n\times n}) \quad (7)$$

where $\mathbf{A}_B$ is the blockmatrix containing all models.

$$\mathbf{A}_B = \begin{pmatrix} \mathbf{A}_{11} & \mathbf{A}_{12} & \mathbf{A}_{13} \\ \mathbf{A}_{21} & \mathbf{A}_{22} & \mathbf{A}_{23} \\ \mathbf{A}_{31} & \mathbf{A}_{32} & \mathbf{A}_{33} \end{pmatrix} \quad (8)$$

Now, given a data vector $\mathbf{b}$, what source distribution $\mathbf{x}$ and model $\mathbf{A}$ gave rise to the data? This question denotes the Blind Deconvolution problem.

2.3. Inverse Problem

Using the forward model we have developed, we now wish to infer the spatial particle distribution from the measured magnetic fields—we are looking for ways to invert the forward model—i.e., we wish to solve the inverse problem. Assuming that the model is accurate, the solution involves computing an estimate that is closely related to the data. For this, we need to minimize the difference between estimate $\mathbf{A}\mathbf{x}$ and data $\mathbf{b}$ via some form of metric. Commonly chosen is the L^2 norm (Euclidean norm) as the distance measure between the two vectors $\mathbf{A}\mathbf{x}$ and $\mathbf{b}$. This choice stems from the differentiability of this distance, because the norm is induced by the inner product such that an analytic expression can be found. Differentiability is useful because the solutions can be iterated via gradient descent procedures. However, because we are dealing with an inverse problem, the least square minimum is not the ideal choice due to numerical instabilities and amplifications of noise in the measurements [1,2,32].

To combat these issues, one needs to regularize the solution. Regularization refers to the addition of constraints to the original problem that limit the size of the solution vector **x**—we are looking to both be close to data **b** and still have physically meaningful results. For this, we add another term to the cost function called the regularizer R.

$$\Phi(\mathbf{x}) = \|\mathbf{A}\mathbf{x} - \mathbf{b}\|_2^2 + \lambda R(\mathbf{x}) \quad (9)$$

A possible geometric meaning of the regularizer can be imagined as the description of a set in which the solution has to lie. Often, one can show that this description is the same as for constrained optimization via the Lagrange multiplier λ. Thus, the role of the regularization parameter λ is to set the solution size (size as in norm of a vector). There are many types of regularizations, most notably Tikhonov regularization (also called ridge regression) and L^1 regularization that promotes sparsity in the solution. The latter is also of interest, because L^1 regularization in combination with a *nonnegativity constraint* sets the total amount of MNP in the system that has to explain the data and, thus, has physical meaning.

Due to the fact that the L^1 regularizer is convex but not differentiable, one can employ iterative solutions schemes for solving the inverse problem. A straightforward technique is the projected gradient method (though many different names exists, such as Projected Landweber Iteration [2]). Here, one performs a simple gradient descent step and then projects back into the feasible set defined for the problem, such as projection into the non-negative orthant for the non-negativity constraint and projection onto the L^1 ball (both can also be described as the projection onto the scaled standard simplex). A visual representation for this procedure can be exemplary observed in Figure 1.

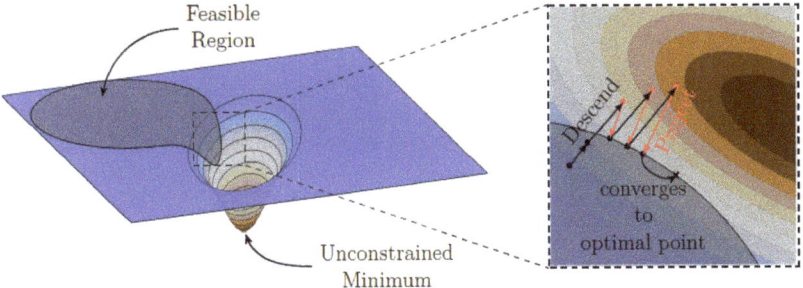

Figure 1. Projected gradient method. Each gradient descent step is projected back into a feasible set. In this manner, the solution becomes regularized, as the solution size cannot lie outside the feasible set. If the feasible set is convex, the solution will converge to an optimal point within the set. Image adapted from [33] (CC BY).

Furthermore, if the magnetic moment direction and sensor sensitive axis also need to be estimated, the objective for solving the inverse problem can be formulated as follows.

$$\min_{\mathbf{x},\mathbf{m},\mathbf{s}} \left\{ \left\| \left(\mathbf{m}^T \otimes \mathbf{I}_{m \times m}\right) \mathbf{A}_B (\mathbf{s} \otimes \mathbf{I}_{n \times n}) \mathbf{x} - \mathbf{b} \right\|_2^2 + \lambda R(\mathbf{x}) \right\} \quad \text{s.t.} \quad \mathbf{x} \geq 0. \quad (10)$$

2.4. Algorithm

To find the correct spatial MNP distribution as well as the correct model, we propose a two step iterative scheme, which successively updates the MNP distribution and the model of the imaging system via restricting entities to their respective feasible domain. For the MNP distribution, this is mainly performed via the non-negativity constraint, and for the model matrices, only a linear combination of nine models (or rather six due to the symmetries; compare Equation (2)) is allowed, thus defining a subspace of possible models. These restrictions allow for the correct estimation of the MNP distribution and model.

First, an algorithm will be investigated for a system where magnetic moment direction **m** is perfectly known and sensitive axis **s** needs to be found in addition to the correct particle distribution, **x**. Then, the general case for unknown magnetic moment direction **m**, unknown sensitive axis **s** and unknown particle distribution **x** will be investigated and important insights will be highlighted.

2.4.1. Estimating Sensor Sensitive Axis

In this section, the system for a fixed magnetic moment direction $\hat{\mathbf{m}}$ with unknown projection axis $\hat{\mathbf{s}}$ from the sensor will be described. We proceed by creating a model matrix that can be described as a superposition of the forward operator in the corresponding axes x, y, z for a fixed magnetic moment direction **m** in the z axis direction, i.e., as shown in Figure 2. We could take any row or column of the block matrix $\mathbf{A}_B$ to construct a forward operator for the subsequent discussion. The chosen case was taken because it is the projection axis and magnetic moment direction that will be investigated experimentally. Model matrix **A** can be written for an unknown projection axis as follows.

$$\mathbf{A} = s_1 \mathbf{A}_{31} + s_2 \mathbf{A}_{32} + s_3 \mathbf{A}_{33} \tag{11}$$

Under these circumstances, the gradient for the cost function can be rewritten as follows:

$$\nabla_\mathbf{s} \Phi = \mathbf{x}^\mathsf{T} \frac{\partial \mathbf{A}^\mathsf{T}}{\partial \mathbf{s}} (\mathbf{A}\mathbf{x} - \mathbf{b})$$
$$= \mathbf{D}^\mathsf{T}(\mathbf{D}\mathbf{s} - \mathbf{b}) \tag{12}$$

with the following.

$$\mathbf{D} = [\mathbf{A}_{31}\mathbf{x}, \mathbf{A}_{32}\mathbf{x}, \mathbf{A}_{33}\mathbf{x}] \tag{13}$$
$$\mathbf{s} = [s_1; s_2; s_3]. \tag{14}$$

As a result, the algorithm to compute particle distribution **x** and model parameters **s** can be written as shown in Algorithm 1. The projection operator $\mathbf{P}_+$ denotes the projection into the non-negative orthant and $\mathbf{P}_C$ denotes projection onto feasible set C. In this case, $\mathbf{P}_C$ also includes a projection onto the unit sphere, as to denote only the direction of the sensitive axis. As a stopping criterion, the discrepancy principle is used [2].

Algorithm 1 Model Estimation For Sensitive Axis

1: Given: iterations K, data **b**, set of possible model parameters C, estimate of noise standard deviation δ, stopping term for discrepancy principle η, projection operator $\mathbf{P}_C$, $\mathbf{P}_+$.
2: Initialize $\mathbf{A}_{31}, \mathbf{A}_{32}, \mathbf{A}_{33}$ possible forward operators, particle distribution **x**, projection estimate **s**.
3: **for** $k = 1$ to K **do**
4:     $\mathbf{A} = s_{1,k}\mathbf{A}_{31} + s_{2,k}\mathbf{A}_{32} + s_{3,k}\mathbf{A}_{33}$.
5:     $\mathbf{x}_{k+1} = \mathbf{P}_+\left(\mathbf{x}_k + \|\mathbf{A}\|_2^{-2}\mathbf{A}^\mathsf{T}(\mathbf{b} - \mathbf{A}\mathbf{x}_k)\right)$
6:     $\mathbf{D} = [\mathbf{A}_{31}\mathbf{x}_{k+1}, \mathbf{A}_{32}\mathbf{x}_{k+1}, \mathbf{A}_{33}\mathbf{x}_{k+1}]$
7:     $\mathbf{s}_{k+1} = \mathbf{P}_C\left(\mathbf{s}_k + \|\mathbf{D}\|_2^{-2}\mathbf{D}^\mathsf{T}(\mathbf{b} - \mathbf{D}\mathbf{s}_k)\right)$
8:     **if** $\|\mathbf{A}\mathbf{x}_{k+1} - \mathbf{b}\|_2 / \delta \leq \eta$ **then**
9:         **return** $\mathbf{x}_{k+1}, \mathbf{s}_{k+1}$
10:    **end if**
11: **end for**

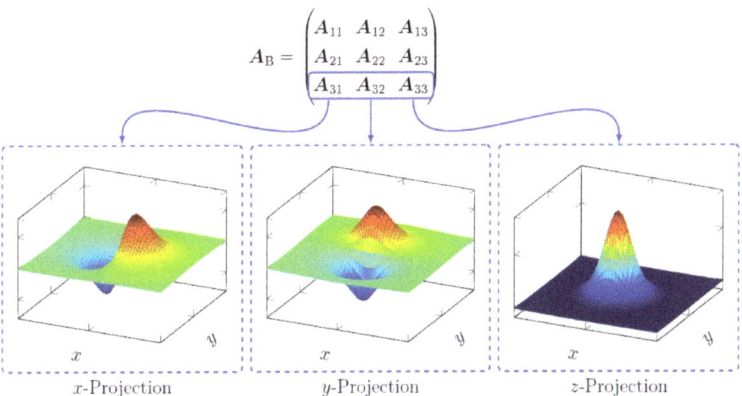

Figure 2. The individual model matrices corresponding to the different combinations of magnetic moment direction and sensitive axis direction refer to different PSFs for the imaging system. In this section, we take the projections for the magnetic moment in the z direction, which is indicated by the model matrices enclosed by the blue borders. The dashed boxes show the corresponding PSFs for the model matrices.

Under the condition that the correct model can be expressed as the linear combination of model matrices corresponding to the x, y and z-projection of the magnetic field, we write the following cost function.

$$\Phi(\mathbf{x}) = \frac{1}{2}\left\|\left(\sum_i^N s_i \mathbf{A}_{3i}\right)\mathbf{x} - \mathbf{b}\right\|_2^2. \tag{15}$$

In this case, we would have $N = 3$ matrices for the projections in the x, y and z directions. Recall the dipole functions as depicted in Figure 2. Take, e.g., the orientation of the dipole in z-direction and take the product of any two different projections of the dipole fields that correspond to the PSFs in Figure 2. The result will yield equal positive and negative parts (given that we have a fine Cartesian discretization and a spatially large enough domain). This will be important in the following step. If we expand the expression above, we have the following:

$$\Phi(\mathbf{x}) = \frac{1}{2}(\mathbf{x}^\mathsf{T}\mathbf{C}^\mathsf{T}\mathbf{C}\mathbf{x} + \mathbf{b}^\mathsf{T}\mathbf{b} - \mathbf{b}^\mathsf{T}\mathbf{C}\mathbf{x} - \mathbf{x}^\mathsf{T}\mathbf{C}^\mathsf{T}\mathbf{b}) \tag{16}$$

with $\mathbf{C} = \sum_i^N s_i \mathbf{A}_{3i}$. Now, the cross terms (i.e., $i \neq j$) in the quadratic forms are equal to zero (considering equidistant sampling of the x-y plane and that the domain is large enough to capture most magnetic field):

$$\mathbf{x}^\mathsf{T}\mathbf{A}_{3i}^\mathsf{T}\mathbf{A}_{3j}\mathbf{x} = 0 \tag{17}$$

which means that the matrix $\mathbf{D}$ is orthogonal.

$$\mathbf{D} = [\mathbf{A}_{31}\mathbf{x}, \mathbf{A}_{32}\mathbf{x}, \mathbf{A}_{33}\mathbf{x}]$$
$$\mathbf{D}^\mathsf{T}\mathbf{D} = \mathrm{diag}(\mathbf{a}). \tag{18}$$

This follows from the fact that the inner product of the two dipole field projections is zero, since it contains equal amounts of positive and negative parts. We can, thus, obtain the following.

$$\Phi(\mathbf{x}) = \frac{1}{2}(1-N)\|\mathbf{b}\|_2^2 + \frac{1}{2}\sum_i^N \|s_i \mathbf{A}_{3i}\mathbf{x} - \mathbf{b}\|_2^2. \tag{19}$$

Since all terms correspond to strictly convex functions, the problem is uniquely solvable. However, because the number of parameters suffices for any of the matrices A_{ji} to express data **b**, one will still need to enforce a non-negativity constraint, which will result in the correct estimation in the end. The derivative with respect to the parameters for the projection **s** is the following:

$$\nabla_i \Phi = \mathbf{d}_i^T(d_i s_i - \mathbf{b}) \tag{20}$$

$$\Rightarrow s_i = \frac{\mathbf{d}_i^T \mathbf{b}}{\mathbf{d}_i^T \mathbf{d}_i}$$

$$s_i = \frac{\mathbf{x}^T \mathbf{A}_{3i}^T \mathbf{b}}{\mathbf{x}^T \mathbf{A}_{3i}^T \mathbf{A}_{3i} \mathbf{x}} \tag{21}$$

which means it can be calculated from the projection of the estimate $\mathbf{A}_{3i}\mathbf{x}$ onto data **b**. The vector **b** can thus be expanded by an N-dimensional subspace that is constructed from the estimated particle distribution **x**.

$$\mathbf{b} = \sum_i^N \frac{\mathbf{b}^T \mathbf{A}_{3i}\mathbf{x}}{\|\mathbf{A}_{3i}\mathbf{x}\|} \frac{\mathbf{A}_{3i}\mathbf{x}}{\|\mathbf{A}_{3i}\mathbf{x}\|}. \tag{22}$$

Combining the results, we obtain the following.

$$\Phi(\mathbf{x}) = \frac{1}{2}(1-N)\|\mathbf{b}\|_2^2 + \frac{1}{2}\sum_i^N \left\|\left(\frac{\mathbf{A}_{3i}\mathbf{x}\mathbf{x}^T\mathbf{A}_{3i}^T}{\mathbf{x}^T\mathbf{A}_{3i}^T\mathbf{A}_{3i}\mathbf{x}} - \mathbf{I}\right)\mathbf{b}\right\|_2^2$$

$$= \frac{1}{2}(1-N)\|\mathbf{b}\|_2^2 + \frac{N}{2}\|\mathbf{b}\|_2^2 - \frac{1}{2}\sum_i^N \frac{(\mathbf{b}^T \mathbf{A}_{3i}\mathbf{x})^2}{\mathbf{x}^T\mathbf{A}_{3i}^T\mathbf{A}_{3i}\mathbf{x}}$$

$$\frac{2}{\|\mathbf{b}\|_2^2}\Phi(\mathbf{x}) = 1 - \sum_i^N \left(\frac{\mathbf{b}^T}{\|\mathbf{b}\|}\frac{\mathbf{A}_{3i}\mathbf{x}}{\|\mathbf{A}_{3i}\mathbf{x}\|}\right)^2$$

$$\frac{2}{\|\mathbf{b}\|_2^2}\Phi(\mathbf{x}) = 1 - \sum_i^N \cos(\theta_i)^2 \tag{23}$$

Here, we see that the cost function is minimized if the projection of estimate $\mathbf{A}_{3i}\mathbf{x}$ onto data **b** is maximized (or that the angle θ_i between data **b** and estimate $\mathbf{A}_{3i}\mathbf{x}$ is minimized). The direction cosines, thus, add up to 1 if we find the minimum of the cost function. This can be imagined as finding a point on a unit sphere, i.e., the decomposition of $\mathbf{b}/\|\mathbf{b}\|$ into orthogonal components related to the projection of the magnetic field (see Figure 3). Our task is, thus, finding the very specific subspace that is able to describe data **b** completely.

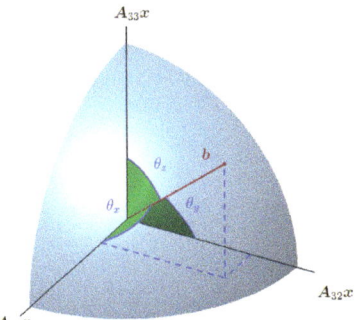

Figure 3. Measurement **b** is estimated to lie in a 3-dimensional subspace spanned by the estimates for the field projections in the x, y and z directions, i.e., $\mathbf{A}_{31}\mathbf{x}$, $\mathbf{A}_{32}\mathbf{x}$ and $\mathbf{A}_{33}\mathbf{x}$. We, thus, need to find a point on a sphere of radius $\|\mathbf{b}\|$.

2.4.2. Estimating Sensor Sensitive Axis and Magnetic Moment Direction

To estimate both the sensitive axis and magnetic moment direction, they need to be updated within each iteration of the algorithm. For this, the general derivatives with respect to **x**, **s** and **m** need to be computed. The derivatives of the cost function can be written as follows:

$$\frac{\partial \Phi}{\partial \mathbf{x}} = \mathbf{A}^\mathsf{T}(\mathbf{A}\mathbf{x} - \mathbf{b}) \tag{24}$$

$$\frac{\partial \Phi}{\partial \mathbf{s}} = \mathbf{M}_\mathbf{s}^\mathsf{T}(\mathbf{M}_\mathbf{s}\mathbf{s} - \mathbf{b}) \tag{25}$$

$$\frac{\partial \Phi}{\partial \mathbf{m}} = \mathbf{M}_\mathbf{m}^\mathsf{T}(\mathbf{M}_\mathbf{m}\mathbf{m} - \mathbf{b}). \tag{26}$$

where $\mathbf{M}_\mathbf{m}$ is given by the following:

$$\mathbf{M}_\mathbf{m} = \left((\mathbf{I}_{3\times 3} \otimes \mathbf{s}^\mathsf{T}) \begin{pmatrix} \mathbf{M}_{1j}^\mathsf{T} \\ \mathbf{M}_{2j}^\mathsf{T} \\ \mathbf{M}_{3j}^\mathsf{T} \end{pmatrix} \right)^\mathsf{T} \tag{27}$$

and $\mathbf{M}_\mathbf{s}$ is given by the following:

$$\mathbf{M}_\mathbf{s} = \left((\mathbf{I}_{3\times 3} \otimes \mathbf{m}^\mathsf{T}) \begin{pmatrix} \mathbf{M}_{i1}^\mathsf{T} \\ \mathbf{M}_{i2}^\mathsf{T} \\ \mathbf{M}_{i3}^\mathsf{T} \end{pmatrix} \right)^\mathsf{T} \tag{28}$$

with the following being the case.

$$\begin{aligned} \mathbf{M}_{ij} &= \begin{pmatrix} \mathbf{A}_{i1} & \mathbf{A}_{i2} & \mathbf{A}_{i3} \end{pmatrix} (\mathbf{I}_{3\times 3} \otimes \mathbf{x}) \\ &= \begin{pmatrix} \mathbf{A}_{i1}\mathbf{x} & \mathbf{A}_{i2}\mathbf{x} & \mathbf{A}_{i3}\mathbf{x} \end{pmatrix}. \end{aligned} \tag{29}$$

Important to note in the gradients is the matrix containing all quadratic forms.

$$\begin{pmatrix} \mathbf{M}_{1j}^\mathsf{T} \\ \mathbf{M}_{2j}^\mathsf{T} \\ \mathbf{M}_{3j}^\mathsf{T} \end{pmatrix} \begin{pmatrix} \mathbf{M}_{1j}^\mathsf{T} \\ \mathbf{M}_{2j}^\mathsf{T} \\ \mathbf{M}_{3j}^\mathsf{T} \end{pmatrix}^\mathsf{T} = \begin{pmatrix} \mathbf{M}_{1j}^\mathsf{T}\mathbf{M}_{1j} & \mathbf{M}_{1j}^\mathsf{T}\mathbf{M}_{2j} & \mathbf{M}_{1j}^\mathsf{T}\mathbf{M}_{3j} \\ \mathbf{M}_{2j}^\mathsf{T}\mathbf{M}_{1j} & \mathbf{M}_{2j}^\mathsf{T}\mathbf{M}_{2j} & \mathbf{M}_{2j}^\mathsf{T}\mathbf{M}_{3j} \\ \mathbf{M}_{3j}^\mathsf{T}\mathbf{M}_{1j} & \mathbf{M}_{3j}^\mathsf{T}\mathbf{M}_{2j} & \mathbf{M}_{3j}^\mathsf{T}\mathbf{M}_{3j} \end{pmatrix}. \tag{30}$$

By plotting this matrix, one can gain an idea about the correlation between possible magnetic field components for different dipole orientations, see Figure 4. We see that because we have off-diagonal elements not equal to zero, the fields for different dipole orientations are correlated. On the other hand, we see that the 3 × 3 block diagonal elements are diagonal sub-matrices, implying that the magnetic dipole fields are orthogonal if the orientation of the dipole lies on the (orthogonal) coordinate system axes. An algorithm that estimates the MNP distribution **x**, sensitive sensor axis **s** and magnetic dipole direction **m** can be seen in Algorithm 2.

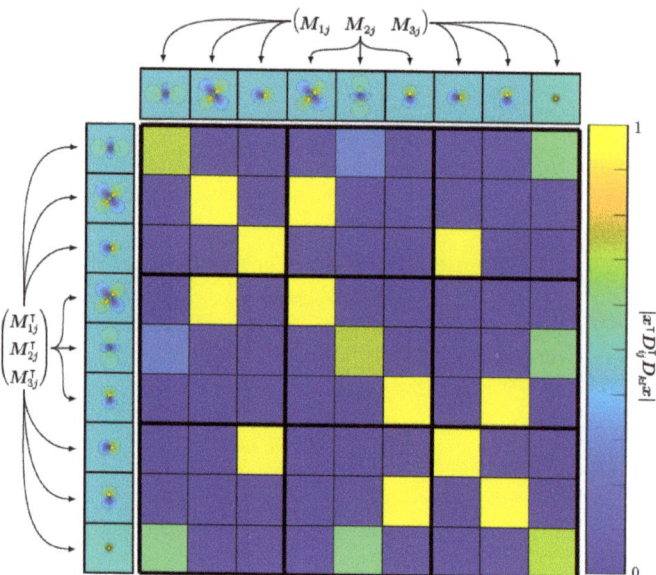

Figure 4. The matrix consisting of all quadratic forms shows the correlation between different magnetic field components for different moment directions. We see that most of the field components are orthogonal to each other, while there are some off-diagonal elements, meaning that there exists a correlation between the corresponding fields. However, they are still linearly independent.

Algorithm 2 Model Estimation

1: Given: iterations K, data **b**, estimate of noise standard deviation δ, stopping term for discrepancy principle η, projection operator $\mathbf{P}_+$, $\mathbf{P}_C^{(1)}$, $\mathbf{P}_C^{(2)}$.
2: Initialize estimate of model parameters **s**, **m**, MNP distribution **x**.
3: **for** $k = 1$ to K **do**
4: $\quad \mathbf{A} = (\mathbf{m}^T \otimes \mathbf{I}_{m \times m}) \mathbf{A}_B (\mathbf{s} \otimes \mathbf{I}_{n \times n})$
5: $\quad \mathbf{x}_{k+1} = \mathbf{P}_+ \left(\mathbf{x}_k - \|\mathbf{A}\|_2^{-2} \mathbf{A}^T (\mathbf{A}\mathbf{x}_k - \mathbf{b}) \right)$
6: $\quad \mathbf{m}_{k+1} = \mathbf{P}_C^{(1)} \left(\mathbf{m}_k - \|\mathbf{M}_\mathbf{m}\|_2^{-2} \mathbf{M}_\mathbf{m}^T (\mathbf{M}_\mathbf{m} \mathbf{m}_k - \mathbf{b}) \right)$
7: $\quad \mathbf{s}_{k+1} = \mathbf{P}_C^{(2)} \left(\mathbf{s}_k - \|\mathbf{M}_\mathbf{s}\|_2^{-2} \mathbf{M}_\mathbf{s}^T (\mathbf{M}_\mathbf{s} \mathbf{s}_k - \mathbf{b}) \right)$
8: $\quad$ **if** $\|\mathbf{A}\mathbf{x}_{k+1} - \mathbf{b}\|_2 / \delta \leq \eta$ **then**
9: $\quad\quad$ **return** $\mathbf{x}_{k+1}, \mathbf{s}_{k+1}, \mathbf{m}_{k+1}$
10: $\quad$ **end if**
11: **end for**

2.4.3. Measurement

The experimental MPM setup can be observed in Figure 5. Shown are the translation stages for position, the sensor and sample and sets of Helmholtz coils for magnetic field generation. One set of Helmholtz coils is used to excite the MNP into the nonlinear magnetic regime, and another set is used for compensation purposes, as will be explained further below. Not shown are the electric appliances, which include an audio amplifier for signal amplification, a charge amplifier for sensor signal amplification and an audio interface to generate the excitation signal and measure the sensor signal. The magnetoelectric sensor used for the experiment is exchange biased [34] in order to avoid an external biasing field, and the fabrication steps can be read up in [6]. The sensor exhibits its first mechanical resonance at about 7.5 kHz. Sensor sensitivity is about 20 kV/T, and the equivalent noise density at resonance is 15 pT/Hz$^{0.5}$. The excitation signal is generated by an RME FireFace UC with a sampling frequency of 192 kHz at $1/3 f_r$ with f_r being the resonant frequency of the sensor. The excitation signal is amplified using a PAS2002 audio amplifier and connected to the Helmholtz coils with additional impedance matching. The AC magnetic field generated is about 10 mT. The sensor is aligned using the manual tip, tilt, rotation and translation stages such that its shape anisotropy is used to attenuate some of the influence of the applied excitation field. Additionally, another magnetic AC field is applied with low amplitude at sensor resonance, for which its amplitude and phase are tuned to destructively interfere with the background signal. Then, the sample containing MNP can be inserted and measured. For measurements, equidistant points (40 × 40) are sampled in space, which correspond to an area of 20 × 20 mm^2, and a signal is measured with a sample rate of 32 kHz and a frame size of 4096 samples, yielding a spectral resolution of 7.8 Hz. In the spectrum, phase and amplitude at resonance are captured, which should ideally only contain the responses to MNPs.

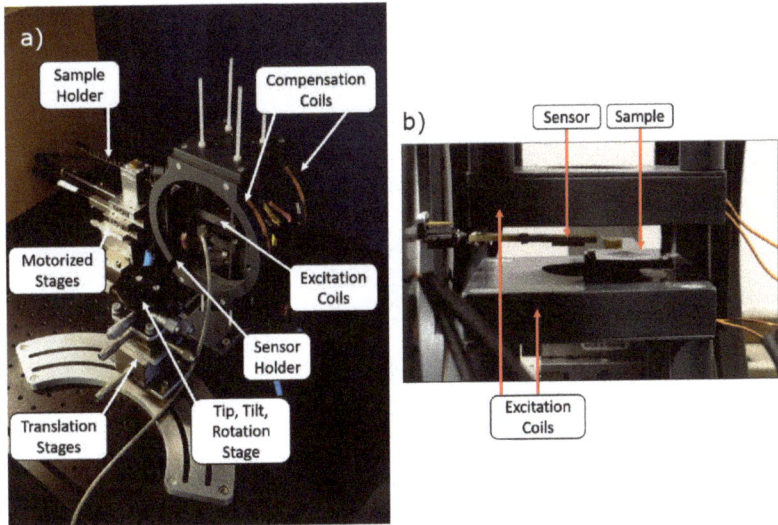

Figure 5. (a) Experimental Magnetic Particle Mapping setup. Manual tip, tilt, rotation and translation stage are used to position the sensor with high precision. Motorized stages are used to move the sample with respect to the sensor. (b) Close-Up of sensor near the sample between the excitation coils.

An image of the MPM sample and the measured magnetic field can be seen in Figure 6. For this, MNP CT100s (fluidMAG, Chemicell, Berlin, Germany) were placed into parallel trenches of a sample holder. The sample has an area of 20 × 20 mm^2. The trenches are 0.5 mm deep and have a length of 1 mm. The filled trenches are 3 mm apart. The total amount of MNP roughly amounts to 300 µg. The magnetoelectric sensor was placed at a distance of circa 2 mm above the sample. Additionally, the associated measured magnetic field can be seen next to it. There exists a translational offset in the origins of both images; hence, the field is not directly above the trenches in direct comparison. This was no influence on the reconstruction.

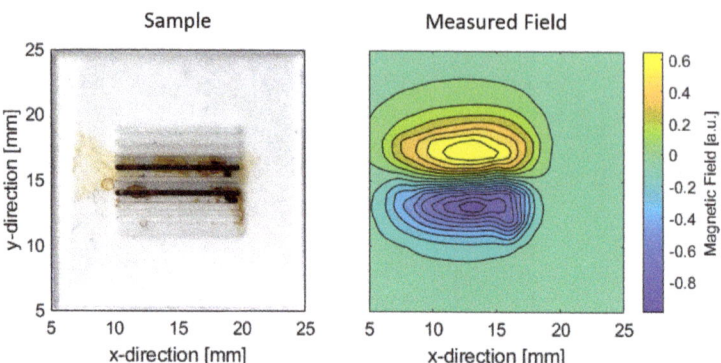

Figure 6. (**Left**): image of the sample with MNP in trenches. (**Right**): measured magnetic field of the sample.

3. Results and Discussion

3.1. Simulation

In the following section, two cases of the blind deconvolution algorithm are investigated, which are listed in Table 1 indicated as Case I and Case II. The cases correspond to the unknown parameters in the model matrix **A**, i.e., sensitive axis **s** and magnetic moment direction **m**. These parameters are either known or unknown and, hence, need to be estimated. Case I for **s** known and **m** is unknown (○), and vice versa (×) they are equivalent due to the bilinear relationship in Equation (2). Case 0 refers to the normal deconvolution when the model is correctly known and will not be treated. The simulations are performed without noise for a maximum number of 500 iterations if not stated otherwise.

Table 1. Investigated cases.

	s Known	s Unknown
m known	Case 0	Case I
m unknown	Case I	Case II

3.1.1. Case I

In the following example, an unknown sensitive axis of the sensor is taken and the magnetic moment direction is known. Even though the cost function is strictly convex (refer to Equation (19)), the reconstruction for any model combination (set of s_i in Equation (19)) could explain the data. What is important is then the non-negativity constraint, such that it acts as a guide to find the correct model. For the case where the magnetic moment direction is known and the sensitive axis has to be estimated while computing the MNP distribution, Algorithm 1 will be used. For each MNP distribution update, the estimated sensitive axis is updated.

An overview of the iterations of the inversion can be seen in Figure 7. As ground truth, the letters "B7" were used. The dimensions were 20×20 mm^2 with 50×50 equidistant points at a z-distance of 1 mm. The data for the reconstruction were computed via a sensitive axis that lies in the x-y plane at an angle of $0°$ (x-direction). The algorithm is initiated via a sensitive axis direction of $90°$. It can be observed that the angle approaches a value of $1°$ for the 50th iteration and that the MNP reconstruction yields the letters "B7", which indicates that the algorithm is able to simultaneously find the MNP distribution and the sensitive axis direction.

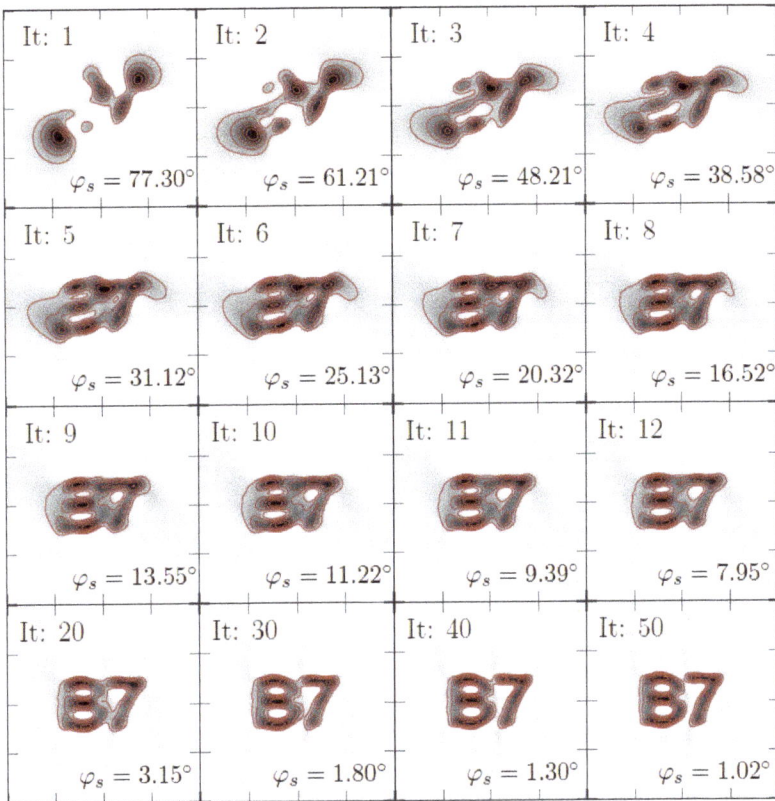

Figure 7. Adaptive reconstruction to iterate the MNP distribution **x** and the sensitive axis **s** simultaneously. Shown are individual iterations as denoted in the upper left corner of each subplot. In the lower right corner of each subplot, the estimated sensitive axis direction (polar) is shown. The ground truth for this axis has an angle of $\varphi_m = 0°$. The black color indicates MNPs, and the red lines are level sets to guide the eye. Dimensions of each square correspond to an area of 20×20 mm^2.

To further investigate the algorithm, a true projection axis for the data is chosen in the x-direction. The algorithm is initiated using a projection axis for all spatial directions that lie on the unit sphere. To quantify the ability to correctly estimate the MNP distribution, the correlation coefficient is taken of the final iteration. Pearson's correlation coefficient (CC) acts as a measure of spatial accuracy of the reconstructed particle distribution compared to the ground truth. The absolute value of the correlation coefficient from the MNP reconstruction for the *initial* sensitive axis direction is used as a radius for that direction. In the case of the correct estimation of the projection axis for all initial directions, the result would be a sphere. Figure 8 shows the results. It can be seen that, for the initial sensitive axes in the upper half plane ($z \geq 0$), the algorithm converges to the correct MNP distribution, indicated by the large correlation coefficient. For some regions in the lower half plane, the algorithm fails to converge to the correct MNP distribution. However, if a good guess is taken for the true sensitive axis, the algorithm will simultaneously find the sensitive axis and the MNP distribution.

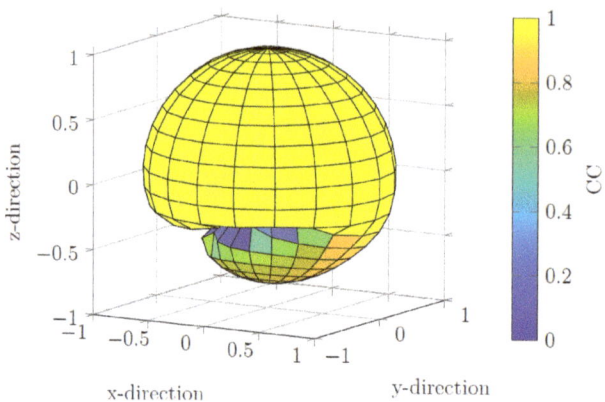

Figure 8. Correlation coefficient for different initial sensitive axis directions. The true sensitive axis lies in the x direction. For the upper half space ($z \geq 0$), the correlation coefficients form a sphere, indicating that the algorithm converges to the true MNP distribution. For the lower half plane, there are regions where the right MNP distribution is not found via the algorithm.

3.1.2. Case II

Next, the case when both the projection axis and the magnetic moment orientation are unknown is considered. In this case, the estimation procedure involves updating the particle distribution, followed by an update of the projection axis of the sensor and followed by an update on the magnetic moment direction. Whether the projection axis or the magnetic moment direction is updated first is irrelevant, because there exists an ambiguity between the magnetic moment direction and the projection axis, as is apparent from the bilinear form that dictates the impulse response of the system and the fact that the matrix of the bilinear form is symmetric (refer to Equation (2)).

To investigate the reliability of the proposed algorithm (see Algorithm 2), one sweeps the parameter space for all orientations of the projection axis and magnetic moment direction. We chose the correct (i.e., belonging to the model that gave rise to the data) projection axis and magnetic moment direction in the x and z directions, respectively. In addition, a box constraint on the projection axis **s** and magnetic moment direction **m** in the form of a predetermined half-space is imposed. That is, the x component of the projection axis cannot be negative, and the z component of the magnetic moment direction cannot be negative. Furthermore, it is imposed that the projection axis and magnetic moment direction is of unit length, which is implemented via projection operators $\mathbf{P}_C^{(1)}$ and $\mathbf{P}_C^{(2)}$. A non-negativity constraint on the particle distribution **x** is imposed as well.

In Figure 9, one can observe the correlation coefficient of the estimated MNP distribution to the ground truth as well as the angle differences of the estimated projection axis of the sensor and the angle difference of the estimated magnetic moment direction (with respect to the ground truths, respectively). It can be seen that correlation coefficient CC is high for most parameter combinations of the directions of sensitive axis **s** and dipole orientation **m**, as indicated by their direction angles in spherical coordinates. For the dipole moment **m**, angles θ_m (polar) and φ_m (elevation) describe the direction, and for the sensitive axis **s**, angles θ_s (polar) and φ_s (elevation) describe the direction. Furthermore, it can also be seen that, for initial parameter combinations where the correlation coefficient is large, the angles between the true and estimated directions of **s** and **m** are small. This indicates that the algorithm is capable of finding the correct MNP distribution while also estimating the directions of **s** and **m** correctly.

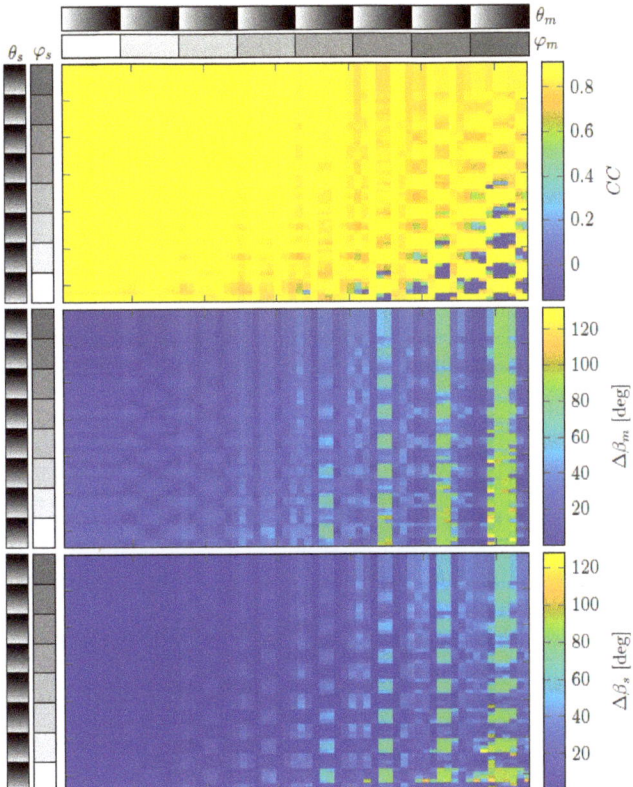

Figure 9. (**Top**): Correlation coefficient of the reconstruction to the ground truth. (**Middle**): Angle error of the magnetic dipole direction. (**Bottom**): Angle error of the sensitive axis direction. All points in the graph are results for a different combination of an initial dipole field direction, indicated by angles θ_m (polar) and φ_m (elevation), and sensitive axis directions, indicated by angles θ_s (polar) and φ_s (elevation). The grey colorbars denote the angle (white is 0° and black is 360°).

3.2. Experiment

Reconstruction

The reconstruction is performed via using an orientation of the sensitive axis that is 20° off the true axis. Not knowing this prior to measurement results in the formation of artifacts in the reconstruction, as can be seen in Figure 10. In direct comparison, the reconstruction using Algorithm 2 results in the reconstruction having significantly less artifacts, being more localized and having a better resolution, because the trenches can roughly be imaged individually. We suspect that the sensor geometry has to be considered in the model for a more accurate reconstruction and better resolution. As of now, the sensor is regarded as point-like. A discussion of the notion of resolution in inverse problems can be found in Appendix A. Additionally, for further enhancement on the image, since fringes are still present around the reconstruction, shrinkage (soft thresholding) can be applied via a projection onto the scaled L^1 ball in addition to the projection into the non-negative orthant. As can be seen from the Figure, fringes are suppressed, and a clearer reconstruction is formed. The choice of the magnitude of the L^1 ball is not chosen arbitrarily but can be roughly estimated from the measurement itself. The discussion on this is shown in Appendix B.

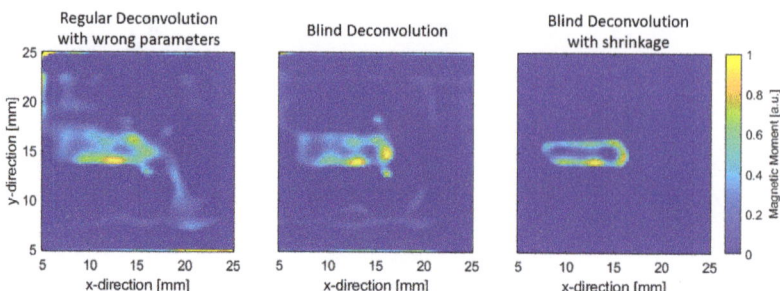

Figure 10. (**Left**): regular deconvolution with sensitive axis off by 20°; (**middle**): Blind Deconvolution as outlined in Algorithm 2; (**right**): Blind Deconvolution with additional shrinkage by soft thresholding.

4. Conclusions

It was shown that the proposed adaptive inversion scheme is able to estimate both the model parameters and MNP distribution simultaneously. The approach shown is able to overcome unknown initial information, such as sensor sensitivity direction, and estimate it correctly. Thus, the generalized model can be used in circumstances where the sensor sensitive axis is not known exactly or incorrectly measured; thereby, it reduces sources of error in the model for better reconstructions, which ultimately improves imaging applications.

Author Contributions: R.-M.F. developed the model and algorithms and carried out MPM measurements and wrote the script. F.F. is the supervisor of R.-M.F. and is the principal investigator of the project. All authors have read and agreed to the published version of the manuscript.

Funding: This work was funded by the German Research Foundation (Deutsche Forschungsgemeinschaft, DFG) through the Collaborative Research Centre CRC 1261 "Magnetoelectric Sensors: From Composite Materials to Biomagnetic Diagnostics" via project B7.

Institutional Review Board Statement: Not applicable.

Informed Consent Statement: Not applicable.

Data Availability Statement: The data that support the findings of this study are available from the corresponding author upon request.

Conflicts of Interest: The authors declare no conflict of interest. The funders had no role in the design of the study; in the collection, analyses or interpretation of data; in the writing of the manuscript; or in the decision to publish the results.

Abbreviations

The following abbreviations are used in this manuscript:

MPM Magnetic Particle Mapping;
MNP Magnetic Nanoparticle;
PSF Point-Spread-Function.

Appendix A. On Notion of Resolution in Inverse Problems

Varieties in approaches of solving inverse problems can be roughly categorized into two types: linear and nonlinear inversion schemes. This refers to the mapping of the measurement data back to the underlying distribution that gave rise to the data. If the inversion is linear, a linear inversion operator can be constructed (i.e., a matrix), and the inversion is independent on the distribution (e.g., MNP) and, hence, depends on the model only. The construction of an inverse operator is only performed to an approximate degree—no full inverse operator will be applied since overfitting would result in erroneous reconstructions that do not reflect reality. Here, statements can be made about the effectiveness of the imaging technique on a general basis. A nonlinear inversion scheme is dependent on the underlying distribution; hence, no general statements can be made. In the following, these statements will be elucidated more in depth.

Resolution in an imaging system refers to its ability to resolve features that are spatially close. The resolution for an imaging system such as a microscope normally uses the Point-Spread Function as a figure of merit. This quantity reflects the spatial spread of a point source input to the imaging system. With the use of deconvolution techniques, the resolution can be enhanced, and the spread of the Point-Spread Function can be reduced.

The proper assessment of the enhancement of a linear imaging system can be performed via the so-called resolution matrix $\mathbf{R}$ [35,36]. The meaning of this operator is that no features can be reconstructed that are sharper than given by the columns, which in turn makes using inverse operators' intrinsic resolution limited [37,38]. To introduce this matrix, we start off with the system of equations that relate particle distribution $\mathbf{x}$ with magnetic field $\mathbf{b}$ via model matrix $\mathbf{A}$.

$$\mathbf{Ax} = \mathbf{b}. \tag{A1}$$

The approach for solving this system of equations uses an inverse operator that, when applied to the left hand side of the equation, yields just the particle distribution $\mathbf{x}$.

$$\mathbf{A}^\dagger \mathbf{Ax} = \mathbf{x}. \tag{A2}$$

Here, $\mathbf{A}^\dagger$ denotes the pseudoinverse of model matrix $\mathbf{A}$. However, under normal circumstances, only an approximate inverse operator is used, which we will denote as $\mathbf{A}_k^\dagger$. Subscript k refers to an iteration number for different approaches of computing the pseudoinverse, which will be shown further below. The reason that only an approximate inverse operator is used is that, otherwise, noise would be amplified in the measurement vector $\mathbf{b}$, which would yield unreliable/unphysical results for the particle distribution $\mathbf{x}$. We can, thus, write the following.

$$\mathbf{A}_k^\dagger \mathbf{Ax} = \mathbf{A}_k^\dagger \mathbf{b}$$
$$\mathbf{R}_k \mathbf{x} = \mathbf{A}_k^\dagger \mathbf{b}. \tag{A3}$$

Operator $\mathbf{R}_k = \mathbf{A}_k^\dagger \mathbf{A}$ refers to the resolution matrix for a given approximate pseudoinverse. To compute the resolution matrix, we choose two different approaches: the singular value decomposition (SVD) and gradient descent (GD). The resolution matrices associated with these approaches are as follows.

$$\mathbf{R}_{\text{GD},k} = \alpha \mathbf{V} \left(\sum_{n=0}^{k-1} \left(\mathbf{I} - \alpha \mathbf{S}^2 \right)^n \right) \mathbf{S}^2 \mathbf{V}^\mathsf{T} \tag{A4}$$

$$\mathbf{R}_{\text{SVD},k} = \mathbf{V}_k \mathbf{S}_k^{-1} \mathbf{U}_k^\mathsf{T} \mathbf{A} \tag{A5}$$

Here, $\mathbf{A} = \mathbf{U}\mathbf{S}\mathbf{V}^\mathsf{T}$, $\alpha = \|\mathbf{A}^\mathsf{T}\mathbf{A}\|^{-1}$ and subscript k either refers to the number of iterations for gradient descent or the number of left and right singular vectors corresponding to the first k singular values (i.e., such that $\mathbf{A} \approx \mathbf{U}_k \mathbf{S}_k \mathbf{V}_k^\mathsf{T}$).

Let us look into the meaning of these matrices in more detail. The resolution matrix can be thought of in two ways:

- The columns can be understood as the spatial spread of a delta-like input, i.e., the Point-Spread Function.
- The rows can be understood as the convolution of the MNP distribution with the Point-Spread Function. It is the (scaled) averaging function of the system.

As observed in Figure A1, the columns (blue enclosure of the matrix) represent the spatial spread of a delta-like input at particle position j. The red enclosure in the matrix refers to the weighted (the weights can be negative) sum of all field contributions from all

MNP, such that it refers to an *averaging function* of the MNP responses. Due to the fact that the system is a linear space-invariant system, the resolution is the same for all positions. *Resolution* could heuristically be chosen to be the full width half maximum of the peak of the columns. It has to be noted that the resolution may be spatially anisotropic depending on the shape of the impulse response of the imaging system.

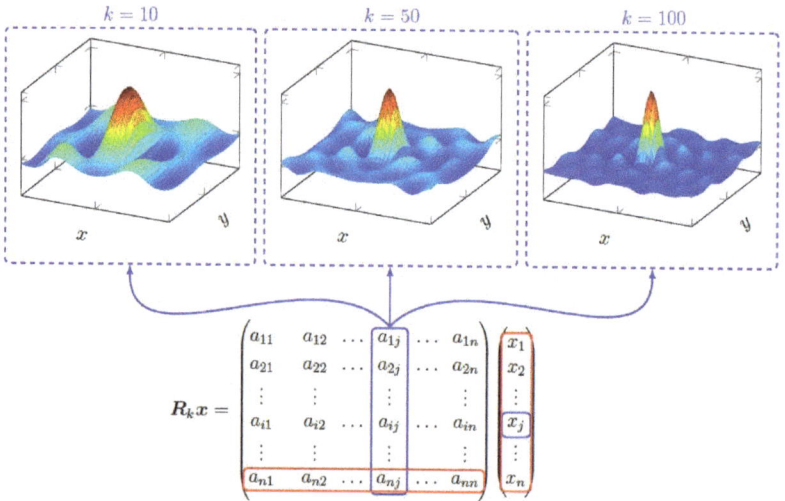

Figure A1. The columns of the resolution matrix $\mathbf{R}_k$ reflects the spread of a delta-like input, which can be thought of as the Point-Spread Function of the system. For the inverse operator for different iterations k, the resolution matrix attains a more delta-like response, which means that the spread is reduced and that resolution is enhanced

The peaks as shown in Figure A1 become narrower with increasing iteration numbers. This means that the resulting overall resolution is given by the last iteration we were able to perform without fitting noise. For this, there are several criteria that can be used for the termination of the reconstruction [2]. It has to be emphasized that the resolution matrix relies on the analytic expressions presented here such that procedures such as the proximal gradient algorithm for the L^1 projection cannot meaningfully produce a resolution matrix. The only cases where an analytic expression can be found is if the regularizer is differentiable (e.g., Tikhonov regularization), semiconvergent properties of gradient descent are used [2] or truncation via a singular value decomposition is applied.

A more general perspective on the resolution matrix (or rather, resolution map) is that it can be viewed as the effect of a single input on the vector **b**, which subsequently needs to be inverted [39]. For this, we look at the change of reconstruction $\delta \mathbf{x}_i$ from unit change $\hat{\mathbf{e}}_i$ in **x** via the inversion operation $\mathcal{G}$.

$$\delta \mathbf{x}_i = \mathcal{G}(\mathbf{A}(\mathbf{x} + \hat{\mathbf{e}}_i)) - \mathcal{G}(\mathbf{A}\mathbf{x}) \tag{A6}$$

In case of linearity, that is, if an analytic expression is found via an approximate (generalized) inverse $\mathbf{A}_k^\dagger$, then we have the following.

$$\begin{aligned}
\delta \mathbf{x}_i &= \mathcal{G}(\mathbf{A}(\mathbf{x} + \hat{\mathbf{e}}_i)) - \mathcal{G}(\mathbf{A}\mathbf{x}) \\
&= \mathbf{A}_k^\dagger \mathbf{A} \hat{\mathbf{e}}_i + \mathbf{A}_k^\dagger \mathbf{A}\mathbf{x} - \mathbf{A}_k^\dagger \mathbf{A}\mathbf{x} \\
&= \mathbf{R}_k \hat{\mathbf{e}}_i.
\end{aligned} \tag{A7}$$

Thus, in the linear case, output $\delta \mathbf{x}_i$ from a unit change directly yields a column of the resolution matrix, which tells us about the Point-Spread Function/spatial spread of the parameter x_i of particle distribution **x**. Important to point out is that this inversion process is only dependent on the model itself and not on the particle distribution. This is a direct result of the linearity of the inversion process.

On the other hand, if the inverse map does not obey linearity, which is the case for, e.g., L^1 regularization, we have the following.

$$\mathcal{G}(\mathbf{A}(\mathbf{x} + \hat{\mathbf{e}}_i)) \neq \mathcal{G}(\mathbf{A}\mathbf{x}) + \mathcal{G}(\mathbf{A}\hat{\mathbf{e}}_i). \tag{A8}$$

As a result, spatial spread δx_i of parameter x_i depends on particle distribution **x** itself, and a general statement of the resolution for a model inversion procedure, when the inverse map is not linear, cannot be given. Therefore, assessment of resolution is in this case is only beneficial if a standard reference is used such that the results can be comparable.

Appendix B. Estimation of Magnetic Content

In the following, it will be shown that the magnetic content of an imaging experiment can be estimated from the measurement alone. For this, we consider a measurement in an x–y plane with equidistant measurement positions. We further set the magnetic moment direction into the z direction and sensitive axis direction in the x direction. The MNP distribution is distributed in an x–y plane below the measurement plane. We, thus, write for magnetic moment **m** as follows:

$$\mathbf{m} = \|\mathbf{m}\| \begin{pmatrix} 0 \\ 0 \\ 1 \end{pmatrix} \tag{A9}$$

and for the sensitive axis **s**, we have the following.

$$\mathbf{s} = \|\mathbf{s}\| \begin{pmatrix} 1 \\ 0 \\ 0 \end{pmatrix} \tag{A10}$$

The associated magnetic field of a point dipole is then the following.

$$B_x = \frac{\mu_0 \|\mathbf{m}\| \|\mathbf{s}\|}{4\pi} \frac{3xz}{(x^2 + y^2 + z^2)^{5/2}}. \tag{A11}$$

The projection above is antisymmetric with respect to the x direction. Integration of this projection in the x-direction yields the following.

$$\int_{-\infty}^{x} B_x dx = -\frac{\mu_0 \|\mathbf{m}\| \|\mathbf{s}\|}{4\pi} \frac{z}{(x^2 + y^2 + z^2)^{3/2}}. \tag{A12}$$

The function is of equal sign from $-\infty$ to $+\infty$ for the x and y variables. The next step involves integrating over the whole $x - y$ plane. The result is the following.

$$\int_{-\infty}^{\infty} \int_{-\infty}^{\infty} -\frac{\mu_0 \|\mathbf{m}\| \|\mathbf{s}\|}{4\pi} \frac{z}{(x^2 + y^2 + z^2)^{3/2}} dx dy = -\frac{\mu_0 \|\mathbf{m}\| \|\mathbf{s}\|}{2}. \tag{A13}$$

Thus, the given procedure is proportional to the magnetic moment, which is, mathematically speaking, the L^1 norm of the negative integrated dipole field projection. Therefore, one can calculate the magnetic moment directly from the data, given that a sufficiently large region of the magnetic field is measured to approximate the procedure given above. It is important to point out that the result is independent of the z-distance. On one hand, this means that one can estimate the magnetic moment without knowing the z-distance, while on the other hand it means that there is no depth information from this procedure. To generalize this approach, given an arbitrary spatial distribution of magnetic moments and given that the measured projection is equal to the example given above, we denote the magnetic field of a magnetic moment as B_D. We can write the magnetic field associated with a moment distribution, given as B_L, as follows.

$$B_L(\mathbf{r}) = \int_\Omega B_D(\mathbf{r}-\mathbf{r}',\mathbf{m})\rho(\mathbf{r}')d^3\mathbf{r}'. \tag{A14}$$

As above, we assume the same direction of magnetic moment $\mathbf{m}$ and integrate in the x-direction and then integrate over the $x-y$ plane. Since the integral is not dependent on the z-distance (as explained above), one obtains the following result.

$$-\frac{2}{\mu_0\|\mathbf{m}\|\|\mathbf{s}\|}\int_{-\infty}^{\infty}\int_{-\infty}^{\infty}\int_{-\infty}^{x} B_L(\mathbf{r})dxdxdy = \int_\Omega \rho(\mathbf{r}')d^3\mathbf{r}'. \tag{A15}$$

Since moment distribution ρ is non-negative (i.e., $\rho \geq 0$), the following statement is obvious.

$$\|\rho\|_1 = \int_\Omega \rho(\mathbf{r}')d^3\mathbf{r}'. \tag{A16}$$

Thus, the total amount of magnetic content can be computed from the magnetic field if the measured projection is the same as in the form above. It would also work for the projection in the y direction. In this way, one can infer the total amount of particles from the measurement itself and can use this for regularization via projection onto the scaled L^1 ball in the projected gradient method. One has to point out that, because the result is independent on the z distance, any reconstruction from *any* height that is able to construct vector $\mathbf{b}$ via estimate $\mathbf{Ax}$ yields the same amount of MNP content.

References

1. Hansen, P.C. Deblurring Images: Matrices, Spectra and Filtering. *J. Electron. Imaging* **2007**, *17*, 019901. [CrossRef]
2. Hansen, P.C. *Discrete Inverse Problems: Insight and Algorithms*; Society for Industrial and Applied Mathematics: Philadelphia, PA, USA, 2010.
3. Yang, Y.; Galatsanos, N.P.; Stark, H. Projection-based blind deconvolution. *J. Opt. Soc. Am. A* **1994**, *11*, 2401. [CrossRef]
4. Kundur, D.; Hatzinakos, D. Blind image deconvolution revisited. *IEEE Signal Process. Mag.* **1996**, *13*, 61–63. [CrossRef]
5. Levin, A.; Weiss, Y.; Durand, F.; Freeman, W.T. Understanding and evaluating blind deconvolution algorithms. In Proceedings of the 2009 IEEE Conference on Computer Vision and Pattern Recognition, CVPR 2009, Miami, FL, USA, 20–25 June 2009; pp. 1964–1971. [CrossRef]
6. Friedrich, R.M.; Zabel, S.; Galka, A.; Lukat, N.; Wagner, J.M.; Kirchhof, C.; Quandt, E.; McCord, J.; Selhuber-Unkel, C.; Siniatchkin, M.; et al. Magnetic particle mapping using magnetoelectric sensors as an imaging modality. *Sci. Rep.* **2019**, *9*, 1–11 . [CrossRef] [PubMed]
7. Durdaut, P. Ausleseverfahren und Rauschmodellierung für Magnetoelektrische und Magnetoelastische Sensorsysteme. Ph.D. Thesis, Christian-Abrechts-Universität zu Kiel, Kiel, Germany, 2019. [CrossRef]
8. Weizenecker, J.; Gleich, B.; Rahmer, J.; Dahnke, H.; Borgert, J. Three-dimensional real-time in vivo magnetic particle imaging. *Phys. Med. Biol.* **2009**, *54*, L1. [CrossRef] [PubMed]
9. Knopp, T.; Buzug, T.M. *Magnetic Particle Imaging: An Introduction to Imaging Principles and Scanner Instrumentation*; Springer: Berlin/Heidelberg, Germany, 2012; pp. 1–204. [CrossRef]
10. Graeser, M.; Knopp, T.; Szwargulski, P.; Friedrich, T.; Von Gladiss, A.; Kaul, M.; Krishnan, K.M.; Ittrich, H.; Adam, G.; Buzug, T.M. Towards Picogram Detection of Superparamagnetic Iron-Oxide Particles Using a Gradiometric Receive Coil. *Sci. Rep.* **2017**, *7*, 1–13. [CrossRef] [PubMed]
11. Zhong, J.; Schilling, M.; Ludwig, F. Spatial and temperature resolutions of magnetic nanoparticle temperature imaging with a scanning magnetic particle spectrometer. *Nanomaterials* **2018**, *8*, 866. [CrossRef] [PubMed]
12. Zhong, J.; Schilling, M.; Ludwig, F. Magnetic nanoparticle temperature imaging with a scanning magnetic particle spectrometer. *Meas. Sci. Technol.* **2018**, *29*, 115903. [CrossRef]
13. Zhong, J.; Schilling, M.; Ludwig, F. Excitation frequency dependence of temperature resolution in magnetic nanoparticle temperature imaging with a scanning magnetic particle spectrometer. *J. Magn. Magn. Mater.* **2019**, *471*, 340–345. [CrossRef]
14. Zhong, J.; Schilling, M.; Ludwig, F. Magnetic nanoparticle-based biomolecule imaging with a scanning magnetic particle spectrometer. *Nanotechnology* **2020**, *31*, 225101. [CrossRef]
15. Liebl, M.; Steinhoff, U.; Wiekhorst, F.; Haueisen, J.; Trahms, L. Quantitative imaging of magnetic nanoparticles by magnetorelaxometry with multiple excitation coils. *Phys. Med. Biol.* **2014**, *59*, 6607–6620. [CrossRef] [PubMed]

16. Rühmer, D.P. Zweidimensionale Scanning-Magnetrelaxometrie mit Fluxgate-Sensoren. Ph.D. Thesis, Technische Universität Braunschweig, Braunschweig, Germany, 2012. [CrossRef]
17. Jaufenthaler, A.; Schier, P.; Middelmann, T.; Liebl, M.; Wiekhorst, F.; Baumgarten, D. Quantitative 2D magnetorelaxometry imaging of magnetic nanoparticles using optically pumped magnetometers. *Sensors* **2020**, *20*, 753. [CrossRef] [PubMed]
18. Schier, P.; Liebl, M.; Steinhoff, U.; Handler, M.; Wiekhorst, F.; Baumgarten, D. Optimizing Excitation Coil Currents for Advanced Magnetorelaxometry Imaging. *J. Math. Imaging Vis.* **2020**, *62*, 238–252. [CrossRef]
19. Oh, J.; Feldman, M.D.; Kim, J.; Condit, C.; Emelianov, S.; Milner, T.E. Detection of magnetic nanoparticles in tissue using magneto-motive ultrasound. *Nanotechnology* **2006**, *17*, 4183–4190. [CrossRef]
20. Hu, G.; He, B. Magnetoacoustic imaging of magnetic iron oxide nanoparticles embedded in biological tissues with microsecond magnetic stimulation. *Appl. Phys. Lett.* **2012**, *100*, 3–5. [CrossRef]
21. Mehrmohammadi, M.; Qu, M.; Ma, L.L.; Romanovicz, D.K.; Johnston, K.P.; Sokolov, K.V.; Emelianov, S.Y. Pulsed magneto-motive ultrasound imaging to detect intracellular accumulation of magnetic nanoparticles. *Nanotechnology* **2011**, *22*, 415105. [CrossRef]
22. Sepúlveda, N.G.; Thomas, I.M.; Wikswo, J.P. Magnetic Susceptibility Tomography for Three-Dimensional Imaging of Diamagnetic and Paramagnetic Objectcts. *IEEE Trans. Magn.* **1994**, *30*, 5062–5069. [CrossRef]
23. Ficko, B.W.; Nadar, P.M.; Hoopes, P.J.; Diamond, S.G. Development of a magnetic nanoparticle susceptibility magnitude imaging array. *Phys. Med. Biol.* **2014**, *59*, 1047–1071. [CrossRef]
24. Ficko, B.W.; Nadar, P.M.; Diamond, S.G. Spectroscopic AC susceptibility imaging (sASI) of magnetic nanoparticles. *J. Magn. Magn. Mater.* **2015**, *375*, 164–176. [CrossRef]
25. Ficko, B.W.; Giacometti, P.; Diamond, S.G. Extended arrays for nonlinear susceptibility magnitude imaging. *Biomed. Tech.* **2015**, *60*, 457–463. [CrossRef]
26. Van Durme, R.; Coene, A.; Crevecoeur, G.; Dupre, L. Model-based optimal design of a magnetic nanoparticle tomographic imaging setup. In Proceedings of the International Symposium on Biomedical Imaging, Washington, DC, USA, 4–7 April 2018; pp. 369–372. [CrossRef]
27. Yorulmaz, O.; Demirel, O.B.; Çukur, T.; Saritas, E.U.; Çetin, A.E. A Blind Deconvolution Technique Based on Projection Onto Convex Sets for Magnetic Particle Imaging. *arXiv* **2017**, arXiv:1705.07506.
28. Huong Giang, D.T.; Dang, D.X.; Toan, N.X.; Tuan, N.V.; Phung, A.T.; Duc, N.H. Distance magnetic nanoparticle detection using a magnetoelectric sensor for clinical interventions. *Rev. Sci. Instrum.* **2017**, *88*, 015004. [CrossRef]
29. Xi, H.; Qian, X.; Lu, M.C.; Mei, L.; Rupprecht, S.; Yang, Q.X.; Zhang, Q.M. A Room Temperature Ultrasensitive Magnetoelectric Susceptometer for Quantitative Tissue Iron Detection. *Sci. Rep.* **2016**, *6*, 1–7. [CrossRef]
30. Lukat, N.; Friedrich, R.M.; Spetzler, B.; Kirchhof, C.; Arndt, C.; Thormählen, L.; Faupel, F.; Selhuber-Unkel, C. Mapping of magnetic nanoparticles and cells using thin film magnetoelectric sensors based on the delta-E effect. *Sens. Actuators A Phys.* **2020**, *309*, 112023. [CrossRef]
31. Wing, G.M. *A Primer on Integral Equations of the First Kind*; Society for Industrial and Applied Mathematics: Philadelphia, PA, USA, 1991. [CrossRef]
32. Nakamura, G.; Potthast, R. *Inverse Modeling: An Introduction to the Theory and Methods of Inverse Problems and Data Assimilation*; Society for Industrial and Applied Mathematics: Philadelphia, PA, USA, 2015; pp. 1–509. [CrossRef]
33. Bolduc, E.; Knee, G.C.; Gauger, E.M.; Leach, J. Projected gradient descent algorithms for quantum state tomography. *Npj Quantum Inf.* **2017**, *3*, 1–9. [CrossRef]
34. Lage, E.; Kirchhof, C.; Hrkac, V.; Kienle, L.; Jahns, R.; Knöchel, R.; Quandt, E.; Meyners, D. Biasing of magnetoelectric composites. *Nat. Mater.* **2012**, *11*, 523–529. [CrossRef]
35. De Peralta Menendez, R.G.; Gonzalez Andino, S.L.; Lütkenhöner, B. Figures of merit to compare distributed linear inverse solutions. *Brain Topogr.* **1996**, *9*, 117–124. [CrossRef]
36. An, M. A simple method for determining the spatial resolution of a general inverse problem. *Geophys. J. Int.* **2012**, *191*, 849–864. [CrossRef]
37. Chen, S.S.; Donoho, D.L.; Saunders, M.A. Atomic decomposition by basis pursuit. *SIAM Rev.* **2001**, *43*, 129–159. [CrossRef]
38. Bertero, M.; Boccacci, P. *Introduction to Inverse Problems in Imaging*; IOP Publishing Ltd.: Bristol, UK, 1998; doi:10.1887/0750304359. [CrossRef]
39. Bangerth, W. A Notion of Resolution in Inverse Problems. Computational Science Stack Exchange. Available online: https://scicomp.stackexchange.com/q/36537 (accessed on 18 December 2020).

Article

Processing Chain for Localization of Magnetoelectric Sensors in Real Time

Christin Bald and Gerhard Schmidt *

Digital Signal Processing and System Theory, Institute of Electrical Engineering and Information Technology, Faculty of Engineering, Kiel University, 24143 Kiel, Germany; cbal@tf.uni-kiel.de
* Correspondence: gus@tf.uni-kiel.de; Tel.: +49-431-880-6125

Abstract: The knowledge of the exact position and orientation of a sensor with respect to a source (distribution) is essential for the correct solution of inverse problems. Especially when measuring with magnetic field sensors, the positions and orientations of the sensors are not always fixed during measurements. In this study, we present a processing chain for the localization of magnetic field sensors in real time. This includes preprocessing steps, such as equalizing and matched filtering, an iterative localization approach, and postprocessing steps for smoothing the localization outcomes over time. We show the efficiency of this localization pipeline using an exchange bias magnetoelectric sensor. For the proof of principle, the potential of the proposed algorithm performing the localization in the two-dimensional space is investigated. Nevertheless, the algorithm can be easily extended to the three-dimensional space. Using the proposed pipeline, we achieve average localization errors between 1.12 cm and 6.90 cm in a localization area of size 50 cm × 50 cm.

Keywords: localization; magnetoelectric sensors; real time; pose estimation

1. Introduction

For the correct solution of inverse problems, such as source reconstruction of biomedical sources, it is essential to know the exact position and orientation of the measuring sensors with respect to the source besides measuring the biomedical signals. Especially in magnetic measurements the sensors do not necessarily have a fixed position and orientation. Thus, a determination of the position and orientation at the beginning of a measurement is not sufficient. Much more desirable is a continuous estimation of the sensor's position and orientation simultaneously with the measurement [1,2].

Magnetic tracking systems are used in many applications, e.g., in indoor positioning systems [3,4] or to locate medical devices inside the body [5,6]. Moreover, magnetic localization approaches are used to determine the position of the subject relative to the sensor array in biomagnetic measurements. A procedure for determining the subject relative to the measuring sensor array, either once at the beginning or also simultaneously with the measurement, was presented in [7]. In [1,2] a method for determining the positions of the individual sensors in a flexible on-scalp MEG system relative to the subject was investigated, which can also be applied during measurement.

Until now, mainly SQUIDs (Super Conducting Quantum Interference Devices) [8] and recently OPMs (Optically Pumped Magnetometers) [9,10] are used for the measurement of biomagnetic signals. Unfortunately, these sensors require a magnetically well shielded environment and are therefore inconvenient in operation. Magnetoelectric sensors, on the other hand, do not require any shielding, no expensive cooling system, and are also very small in size, which makes them ideal for array applications. The sensors are composed of a magnetostrictive and a piezoelectric layer and use the resonant structure of a cantilever [11]. Detection limits in the sub-nT regime have been reached recently [12–15] using modulation techniques such as the ΔE-effect [16] for the detection of low-frequency magnetic fields.

Citation: Bald, C.; Schmidt, G. Processing Chain for Localization of Magnetoelectric Sensors in Real Time. *Sensors* **2021**, *21*, 5675. https://doi.org/10.3390/s21165675

Academic Editor: Daniel Ramos

Received: 30 July 2021
Accepted: 20 August 2021
Published: 23 August 2021

Publisher's Note: MDPI stays neutral with regard to jurisdictional claims in published maps and institutional affiliations.

Copyright: © 2021 by the authors. Licensee MDPI, Basel, Switzerland. This article is an open access article distributed under the terms and conditions of the Creative Commons Attribution (CC BY) license (https://creativecommons.org/licenses/by/4.0/).

Consequently, magnetoelectric sensors could be a promising alternative for biomagnetic field measurements in the near future [17].

Since the feasibility of measurements with simultaneous localization using magnetoelectric sensors has been shown in a previous work [15], this work will focus on the processing chain for an enhanced localization of magnetoelectric sensors in real time. However, by adapting the coil excitation signals the method can be used for arbitrary magnetic field sensors. This contribution will step through the general processing overview shown in Figure 1.

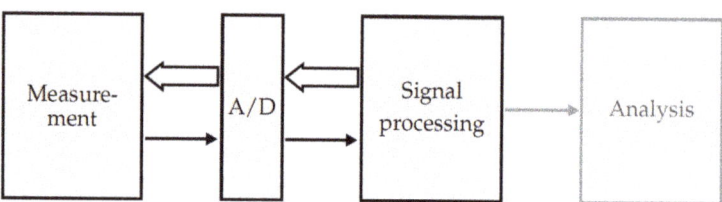

Figure 1. General overview of a medical system operating with magnetic sensors. The measurements are performed simultaneously with localizing the sensors. After transforming the signals into the digital domain, the signals are processed and analyzed. Since the analysis of the measured magnetic signals is not in the focus of this contribution, the corresponding box is depicted in gray.

In Section 2, the magnetoelectric sensor used in this paper will be presented and characterized. After explaining the so-called forward problem in Section 3, the real-time localization approach will be presented in Section 4. The presented localization processing chain will be verified by measurements presented in Section 5. The paper closes with a conclusion and an outlook in Section 6.

2. Magnetoelectric Sensor

For the proof of principle an exchange bias ΔE-effect sensor as depicted in Figure 2a will be used in this contribution. A sensor of the same type has already been used in a previous work [15]. The sensor is based on a polysilicon cantilever with a size of 1 mm width, 3 mm length, and 50 µm thickness. The cantilever is covered by a 4 µm thick magnetic multilayer (20 × Ta/Cu/MnIr/FeCoSiB) and a 2 µm thick piezoelectric layer (AlN). Further details about the fabrication process of the sensor can be found in [15]. The magnetic multilayer consists of ferromagnetic and antiferromagnetic layers, which ensure the self-biasing of the sensor and thus lead to a shift of the magnetization curve of the sensor [18]. Hence, the sensor can be operated without applying an external bias field, which is especially favorable regarding array applications. The sensor is connected to a low-noise JFET charge amplifier [19] and placed on a printed circuit board. The whole sensor system is encapsulated in a 2.1 mm thick brass cylinder for electrical shielding.

As shown in [15], the localization of the magnetoelectric sensor can be performed simultaneously with a measurement without loss of information or degradation of the signals. The first bending mode was used to localize the sensor, while an artificial heart signal was measured in the second mode using the ΔE-effect. Hence, also in this contribution only frequencies around the first bending mode will be used for the transmission of the localization signals. The amplitude and phase response around the first bending mode of the magnetoelectric sensor used are shown in Figure 2c. The characterization measurements have been performed in a magnetically, electrically, and acoustically shielded environment [11]. The magnitude and phase response of the sensor have been measured applying a magnetic field of $b_{ac} = 1$ µT on the sensor's long axis. The performance of the sensor can be determined as described in [20]. The sensor has a resonance frequency of $f_r = 7.712$ kHz and a -3 dB bandwidth of $bw_{-3dB} = 10.2$ Hz. Since the brass cylinder acts as a low-pass filter with a cut-off frequency of approximately $f_c \approx 1.5$ kHz [15], the sensor's performance will be improved when removing the brass cylinder. Nevertheless, the encapsulation is necessary for electrical shielding and acts as a mechanical protection. Moreover, the limit-of-detection in the first bending mode with brass cylinder is still sufficient, because the coils can simply emit higher field amplitudes. The maximum sensitivity of the sensor is reached when a magnetic field is applied on the sensitive axis of the sensor. However, the sensitive axis of the sensor is not necessarily equal to the long axis of the sensor. There can be a tilt γ between these two axes [15,21], which is visualized in Figure 2b.

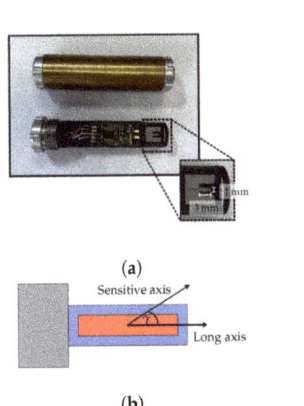
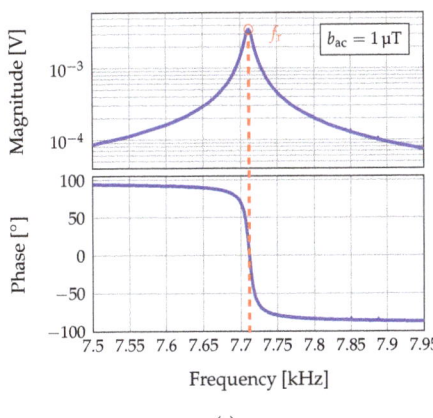

Figure 2. (a) Exchange bias ΔE-effect sensor used in this study. The sensor is based on a cantilever of size 3 mm × 1 mm. The cantilever is placed on a printed circuit board and connected to a low-noise JFET charge amplifier [19]. The sensor is encapsulated by a brass cylinder for electrical shielding and mechanical protection. (b) Visualization of the relationship between the sensitive and the long axis of the sensor. (c) Magnitude and phase response of the sensor in the first bending mode applying a magnetic field of $b_{ac} = 1$ µT. The sensor has a resonance frequency of $f_r = 7.712$ kHz and a bandwidth of $bw_{-3dB} = 10.2$ Hz.

3. Forward Problem

For the localization of the magnetoelectric sensor, coils are placed outside the localization area as shown in Figure 3. If the distance between the sensor and the coil is large enough, the magnetic field of the coil i at the sensor position $\vec{r}_s(t)$ can be approximated by the field of a magnetic dipole [22]:

$$\vec{b}_i(t, \vec{r}_s(t)) = \frac{\mu_0}{4\pi} \frac{3\vec{r}_{cs,i}(t)\left(\vec{m}_{c,i}(t)^T \vec{r}_{cs,i}(t)\right) - \vec{m}_{c,i}(t)\|\vec{r}_{cs,i}(t)\|_2^2}{\|\vec{r}_{cs,i}(t)\|_2^5} \quad (1)$$

Here, μ_0 is the permeability of vacuum, $\vec{m}_{c,i}(t)$ the magnetic dipole moment of the coil i, and $\vec{r}_{cs,i}(t) = \vec{r}_s(t) - \vec{r}_{c,i}$ the distance vector between the sensor at position $\vec{r}_s(t)$ and the coil i at position $\vec{r}_{c,i}$. The superscript T denotes the transpose of the vector. The positions of the coils $\vec{r}_{c,i}$ are fixed during the measurement and therefore time independent. In this study, $N_c = 6$ coils have been used. The coils have an effective diameter of 2.6 cm and consist of about 350 turns of enameled copper wire with a wire cross section of 0.13 mm². The impedances of the six coils, separated into magnitude and phase, are shown in Figure 4.

The signal measured by the sensor

$$u_{in}(t) = h_s(t) * \left(\vec{d}_s^T(t) \sum_{i=1}^{N_c} \vec{b}_i(t, \vec{r}_s(t))\right) \quad (2)$$

can be described as a voltage at the output of the sensor system. At the location of the sensor a superposition of the magnetic fields of the coils is present. Due to the directional characteristic of the sensor $\vec{d}_s(t)$, only a part of the applied magnetic field is picked up. The conversion of magnetic field into voltage by the sensor system (including the charge amplifier) is described by the impulse response $h_s(t)$. Equation (2) is valid at least for the frequencies around the first bending mode [15]. For simplification, no noise sources are considered here.

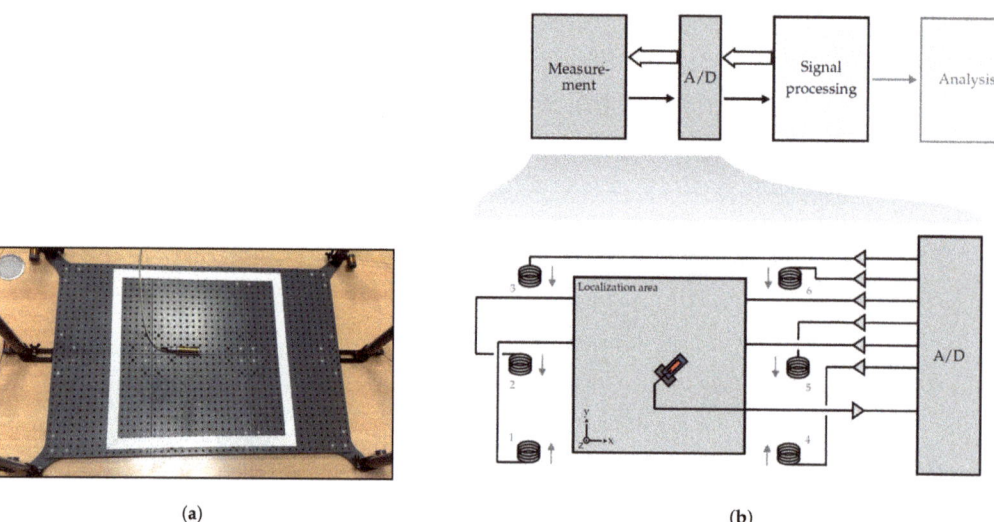

Figure 3. Real (**a**) and schematic (**b**) measurement setup for the localization of magnetoelectric sensors. The coils are placed outside of the localization area and transmit orthogonal signals, which are measured by the sensor. The localization area (box bounded by white stripes in (**a**)) is of size 50 cm × 50 cm.

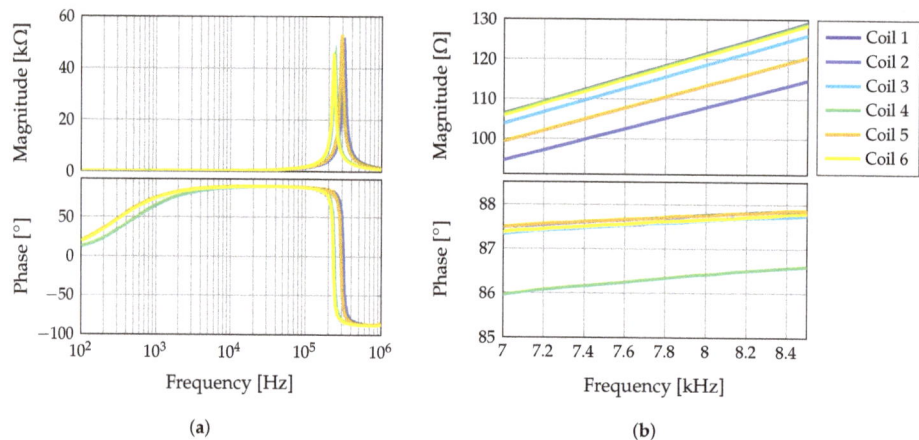

Figure 4. Coil impedances separated into magnitude and phase. In (**a**) the whole spectrum from 100 Hz up to 1 MHz is shown, so that the resonance of the coils can be seen. In (**b**) the frequency range is scaled to the frequency range of the excitation signals.

4. Localization Processing Chain

For the estimation of the sensor's position and orientation an inverse problem must be solved. The processing chain for solving this inverse problem is shown in Figure 5.

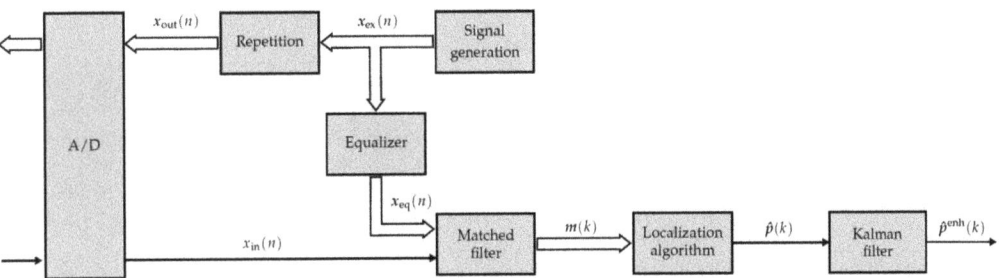

Figure 5. Processing chain for the localization of magnetoelectric sensors in real time. The input signal of the sensor is matched filtered with the equalized coil signals. The matched filter outputs at time lag zero are compared with the lead-field matrix entries. A first order Kalman filter smooths the estimated position-orientation-pairs over time to mitigate possible outliers.

For this purpose, the localization area is first divided into a discrete grid containing N_p different position-orientation-pairs $P = [p_1, \ldots, p_j, \ldots, p_{N_p}]$, with $p_j = [\vec{r}_{p,j}^T, \vec{d}_{p,j}^T]^T$ consisting of a position vector $\vec{r}_{p,j}$ (containing x, y, and z components) and an orientation vector $\vec{d}_{p,j}$ (directivity described by roll φ, pitch θ, and yaw ψ). The lead-field matrix

$$A = [a_1, \ldots, a_j, \ldots, a_{N_p}] = \begin{bmatrix} a_{11} & \cdots & a_{1j} & \cdots & a_{1N_p} \\ \vdots & \ddots & \vdots & \ddots & \vdots \\ a_{i1} & \cdots & a_{ij} & \cdots & a_{iN_p} \\ \vdots & \ddots & \vdots & \ddots & \vdots \\ a_{N_c1} & \cdots & a_{N_cj} & \cdots & a_{N_cN_p} \end{bmatrix} \quad (3)$$

describes the forward problem for the defined position-orientation-pairs in P. That means, the lead-field matrix entry of row i and column j is defined as

$$a_{ij} = \vec{d}_{p,j}^T \frac{3\vec{r}_{cp,ij}\left(\vec{d}_{c,i}^T \vec{r}_{cp,ij}\right) - \vec{d}_{c,i}\|\vec{r}_{cp,ij}\|_2^2}{\|\vec{r}_{cp,ij}\|_2^5} \quad (4)$$

and thus describes the influence of coil i on the sensor, if the sensor would have the position and orientation described by p_j. The distance vector is defined as $\vec{r}_{cp,ij} = \vec{r}_{p,j} - \vec{r}_{c,i}$ and the orientation vector $\vec{d}_{p,j}$ can be described by the angles θ and ψ (φ is always zero here) using [23]:

$$\vec{d}_{p,j} = \begin{pmatrix} \cos(\theta)\cos(\psi) \\ \cos(\theta)\sin(\psi) \\ -\sin(\theta) \end{pmatrix} \tag{5}$$

Equation (4) is a reduced magnetic dipole equation. The prefactor of Equation (1) is neglected and the magnetic dipole moment of the coil is reduced to the orientation of the coil.

4.1. Signal Generation and Equalizer

To separate the mixed signals received by the sensor, the signals of the coils must be orthogonal. Two signals $x_i(n)$ and $x_j(n)$ are orthogonal for $n \in \{L_1, \ldots, L_2\}$, if the following condition

$$\sum_{n=-L_1}^{L_2} x_i(n)\, x_j^*(n) = \begin{cases} E_i, & i = j \\ 0, & i \neq j \end{cases} \tag{6}$$

is fulfilled [24]. Extending this equation to the constant E_i being 1, the signals are called orthonormal [24]. This is necessary to extract the individual coil amplitudes from the sensor signal and make them comparable. Different approaches can be used for the generation of orthogonal signals, e.g., using a TDMA (Time Division Multiple Access), an FDMA (Frequency Division Multiple Access) or a CDMA (Code Division Multiple Access) approach [25]. Due to the small bandwidth of the sensor, a TDMA approach is used in this contribution. Thus, the excitation signals

$$x_{\text{ex},i}(n) = \cos(2\pi f_r(n - \kappa_i))\, w(n - \kappa_i) \quad \text{with } \kappa_i = (i-1)L_{\text{mf}} \quad \forall i \in \{1, \ldots, N_c\} \tag{7}$$

are cosine signals at the resonance of the magnetoelectric sensor [15]. The signals are weighted with a Hann window $w(n)$ [26] of length L_{sig}, so that a smoothed in- and out-fading of the signals is ensured. Additionally, the condition $L_{\text{mf}} \geq L_{\text{sig}}$ must be fulfilled. If $L_{\text{mf}} > L_{\text{sig}}$, there is a pausing time between two consecutive coil signals. This is important when considering the impulse response of the sensor. The excitation signals are repeated every $L_r = N_c L_{\text{mf}}$ samples

$$x_{\text{out},i}(n) = x_{\text{ex},i}(n - \lambda L_r) \quad \text{with } \lambda \in \mathbb{Z} \tag{8}$$

and transmitted by the coils after D/A conversion. It should be noted that $L_r = L_{\text{mf}}$ if an FDMA or a CDMA approach is chosen, because the signals are transmitted simultaneously by all coils.

As can be seen from Equation (1) the magnetic field of a coil is proportional to the driving current. Since the output of the D/A converter is proportional to a voltage, the excitation signals $x_{\text{ex}}(n) = [x_{\text{ex},1}(n), \ldots, x_{\text{ex},N_c}(n)]^T$ are linearly deformed in amplitude and phase. This can be described by the impulse response of the coil $h_{c,i}(n)$, denoting the relationship between the voltage and the current of the coil i. Additionally, the signals are modified by the impulse response of the magnetoelectric sensor $h_s(n)$, as can be seen from Equation (2). Thus, the signals $x_{\text{ex}}(n)$ must be equalized either prior to the deformation due to the coil and sensor impulse response or before they are forwarded to the matched filter. In this contribution, the matched filter impulse responses are adapted to match with the modified transmitted signals, so that they are again comparable to the coil signals measured by the sensor. Each coil excitation signal is adjusted individually by the equalizer

$$h_{\text{eq},i}(n) = \hat{g}_{c,i}\, \hat{h}_{c,i}(n) * \hat{h}_s(n), \tag{9}$$

with $\hat{h}_{c,i}(n)$ and $\hat{h}_s(n)$ denoting the approximated impulse responses of the coil i and the magnetoelectric sensor, respectively. The factor $\hat{g}_{c,i}$ describes the influence of other components in the measurement setup. This includes for example different gains of the individual coil amplifier channels and different conversion factors of the coils from current to magnetic field. This is, e.g., due to variances in the number of windings. These values are approximated by a constant for the considered frequency range. The values for the six coils and amplifier channels used in this study are given in Table 1.

Table 1. Parameter of the coils and amplifier channels used in this study. The conversion factors of the coils describing the relationship between current and magnetic field are described by the column *conversion*. The gains of the coil amplifier channels are normalized to the maximum value (channel 6). Both values are determined at the resonance frequency of the sensor.

Number	Conversion (mT/A) (@7712 Hz)	Amplifier Gain (Relative, @7712 Hz)
1	13.2	0.9632
2	12.4	0.9823
3	12.8	0.9592
4	12.8	0.9751
5	12.2	0.9549
6	12.5	1.0000

The equalized signals are calculated according to

$$x_{eq,i}(n) = g_{eq,i} \underbrace{x_{ex,i}(n) * h_{eq,i}(n)}_{\tilde{x}_{eq,i}(n)} \tag{10}$$

and forwarded to the matched filter. The weighting factor $g_{eq,i} = \frac{1}{E_i}$ ensures that each equalized signal has the same correlation output value one between the signals $x_{eq,i}(n)$ and $\tilde{x}_{eq,i}(n)$ at lag zero. The factor E_i is the auto-correlation output of $\tilde{x}_{eq,i}(n)$ at lag zero. In Figure 6 the cross correlation of the signals $x_{eq,i}(n)$ and $\tilde{x}_{eq,i}(n)$ at time lag zero is visualized.

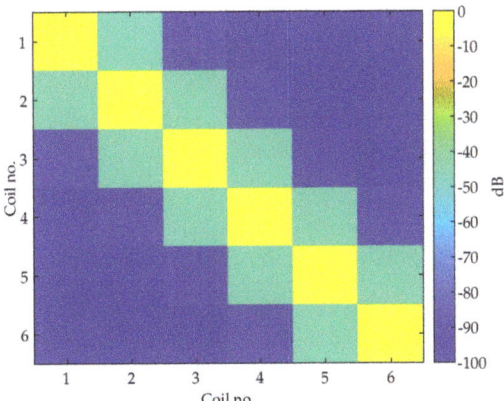

Figure 6. Cross correlation between individual equalized coil excitation signals.

It is obvious that adjacent coils signals have cross correlation values different than zero. This is due to the decay behavior of the sensor impulse response. Nevertheless, the shown values are still sufficient for separating the coil signals.

4.2. Matched Filter

As described in Equation (2), the input signal of the sensor is a superposition of the magnetic coil signals. Additionally, noise sources superimpose the signal. To obtain a high signal-to-noise ratio (SNR) the coil amplitudes can be increased, which leads to a high energy consumption. Alternatively, a matched filter [27] can be used, which increases the SNR and thus can perform well also with lower energy consumption. This additionally makes the algorithm more robust against distortions. Hence, to obtain the amplitudes of the coil signals measured by the magnetoelectric sensor, the sensor input signal $x_{\text{in}}(n)$ is matched filtered with the equalized coil excitation signals

$$x_{\text{mf},i}(n) = x_{\text{in}}(n) * x_{\text{eq},i}(L_{\text{r}} - n). \tag{11}$$

The matched filter output can be evaluated every L_{r} samples, since all coil signals have then been completely transmitted once

$$m_i(k) = x_{\text{mf},i}(kL_{\text{r}}), \tag{12}$$

with the value $m_i(k)$ corresponding to the amplitude of the coil signal i at the sensor. These amplitudes are summarized in the vector $\boldsymbol{m}(k) = [m_1(k), m_2(k), \ldots, m_{N_c}(k)]^{\text{T}}$ and forwarded to the localization algorithm.

4.3. Localization Algorithm

For the estimation of the position and orientation of the sensor, the matched filter output vector $\boldsymbol{m}(k)$ is compared with the columns of the lead-field matrix $\boldsymbol{A}$. As stated in Equation (4), each column $\boldsymbol{a}_j$ describes the coil amplitudes that would be measured by the sensor (after being filtered by the matched filter), if the sensor would occupy the defined position and orientation pair described by $\boldsymbol{p}_j$. To be more robust against gain uncertainties and to ensure comparability between the measured coil amplitudes and the lead-field matrix columns, the vectors are normalized to the respective absolute maximum value beforehand. The values for the cost function $\boldsymbol{c}(k) = [c_1(k), \ldots, c_j(k), \ldots, c_{N_p}(k)]^{\text{T}}$ are calculated by [15]:

$$c_j(k) = \sum_{i=1}^{N_c} \left| \frac{a_{ij}}{\max\{|\boldsymbol{a}_j|\}} - \frac{m_i(k)}{\max\{|\boldsymbol{m}(k)|\}} \right|^2 \quad \forall j \in \{1, \ldots, N_p\} \tag{13}$$

The estimated sensor position and orientation is then given by the position-orientation pair of the forward problem with the minimum cost function value

$$\hat{\boldsymbol{p}}(k) = \boldsymbol{p}_{l(k)} \text{ with } l(k) = \underset{j}{\operatorname{argmin}} \ c_j(k) \tag{14}$$

and can also only be determined every L_{r} samples. It is obvious that localization errors occur due to the discretization of the localization area. If the sensor is not directly located on a defined grid point, the localization error will be at least the distance between the closest grid point and the sensor's location. Unfortunately, even higher localization errors can occur in some forward model configurations, due to the shape of the cost function $\boldsymbol{c}(k)$. Further information can be found in Appendix A. To overcome this problem a higher resolution is required. However, this will increase the computational complexity dramatically and hence can endanger the real-time capability of the system. By increasing the resolution iteratively, the localization can be performed with high accuracy and a moderate increase in computational complexity.

The flow chart for the iterative localization process is shown in Figure 7. The first iteration is calculated as described before. Instead of considering only one estimated position-orientation-pair as stated in Equation (14), the N_{b} best position-orientation-pairs are taken into account. The minima and maxima of the included position-orientation-pairs plus a fraction (one step size) of the previous grid form the boundaries of the new grid. The new grid is again divided into N_{p} position-orientation-pairs and the forward problem as described in Equation (4) is calculated. From then on, the steps are repeated as described in the upper part of this section. Grid refinement stops when either the resolution between two adjacent grid points is less than a specified resolution or the maximum number of iterations N_{it} has been reached. The estimated position-orientation pair is given in the last iteration as described by Equation (14).

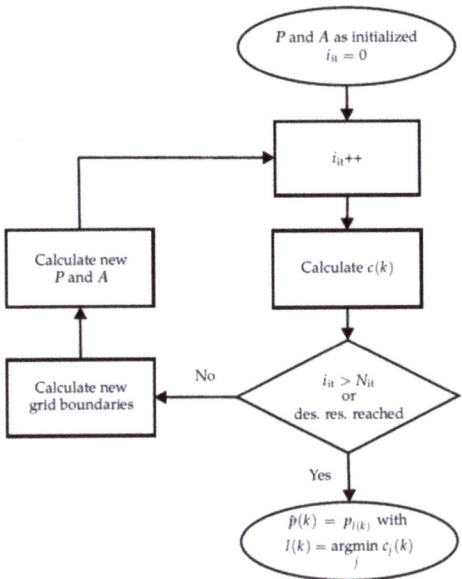

Figure 7. Flow chart of the iterative localization approach for one time step k. As long as neither the maximum number of iterations nor the desired resolution is reached, the algorithm keeps refining the localization grid.

4.4. Postprocessing

To mitigate possible outliers in the localization results, a linear Kalman filter for smoothing the estimated localization outcomes is used. The measurement equation of the system can be described by

$$\hat{p}(k) = Hs(k) + n_m(k), \quad (15)$$

with the state variables $s(k)$, the observation model H transforming the states into the measurement variables, and supposing white Gaussian measurement noise $n_m(k)$ [28]. Assuming a linear system, the state variables are updated via the equation

$$s(k) = Fs(k-1) + n_p(k). \quad (16)$$

The matrix F is the state transition matrix and $n_p(k)$ is white noise with zero mean [28]. Due to the noise processes the measurement variables are subject to errors. The Kalman filter attempts to predict the states and thus reduces the influence of the noises stated in Equations (15) and (16). The calculations are performed according to the descriptions in [28]. Based on the $N_m = 6$ measured variables summarized in $\hat{p}(k)$, there are $N_s = 3 \cdot N_m = 18$ states available, when additionally considering the velocity and acceleration of the measured variables. The initialization of the variables and covariance matrices are described in Appendix B. The enhanced localization output is denoted as $\hat{p}^{enh}(k)$. It is worth noting that the Kalman filter outputs should not lie outside the localization area and are thus restricted to the boundaries of the localization grid.

5. Measurements and Results

For the proof of principle, the measurements were performed in a two-dimensional space, i.e., only considering the x and y components of the sensor position and only considering the angle yaw ψ for the orientation of the sensor. The z component of the sensor's position, as well as the orientation angles roll φ and pitch θ are assumed to be zero. Nevertheless, the proposed method can easily be performed in the three-dimensional space without any restrictions. More coils should be used for this, positioned in the three-dimensional space. Additionally, a smaller initial grid resolution might be used to keep the computational complexity low. The measurements were performed with a real-time system developed at the chair of Digital Signal Processing in Kiel [29]. A picture of the graphical user interface of the tool is shown in Figure 8.

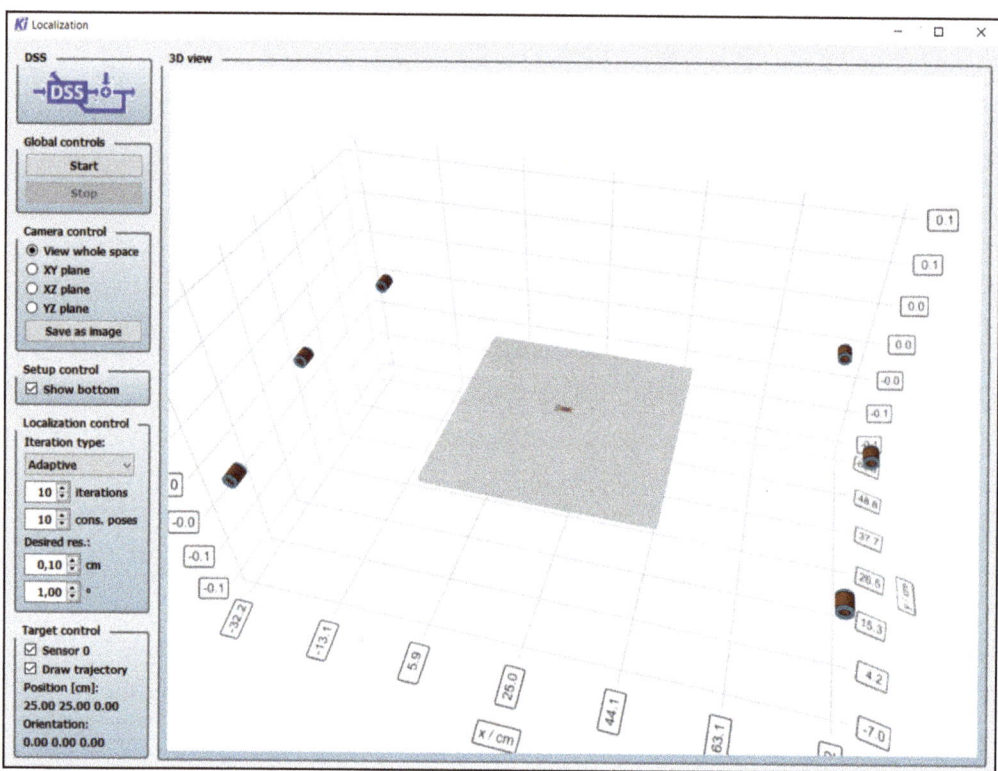

Figure 8. Graphical user interface of the real-time system used for localizing the magnetic sensors. The estimated position and orientation of the sensor is shown graphically in the 3D view and as text in the lower left corner. The number of iterations N_{it}, the number of considered position-orientation pairs for refining the localization grid N_b and the desired resolution in position and orientation are adjustable during runtime.

The parameter used for the measurements (as defined in the sections above) are listed in Table 2. The localization area is limited to values between 0 cm and 50 cm in x and y direction and between $-90°$ and $90°$ for the yaw angle.

Table 2. Parameter of the coil signals used in this study.

Parameter	N_c	L_{sig}	L_{mf}	f_s (kHz)	N_{it}	N_b	N_p
Value	6	2048	28,672	192	10	10	49,419

The waiting time between two consecutive coil signals must be rather high due to the decay of the impulse response of the sensor. Consequently, a total time of $\frac{L_t}{f_s} = 896$ ms, with f_s being the sampling rate, is needed to completely transmit the signals and thus to generate one localization result. This time can be shortened tremendously when using a sensor with a higher bandwidth. The sensor is placed in fixed positions and with fixed orientations to determine the accuracy of the algorithm. The tested position-orientation pairs $p_{s,j}$ were chosen randomly and are shown in Figure 9. The arrow directions represent the sensor's long axis.

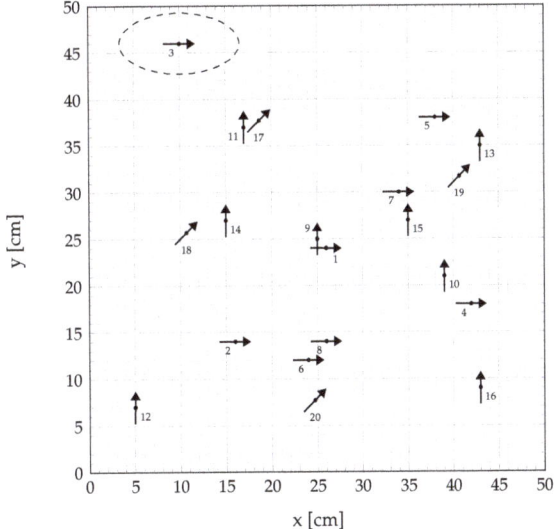

Figure 9. Position-orientation-pairs of the sensor used for testing the localization algorithm.

The tilt between the sensor's sensitive and long axis has been approximately determined by manually rotating the sensor in a Helmholtz coil. The tilt was measured outside a shielded environment and results in $\gamma \approx -45°$. However, the tilt of the sensor depends on various factors, such as the strength of the bias field [30] (e.g., the earth's magnetic field) and is therefore only an approximation.

The results of the localization for the position-orientation pair $p_{s,3}$ over time are shown as an example in Figure 10. Here, $x_{s,j}$, $y_{s,j}$, and $\psi_{s,j}$ denote the x and y component and the yaw angle of the sensor at position $p_{s,j}$, respectively. The localization output without the Kalman filter is described by $\hat{x}_{s,j}(k)$, $\hat{y}_{s,j}(k)$, and $\hat{\psi}_{s,j}(k)$ and after smoothing by the Kalman filter by $\hat{x}_{s,j}^{enh}(k)$, $\hat{y}_{s,j}^{enh}(k)$, and $\hat{\psi}_{s,j}^{enh}(k)$.

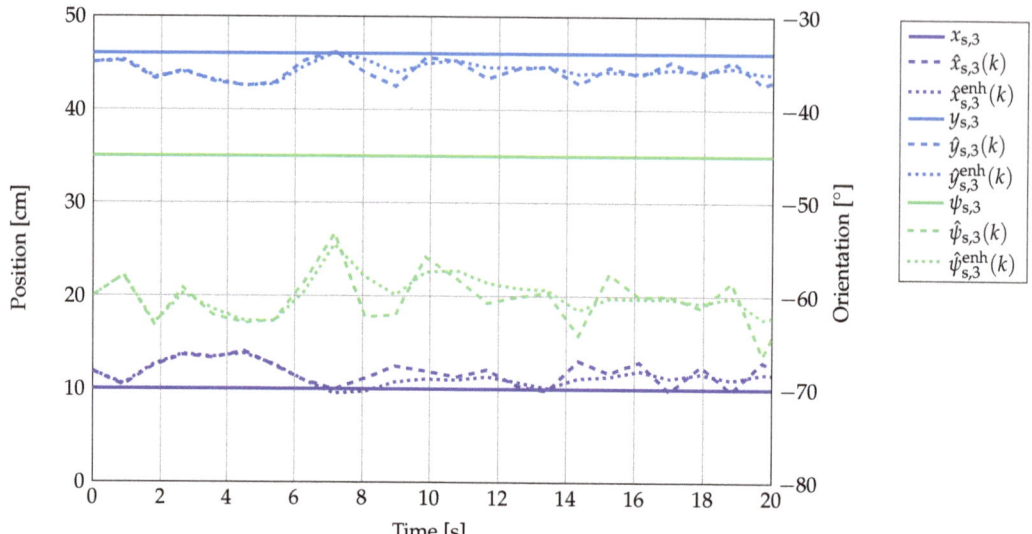

Figure 10. Real and estimated position and orientation of $p_{s,3}$ over time. The variances in the localization result are due to the presence of noise and cross talk in the measurement hardware.

There are some variances of the localization result over time. This is mainly due to the presence of noise, which leads to slightly varying amplitudes at the sensor and thus to small variations in the localization outcomes. The offset error might be due to cross talk between the coil amplifier channels and coupling of the magnetic coil signals into the cables and electronics. Additionally, it can be seen that the Kalman filter smooths the estimation output over time, so that outliers in the localization results do not have such a high impact.

To quantify the accuracy of the algorithm, a localization error is defined according to

$$\bar{e}_{r,j} = \frac{1}{N_{\text{meas}}} \sum_{k=0}^{N_{\text{meas}}-1} \sqrt{\left(\hat{x}_{s,j}^{\text{enh}}(k) - x_{s,j}\right)^2 + \left(\hat{y}_{s,j}^{\text{enh}}(k) - y_{s,j}\right)^2} \quad (17)$$

for the position estimation and

$$\bar{e}_{d,j} = \frac{1}{N_{\text{meas}}} \sum_{k=0}^{N_{\text{meas}}-1} \sqrt{\left(\hat{\psi}_{s,j}^{\text{enh}}(k) - \psi_{s,j}\right)^2} \quad (18)$$

for the orientation estimation. The number of localization outcomes is set to $N_{\text{meas}} = 50$, according to a measurement time of 44.8 s. The accuracy of the localization results for all tested position-orientation-pairs is shown in Figure 11.

The localization error is lying between 1.12 cm and 6.90 cm for the position estimation and results in a mean error of about 3.44 cm. The error for the orientation estimation is between 3.02° and 16.76°. This results in a mean error of about 11.23°. When considering fixed positions and orientations of the sensor, the localization output can be averaged over time and compared to the real sensor position/orientation afterwards. When doing so, the localization accuracy can be slightly improved and results in values between 0.46 cm and 6.52 cm for the position estimation and 1.54° and 15.35° for the orientation estimation. The average error reduces to 3.14 cm and 10.54°, respectively.

The high errors for the estimation of the sensor's position can result from the noise and the cross talk in the measurement system. Due to the different distances between the coils and the sensor the SNR is dependent on the sensor position/orientation. For example, looking at position $p_{s,9}$ an average SNR of about 9.5 dB is obtained at the sensor. Higher coil currents would lead to an improved SNR. Nevertheless, the goal was to localize with a minimum amount of energy. To avoid cross talk, the cables as well as the amplifier channels must be shielded. Additionally, the remaining cross talk can already be considered when setting up the forward problem or with an appropriate initial calibration. However, since the focus of this work is on the real-time localization pipeline and the calibration will be very extensive, it is not the subject of this work, but will be taken into account in our future work. The high error variance of the orientation estimation can partly be due to the change of the sensor's sensitive axis with respect to the bias field. Even a rotation in the earth's magnetic field can tilt the sensitive axis [30]. This problem only occurs with the sensors presented here and not with other types of magnetic field sensors. Furthermore, possible calibration errors do not only influence the position estimation but also the orientation estimation.

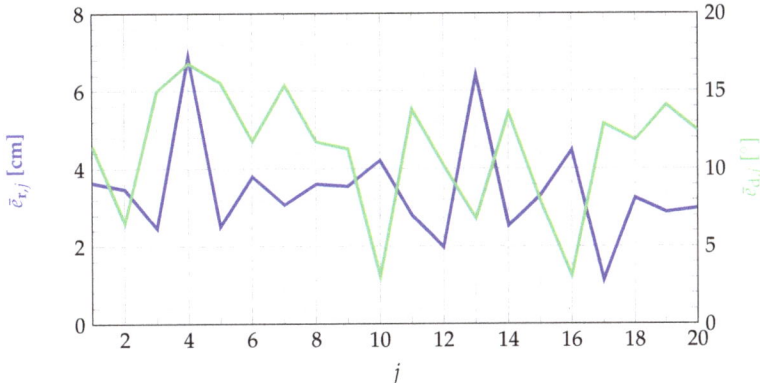

Figure 11. Mean localization errors for all tested position-orientation pairs. The index j depicts the respective position-orientation pair $p_{s,j}$ of the sensor as defined in Figure 9.

6. Conclusions and Outlook

An algorithmic pipeline for localizing magnetic sensors in real time was presented in this contribution. Besides the localization algorithm itself, pre- and postprocessing steps for an enhanced estimation of the sensor's position and orientation have been described. The potential of the proposed algorithm was emphasized by measurements with a magnetoelectric exchange bias ΔE-effect sensor. Nevertheless, the proposed method can be applied to any type of magnetic field sensor. Only the coil excitation signals must be adapted to the properties (frequency range, dynamic range, etc.) of the magnetic sensor used. Using the magnetoelectric sensor, a mean localization error of 3.44 cm has been reached. For the proof of principle the localization of the sensor has been limited to the two-dimensional space. Nevertheless, the localization can be easily extended to the three-dimensional space.

The achieved results are comparable with other magnetic position estimation approaches. In [4] a 3 × 3 m grid was used for localizing, achieving an accuracy of less than 10 cm. Localizing with a 3D sensor in a grid size of 8 × 7 cm an accuracy of 2.6 mm could be reached in [31]. In [2] a localization accuracy of ≤ 2 mm has been achieved, using coils for the localization of the sensors in a flexible MEG system. That shows that our localization method performs well, but can still be improved. However, the localization method investigated in this contribution can be performed in real time and in parallel to magnetic measurements without any degradation. Additionally, the robustness in noisy environments is increased by the usage of a matched filter and the smoothing of the localization outcomes via the linear Kalman filter. Moreover, the usage of a magnetoelectric sensor can be advantageous with respect to later medical applications due to its small size and the low production and operation costs.

Until now, the biggest limiting factor is the hardware used. Moreover, only a simplified model of the magnetoelectric sensor has been used, where the sensor is reduced to a point model. A more detailed model of the sensor, which also considers the dimensions of the sensor as well as a bias field dependent tilting of the sensor's sensitive axis, could improve the localization accuracy. The results can be further improved using multiple sensors included in an array with fixed distances and orientations as described in [1]. To reduce the transmitting time of the coils and thus to increase the rate of localization outcomes, FDMA or CDMA approaches would be beneficial. This would require an adaption of the sensor hardware.

Author Contributions: Conceptualization, C.B. and G.S.; methodology, C.B. and G.S.; software, C.B.; validation, G.S.; formal analysis, C.B.; investigation, C.B.; writing—original draft preparation, C.B.; writing—review and editing, C.B. and G.S.; visualization, C.B.; supervision, G.S.; project administration, G.S.; funding acquisition, G.S. All authors have read and agreed to the published version of the manuscript.

Funding: This work was funded by the German Research Foundation (Deutsche Forschungsgemeinschaft, DFG) via the collaborative research center CRC 1261.

Institutional Review Board Statement: Not applicable.

Informed Consent Statement: Not applicable.

Data Availability Statement: Not applicable.

Acknowledgments: The authors would like to thank Alexander Teplyuk for building the coil amplifier and Christine Kirchhof, Phillip Durdaut, and Jens Reermann for building up the sensor.

Conflicts of Interest: The authors declare no conflict of interest. The funders had no role in the design of the study; in the collection, analyses, or interpretation of data; in the writing of the manuscript, or in the decision to publish the results.

Abbreviations

The following abbreviations are used in this manuscript:

TDMA	Time Division Multiple Access
FDMA	Frequency Division Multiple Access
CDMA	Code Division Multiple Access
SNR	Signal-to-Noise Ratio

Appendix A. Localization Errors

As already mentioned in Section 4.3, localization errors can occur due to the discretization of the localization area. It is obvious that the minimal localization error is defined between the real sensor position/orientation and the closest grid point and is only zero, when the sensor is lying directly on a defined grid point. Besides these obvious errors even higher localization errors can occur, if the grid resolution is not high enough. This highly depends on the coil arrangement, i.e., on the setup of the forward problem. In Figure A1 the cost function of a sensor located at point A (only considering a 2D plane and assuming a fixed orientation) is shown assuming an exemplary (not well set up) forward problem.

Now assuming a localization grid with a coarse resolution, which is visualized by the black lines. The points, where the lines are crossed are the potential grid positions. The minimal localization error would be the distance between point A and point B. However, the position-orientation pair with the smallest cost function is found at point C. This is due to the shape of the cost function and the small grid resolution, whereby the grid points do not lie on the minimum of the cost function. This results in a large localization error. Thus, a higher grid resolution is needed, which leads to a higher computational complexity and hence can endanger the real-time capability of the system. To overcome this, an iterative localization is implemented, which refines the grid resolution progressively.

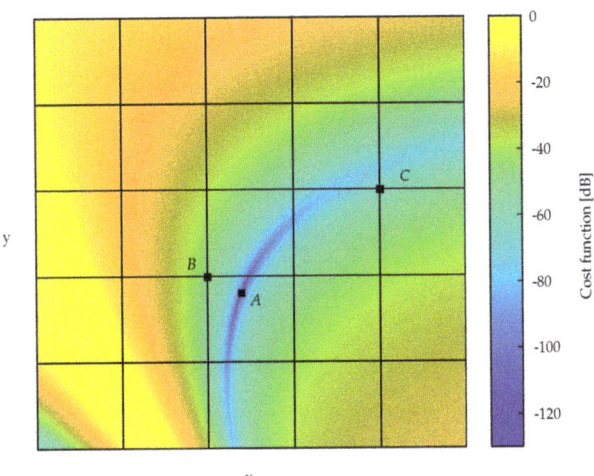

Figure A1. Exemplary cost function. The sensor is located at point A. Due to the relatively coarse grid (black lines), there will be a localization error of at least the distance between the point B and point A. However, due to the shape of the cost function, the minimum that is crossing the grid lines—and thus the localization outcome—is at point C.

Appendix B. Initialization of the Kalman Matrices

As already stated in Section 4.4 the calculations of the Kalman filter are performed as described in [28]. The initialization of the matrices is filled as described for a discrete Wiener process acceleration model [28].

- State transition matrix

$$F = \begin{bmatrix} I_{N_m} & \Delta t I_{N_m} & \frac{1}{2}\Delta t^2 I_{N_m} \\ 0_{N_m \times N_m} & I_{N_m} & \Delta t I_{N_m} \\ 0_{N_m \times N_m} & 0_{N_m \times N_m} & I_{N_m} \end{bmatrix} \tag{A1}$$

- Measurement matrix

$$H = [I_{N_m}, 0_{N_m \times 2N_m}] \tag{A2}$$

- Covariance matrix of the process noise

$$Q = \mathrm{E}\{n_p\, n_p^T\} = \begin{bmatrix} \frac{\Delta t^4}{4} I_{N_m} & \frac{\Delta t^3}{2} I_{N_m} & \frac{\Delta t^2}{2} I_{N_m} \\ \frac{\Delta t^3}{2} I_{N_m} & \frac{\Delta t^2}{2} I_{N_m} & \Delta t I_{N_m} \\ \frac{\Delta t^2}{2} I_{N_m} & \Delta t I_{N_m} & I_{N_m} \end{bmatrix} \hat{\sigma}_p^2 \tag{A3}$$

- Covariance matrix of the measurement noise

$$R = \mathrm{E}\{n_m\, n_m^T\} = I_{N_m} \hat{\sigma}_m^2 \tag{A4}$$

- State covariance matrix

$$S(0) = I_{3N_m} \hat{\sigma}_s^2 \tag{A5}$$

Here, I_{M_1} denotes a unit matrix of size $M_1 \times M_1$ and $0_{M_1 \times M_2}$ a matrix filled with zeros of size $M_1 \times M_2$. $\mathrm{E}\{\dots\}$ is the expected value operator and $\Delta t = \frac{L_r}{f_s}$, with f_s being the sampling rate. Considering the high measurement noise and assuming a slowly moving or fixed sensor (i.e., low process noise), the estimated variances are set to $\hat{\sigma}_p^2 = 0.001$, $\hat{\sigma}_m^2 = 10$, and $\hat{\sigma}_s^2 = 500$. These values were chosen exemplary.

References

1. Pfeiffer, C.; Andersen, L.M.; Lundqvist, D.; Hämäläinen, M.; Schneiderman, J.F.; Oostenveld, R. Localizing on-scalp MEG sensors using an array of magnetic dipole coils. *PLoS ONE* **2018**, *13*, e0191111. [CrossRef] [PubMed]
2. Pfeiffer, C.; Ruffieux, S.; Andersen, L.M.; Kalabukhov, A.; Winkler, D.; Oostenveld, R.; Lundqvist, D.; Schneiderman, J.F. On-scalp MEG sensor localization using magnetic dipole-like coils: A method for highly accurate co-registration. *NeuroImage* **2020**, *212*, 116686. [CrossRef] [PubMed]
3. Blankenbach, J.; Norrdine, A.; Hellmers, H. A robust and precise 3D indoor positioning system for harsh environments. In Proceedings of the 2012 International Conference on Indoor Positioning and Indoor Navigation, Sydney, Australia, 13–15 November 2012.
4. De Angelis, G.; Pasku, V.; De Angelis, A.; Dionigi, M.; Mongiardo, M.; Moschitta, A.; Carbone, P. An indoor AC system. *IEEE Trans. Instrum. Meas.* **2015**, *64*, 1275–1283. [CrossRef]
5. Hu, C.; Meng, M.Q.; Mandal, M. Efficient magnetic localization and orientation technique for capsule endoscopy. In Proceedings of the 2005 IEEE/RSJ International Conference on Intelligent Robots and Systems, Edmonton, AB, Canada, 2–6 August 2005; pp. 628–633.
6. Polhemus. Motion Tracking Overview. Available online: https://polhemus.com/motion-tracking/overview/ (accessed on 18 July 2021).
7. Erné, S.; Narici, L.; Pizzella, V.; Romani, G. The Positioning Problem in Biomagnetic Measurements: A Solution for Arrays of Superconducting Sensors. *IEEE Trans. Magn.* **1987**, *23*, 1319–1322. [CrossRef]
8. Sternickel, K.; Braginski, A.I. Biomagnetism using SQUIDs: Status and perspectives. *Supercond. Sci. Technol.* **2006**, *19*, 3. [CrossRef]
9. Borna, A.; Carter, T.R.; Colombo, A.P.; Jau, Y.Y.; Berry, C.; McKay, J.; Stephen, J.; Weisend, M.; Schwindt, P.D. A 20-channel magnetoencephalography system based on optically pumped magnetometers. *arXiv* **2017**, arXiv:1706.06158.
10. Johnson, C.N.; Schwindt, P.D.; Weisend, M. Multi-sensor magnetoencephalography with atomic magnetometers. *Phys. Med. Biol.* **2013**, *58*, 6065–6077. [CrossRef] [PubMed]
11. Jahns, R.; Knöchel, R.; Greve, H.; Woltermann, E.; Lage, E.; Quandt, E. Magnetoelectric sensors for biomagnetic measurements. In Proceedings of the IEEE International Symposium on Medical Measurements and Applications, Bari, Italy, 30–31 May 2011; pp. 107–110.
12. Reermann, J.; Zabel, S.; Kirchhof, C.; Quandt, E.; Faupel, F.; Schmidt, G. Adaptive Readout Schemes for Thin-Film Magnetoelectric Sensors Based on the delta-E Effect. *IEEE Sensors J.* **2016**, *16*, 4891–4900. [CrossRef]
13. Zabel, S.; Kirchhof, C.; Yarar, E.; Meyners, D.; Quandt, E.; Faupel, F. Phase modulated magnetoelectric delta-E effect sensor for sub-nano tesla magnetic fields. *Appl. Phys. Lett.* **2015**, *107*, 152402. [CrossRef]
14. Zabel, S.; Reermann, J.; Fichtner, S.; Kirchhof, C.; Quandt, E.; Wagner, B.; Schmidt, G.; Faupel, F. Multimode delta-E effect magnetic field sensors with adapted electrodes. *Appl. Phys. Lett.* **2016**, *108*, 222401. [CrossRef]
15. Spetzler, B.; Bald, C.; Durdaut, P.; Reermann, J.; Kirchhof, C.; Teplyuk, A.; Meyners, D.; Quandt, E.; Höft, M.; Schmidt, G.; et al. Exchange biased delta-E effect enables the detection of low frequency pT magnetic fields with simultaneous localization. *Sci. Rep.* **2021**, *11*, 1–14. [CrossRef] [PubMed]
16. Ludwig, A.; Quandt, E. Optimization of the ΔE-effect in thin films and multilayers by magnetic field annealing. In Proceedings of the IEEE International Magnetics Conference, Amsterdam, The Netherlands, 28 April–2 May 2002.
17. Reermann, J.; Durdaut, P.; Salzer, S.; Demming, T.; Piorra, A.; Quandt, E.; Frey, N.; Höft, M.; Schmidt, G. Evaluation of magnetoelectric sensor systems for cardiological applications. *Measurement* **2018**, *116*, 230–238. [CrossRef]
18. Lage, E.; Kirchhof, C.; Hrkac, V.; Kienle, L.; Jahns, R.; Knöchel, R.; Quandt, E.; Meyners, D. Exchange biasing of magnetoelectric composites. *Nat. Mater.* **2012**, *11*, 523–529. [CrossRef] [PubMed]
19. Durdaut, P.; Penner, V.; Kirchhof, C.; Quandt, E.; Knöchel, R.; Höft, M. Noise of a JFET Charge Amplifier for Piezoelectric Sensors. *IEEE Sensors J.* **2017**, *17*, 7364–7371. [CrossRef]
20. Elzenheimer, E.; Bald, C.; Engelhardt, E.; Hoffmann, J.; Bahr, A.; Höft, M.; Schmidt, G. Quantitative Magnetometer System Evaluation for Biomedical Applications. In *Manuscript in Preparation*; Chair of Digital Signal Processing and System Theory: Kiel, Germany, 2021.
21. Jahns, R.; Piorra, A.; Lage, E.; Kirchhof, C.; Meyners, D.; Gugat, J.L.; Krantz, M.; Gerken, M.; Knöchel, R.; Quandt, E. Giant magnetoelectric effect in thin-film composites. *J. Am. Ceram. Soc.* **2013**, *96*, 1673–1681. [CrossRef]
22. Bao, J.; Hu, C.; Lin, W.; Wang, W. On the magnetic field of a current coil and its localization. In Proceedings of the IEEE International Conference on Automation and Logistics, Zhengzhou, China, 15–17 August 2012; pp. 573–577.
23. Ang, M.H.; Tourassis, V.D. Singularities of Euler and Roll-Pitch-Yaw Representations. *IEEE Trans. Aerosp. Electron. Syst.* **1987**, *AES-23*, 317–324. [CrossRef]
24. Oppenheim, A.V.; Willsky, A.S. *Signals & Systems*, 2nd ed.; Prentice-Hall, Inc.: Hoboken, NJ, USA, 1997; p. 280.
25. Faruque, S. Time Division Multiple Access (TDMA). In *Radio Frequency Multiple Access Techniques Made Easy*; Faruque, S., Ed.; Springer International Publishing: Cham, Switzerland, 2019; pp. 35–43.
26. Proakis, J.G.; Manolakis, D.G. *Digital Signal Processing: Principles, Algorithms, and Applications*, 3rd ed.; Prentice-Hall, Inc.: Hoboken, NJ, USA, 1996; pp. 626–627.
27. Turin, G.L. An Introduction to Matched Filters. *Ire Trans. Inf. Theory* **1960**, *6*, 311–329. [CrossRef]

28. Bar-Shalom, Y.; Li, X.R.; Kirubarajan, T. *Estimation with Application to Tracking and Navigation*; John Wiley & Sons, Inc.: Hoboken, NJ, USA, 2001; pp. 200–210. 274.
29. Digital Signal Processing and System Theory. Real-time Framework. Available online: https://dss.tf.uni-kiel.de/index.php/research/realtime-framework (accessed on 18 July 2021).
30. Durdaut, P. Ausleseverfahren und Rauschmodellierung für Magnetoelektrische und Magnetoelastische Sensorsysteme. Ph.D. Thesis, Kiel University, Kiel, Germany, 2019. (In German)
31. Dai, H.; Song, S.; Zeng, X.; Su, S.; Lin, M.; Meng, M.Q. 6-D Electromagnetic Tracking Approach Using Uniaxial Transmitting Coil and Tri-Axial Magneto-Resistive Sensor. *IEEE Sensors J.* **2018**, *18*, 1178–1186. [CrossRef]

Article

Active Magnetoelectric Motion Sensing: Examining Performance Metrics with an Experimental Setup

Johannes Hoffmann [1], Eric Elzenheimer [1], Christin Bald [1], Clint Hansen [2], Walter Maetzler [2] and Gerhard Schmidt [1,*]

[1] Institute of Electrical Engineering and Information Technology, Faculty of Engineering, Kiel University, 24143 Kiel, Germany; jph@tf.uni-kiel.de (J.H.); ee@tf.uni-kiel.de (E.E.); cbal@tf.uni-kiel.de (C.B.)
[2] Department of Neurology, Kiel University, 24105 Kiel, Germany; c.hansen@neurologie.uni-kiel.de (C.H.); w.maetzler@neurologie.uni-kiel.de (W.M.)
* Correspondence: gus@tf.uni-kiel.de; Tel.: +49-431-880-6125

Citation: Hoffmann, J.; Elzenheimer, E.; Bald, C.; Hansen, C.; Maetzler, W.; Schmidt, G. Active Magnetoelectric Motion Sensing: Examining Performance Metrics with an Experimental Setup. *Sensors* **2021**, *21*, 8000. https://doi.org/10.3390/s21238000

Academic Editor: Angelo Maria Sabatini

Received: 26 October 2021
Accepted: 28 November 2021
Published: 30 November 2021

Publisher's Note: MDPI stays neutral with regard to jurisdictional claims in published maps and institutional affiliations.

Copyright: © 2021 by the authors. Licensee MDPI, Basel, Switzerland. This article is an open access article distributed under the terms and conditions of the Creative Commons Attribution (CC BY) license (https:// creativecommons.org/licenses/by/ 4.0/).

Abstract: Magnetoelectric (ME) sensors with a form factor of a few millimeters offer a comparatively low magnetic noise density of a few $pT/\sqrt{Hz}$ in a narrow frequency band near the first bending mode. While a high resonance frequency (kHz range) and limited bandwidth present a challenge to biomagnetic measurements, they can potentially be exploited in indirect sensing of non-magnetic quantities, where artificial magnetic sources are applicable. In this paper, we present the novel concept of an active magnetic motion sensing system optimized for ME sensors. Based on the signal chain, we investigated and quantified key drivers of the signal-to-noise ratio (SNR), which is closely related to sensor noise and bandwidth. These considerations were demonstrated by corresponding measurements in a simplified one-dimensional motion setup. Accordingly, we introduced a customized filter structure that enables a flexible bandwidth selection as well as a frequency-based separation of multiple artificial sources. Both design goals target the prospective application of ME sensors in medical movement analysis, where a multitude of distributed sensors and sources might be applied.

Keywords: motion tracking; magnetoelectric sensors; artificial fields

1. Introduction

Magnetic sensing is well established in movement analysis [1,2] as most inertial measurement units (IMUs) contain 3D magnetometers to determine the unit's orientation in the horizontal plane (compass). Common scientific applications include human movement analysis, where medical doctors use motion tracking systems in both diagnosis and therapy of neurodegenerative disorders with movement-related symptoms: sensors are commonly applied in a lab-based assessment to examine early motor markers of Parkinson's disease [3]. Patient monitoring in the home setting is another emerging approach that relies heavily on wearable sensor technology [4,5].

Active magnetic motion tracking shows the potential to be a supplemental source of reference data [6]. In comparison to the passive magnetic method, it utilizes artificial magnetic fields of excitation coils in combination with tracking algorithms to obtain the relative position and orientation data between source and sensor [7]. Integrated commercial magnetometers based on Hall-effect sensors are well-suited for pure geomagnetic applications (up to 50 µT), as they offer sufficient noise performance (e.g., below 0.3 µT for BMX005, Bosch Sensortec [8]) at low cost and small integration size. For active magnetic motion tracking, sensor performance is of much greater concern, as actuators (excitation coils) consume significant amounts of power to generate magnetic fields, which is noticeable in the large required currents (e.g., 1.5 A [6]). A low power consumption is generally desirable as it improves runtime for battery-powered applications. In essence, the required power to reach a given signal-to-noise ratio (SNR) at a given distance (and orientation) can be significantly decreased by the selection of a performant sensor type in combination with optimized transmission and readout schemes.

In this contribution, we employ magnetoelectric (ME) delta-E-effect sensors [9] in direct detection of the first bending mode (approximately 7.7 kHz) with a magnetic noise density below 10 pT/$\sqrt{\text{Hz}}$ in resonance. Assuming mono-frequent signals at a sensor-optimized frequency, the sensors perform in the same order of magnitude as fluxgate sensors (e.g., FL1-100, Stefan Mayer Instruments: 20 pT/$\sqrt{\text{Hz}}$ at 1 Hz [10]) and slightly inferior to total field optically pumped magnetometers (e.g., QTFM, QuSpin: below 1 pT/$\sqrt{\text{Hz}}$ [11]). However, OPMs require heating power (e.g., total field OPM, Twinleaf: 0.7 W [12]) for laser and cell temperature stabilization, which might be undesirable in body-worn setups. Better performing sensors like superconducting quantum interference devices (SQUIDs) are not considered here, as they bring the significant drawbacks of supercooling (SQUIDs) or magnetically shielded operation (zero field OPMs). Aside from sensor noise, requirements differ heavily between applications as the term "motion sensing" is used interchangeably in indoor tracking/navigation, localization, and motion tracking of humans. Relevant performance metrics include measurement volume, (stationary) spatial accuracy, and update rate/bandwidth.

Indoor tracking or navigation systems as in [13,14] offer a measurement volume of multiple cubic meters aided by IMUs. In contrast, specialized devices for the tracking of medical instruments (e.g., Polhemus Viper [15]) target a much smaller volume and achieve sub-millimeter accuracy for the stationary case. Depending on actuator and sensor, different frequency ranges can be targeted: direct current (dc) systems have to compensate for the geomagnetic field and other stray fields (hard iron) [16]. Alternating current (AC) systems [15] might operate in the very low frequency (VLF) band from 3 kHz to 30 kHz, which is also suitable for the delta-E sensor type. Such systems avoid the geomagnetic field and some DC stray fields but have to deal with eddy currents in nearby conductive materials [17]. Due to their propagation velocity at the speed of light and their wavelengths in the 10 km range, these VLF approaches do not suffer from multi-path propagation and the Doppler effect [7], which results in a straight forward channel model for the stationary case. However, this does not eliminate the impact of relative movement in the context of complex field geometries.

Motion tracking systems commonly (e.g., [13]) use stationary sources in combination with distributed sensors. As most systems use multiple (orthogonal) sources, some form of multiple-access technique has to be employed to divide the source signals at the sensor. The sources might therefore operate sequentially as in [6] (time division multiple access, TDMA), by applying spreading codes (code division multiple access, CDMA) [13] or in separate frequency bands (frequency division multiple access, FDMA), such as the magnetoelectric approach presented here. The operation of multiple interoperable transmitters and receivers potentially enables powerful tracking approaches, as each additional unit results in a multitude of new data points. For a number of L sources and M sensors, this results in $L \times M$ data points.

This article focuses on the sensing of magnetic fields where characteristics of excitation, field, sensor, and basic signal enhancement are considered. This does not include tracking and sensor fusion algorithms [14], which are typically the following processing stages and can contribute to a significantly improved overall performance.

This paper is structured as follows: Section 2 details the characterization of the components and their mathematical description in the system. Section 3 introduces the experimental setup. It also features the acquired signals for the single channel and the FDMA approach as well as noise measurements and derived performance metrics. Section 4 contains a conclusion and an outlook.

2. System Characterization and Simulation

The magnetoelectric motion sensing system (Figure 1) consists of three subsystems:

(1) The transmitter generates a magnetic AC field by feeding a periodic excitation signal $g(n)$ (zero-mean, normalized power) through an amplifier to a coil.
(2) The receiver (1D sensor, fixed orientation) moves on a trajectory $\vec{r}(t)$ relative to the transmitter unit. Thus, it senses the excitation signal $g(t)$ weighted with the magnetic flux density $d(t)$ (desired signal) along the trajectory. This behavior resembles an amplitude modulation.
(3) The signal processing system's task is to estimate an unbiased $\hat{d}(k)$ in a frame-based fashion with the subsampled time index k. Therefore, prior knowledge on the sensor system (signal enhancement) and the excitation sequence (matched filter) is applied.

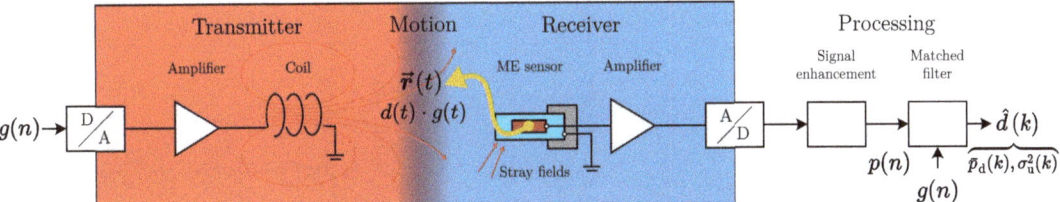

Figure 1. Magnetic motion sensing system overview.

The mathematical description of our approach is based on a distinction between the periodic excitation signal $g(n)$ and the discrete desired signal $d(n)$, which corresponds to a DC magnetic field with time variance induced by motion. The central component of the enhanced signal $p(n)$ is the modulation term (with minor deviations between discrete and continuous signals omitted). Undesired signal components like stray fields and sensor intrinsic noise are represented in the aggregated noise term $u_\mathrm{p}(n)$:

$$p(n) = d(n)\,g(n) + u_\mathrm{p}(n). \tag{1}$$

The matched filter separates the desired signal $d(n)$ from the excitation signal $g(n)$. Therefore, the output estimation is comprised of only the desired signal power $\bar{p}_\mathrm{d}(k)$ and the noise power $\sigma_\mathrm{u}^2(k)$ (zero-mean). Thus, the resulting signal-to-noise ratio (SNR) is independent of the excitation signal waveform:

$$SNR(k) = \frac{\bar{p}_\mathrm{d}(k)}{\sigma_\mathrm{u}^2(k)}. \tag{2}$$

We introduce a position-dependent SNR for spatial performance considerations that only depends on the sensor position $\vec{r}$ at a given moment in time, without prior knowledge of the trajectory. Its definition requires multiple assumptions:

(1) The period of the highest frequency component in the desired signal $T_\mathrm{d}^{\max}$ is much longer than the temporal length of the matched filter T_mf. This leads to a desired signal power $d^2(\vec{r})$ which is approximately constant within a frame.
(2) The noise power is both space- and time-invariant within the measurement area and duration.

$$SNR(\vec{r}) = \frac{d^2(\vec{r})}{\sigma_\mathrm{u}^2} \tag{3}$$

The following subsections focus on the contributions of the source and the sensor system regarding this metric in the system's operating frequency range near the sensor's resonance frequency f_0 of approximately 7.7 kHz.

2.1. Source Selection and Limitations

As a setup with multiple homogeneous or linear gradient fields is difficult to employ in a scenario with distributed moving sources, we used small cylindrical air coils. Manufacturing was done by spooling approximately 80 windings (N_w) of 0.5 mm enameled copper wire on a 3D-printed plastic body. The medium radius R was approximately 6.7 mm.

According to Ampere's circuital law [17], the magnetic flux density of a conductor loop is driven by the current I_{rms}. For a specified input voltage (voltage source) I_{rms} is limited by the electrical impedance of the coil, whose equivalent circuit is assumed here as a copper resistance in series with an inductance. Both parameters of the primarily used coil in the relevant frequency range were measured with a vector network analyzer (Bode 100, Omnicron Labs [18]). The resistance (Figure 2a) is close to 0.24 Ω at f_0, while the inductance (Figure 2b) is almost constant with 24 µH. These comparatively low values have to be considered with regard to accurate current measurements.

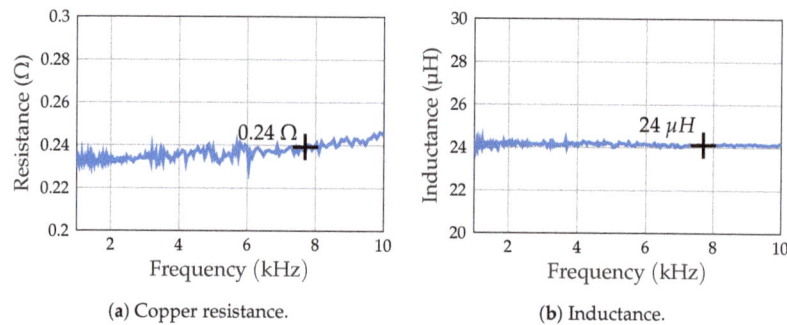

(a) Copper resistance.   (b) Inductance.

Figure 2. Impedance of the primary excitation coil.

The resulting magnetic flux density $\vec{b}$ of the coil is also affected by its physical properties. As a universal approach, the law of Biot–Savart [17] can be applied for the field computation of an arbitrary ensemble of line conductors. While there are analytic solutions for a simple cylindrical helix coil [19], a more complex coil might require a computationally intensive discrete model.

The magnetic dipole equation [20] offers a lightweight alternative, assuming the displacement r is much greater than the medium radius of the coil R. The actual distance requirement is dependent on the coil and the acceptable imitation error and might lay between 4 and $10R$ [21]. The magnetic flux density $\vec{b}(\vec{r})$ at a specified position $\vec{r}$ depends on the vacuum permeability μ_0, and the coil's magnetic dipole moment $\vec{m}$ with the number of windings N. Each vector $(\vec{r}, \vec{m})$ is comprised of the vector norm (r, m) and the corresponding unit vector $(\vec{e}_r, \vec{e}_m)$. This equation describes a time-invariant (DC) magnetic field, which yields an AC field, when modulated with the excitation signal $g(n)$:

$$\vec{b}(\vec{r}) = \frac{\mu_0}{4\pi r^2} \cdot \frac{3\vec{r}(\vec{m} \circ \vec{r}) - \vec{m} r^2}{r^3} \quad \text{with} \quad \vec{m} = N_w \, I_{rms} \, R^2 \, \pi \, \vec{e}_m. \tag{4}$$

The magnetic flux density at the sensor's position is comprised of constant, distance-dependent, and directivity-related contributions. This results in a nominal flux density b_0 with a field constant a_0, a $1/r^3$ decay (path loss), and a corresponding loss due to directivity $\vec{\Theta}(\vec{e}_r)$.

$$\vec{b}(\vec{r}) = \underbrace{a_0 \, I_{rms}}_{b_0} \cdot \frac{1}{r^3} \cdot \vec{\Theta}(\vec{e}_r) \tag{5}$$

The nominal field b_0 determines the feasible magnetic flux density at a specified position. It is constrained by competing geometric and electric parameters, as only a limited number of windings N_w at a radius R fit into a constrained space. A reduced wire cross section enables more windings, but increases the electrical impedance and vice versa. An increase in current is generally possible within the voltage range of the source, but is ultimately limited by thermal constraints.

2.2. Sensor Characterization and Idealization

This motion sensing approach uses exchange bias magnetoelectric delta-E cantilever sensors with a sensor element made of multiple layers of magnetostrictive FeCoSiB material and piezoelectric material on a silicon substrate [22]. A sensor element and a low-noise charge amplifier [23] together form the ME sensor system.

The ME cantilever sensor is a vector magnetometer [24]. Accordingly, it will pick up magnetic fields depending on the orientation of its sensitive axis $\vec{e}_s$. Fabrication might cause a tilt between the physical long axis and the sensitive axis. Based on our observations while rotating the sensor in a Helmholtz coil, the tilt is very small (low single digit degrees maximum) for this specific sensor. Thus, the magnetic flux density $d(\vec{r})$ that the sensor picks up due to modulation might be modeled as the dot product of the sensor orientation and the magnetic vector field:

$$d(\vec{r}) = \vec{e}_s \circ \vec{b}(\vec{r}). \qquad (6)$$

The sensor characterization measurements were done in the optimized conditions of a magnetically shielded measurement setup [25]. Therefore, real-world measurements will show an inferior sensor performance. The selected frequency space is targeted at the first bending mode close to 7.7 kHz, which is specified by the mechanical properties of the cantilever. Delta-E effect readout schemes for much lower frequencies [9] are also possible for this type of sensor, but not used in this application.

Based on the assumption of Linearity and time invariance (LTI), the measured frequency response $\hat{H}(e^{j\Omega})$ (Figure 3a) quantifies the conversion of a magnetic field into a voltage. It is synonymous with the sensor's sensitivity in this context.

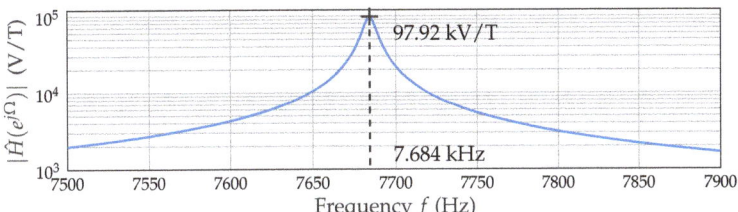

(a) Sensitivity curve with peak sensitivity at the resonance frequency.

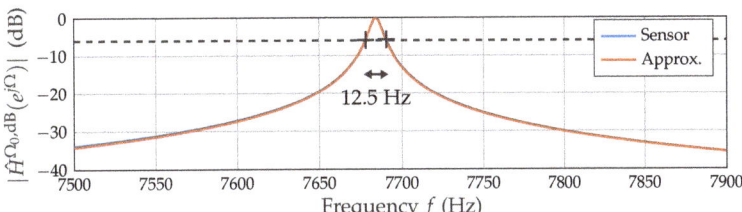

(b) Normalized filter characteristic with IIR peak filter defined by 6 dB bandwidth.

Figure 3. Frequency response of the ME sensor.

The sensitivity curve (Figure 3a) results from multiple discrete measurement points interpolated using a spline method. The data were obtained using a magnetic sine sweep at a peak amplitude of 100 nT. The peak sensitivity (approximately 100 kV/T) is reached at the resonance frequency (approximately 7.684 kHz). The filter characteristic (Figure 3b) of the sensor is obtained from the logarithmic amplitude response normalized to the peak sensitivity $\hat{H}^{\Omega_0,\text{dB}}(e^{j\Omega})$. The terminology for discrete-time signals with the normalized angular frequency Ω and the corresponding resonance frequency Ω_0 leads to:

$$|\hat{H}^{\Omega_0,\text{dB}}(e^{j\Omega})| = 20 \log_{10}\left(\frac{|\hat{H}(e^{j\Omega})|}{|\hat{H}(e^{j\Omega_0})|}\right). \tag{7}$$

The resulting 6 dB bandwidth of 12.5 Hz and the resonance frequency were adopted to design a digital filter that approximates the filter characteristic in the region of interest. Based on a priori knowledge of the mechanical cantilever, an infinite impulse response (IIR) peak (resonator) filter [26] was chosen (Figure 3b).

The voltage noise amplitude spectral density (ASD) $\hat{A}_{vv}(\Omega)$ (Figure 4a) is the other crucial performance metric obtained from the ME sensor characterization. It enables the estimation of the noise power at specific frequencies.

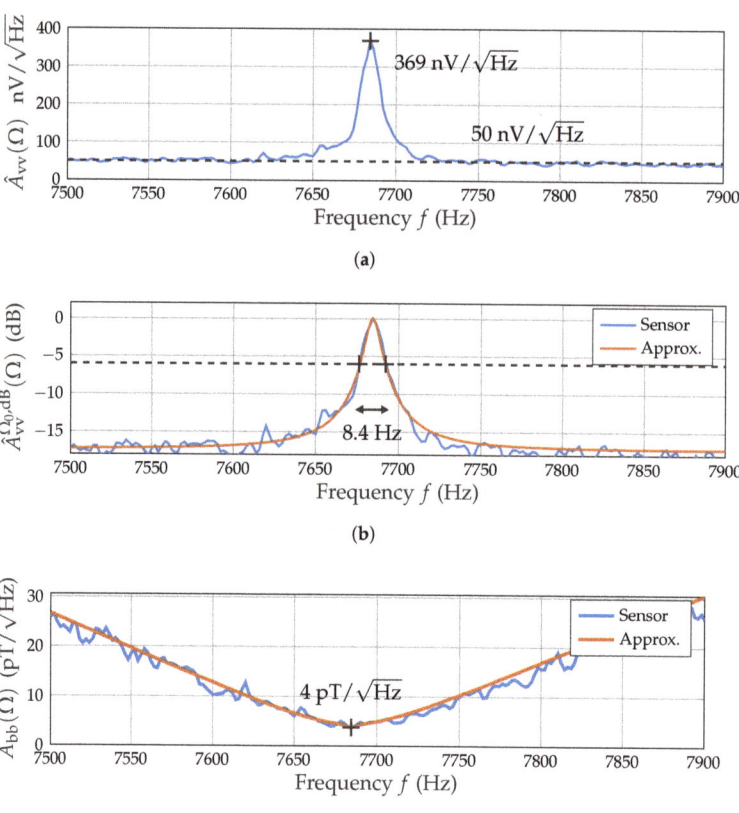

Figure 4. ME sensor noise characteristics. (**a**) Voltage noise spectral density with peak and floor value. (**b**) Normalized noise filter with IIR peak filter approximation defined by 6 dB bandwidth. (**c**) Magnetic noise density with minimum and IIR peak filter approximation.

$\hat{A}_{vv}(\Omega)$ (Figure 4a) reaches its local maximum of 370 nV/$\sqrt{\text{Hz}}$ at the sensor's resonance frequency and approaches a floor level of approximately 50 nV/$\sqrt{\text{Hz}}$. The filter characteristic is again obtained from the logarithmic noise density by normalizing it to the peak noise value in resonance:

$$\hat{A}_{vv}^{\Omega_0,dB}(\Omega) = 20\log_{10}\left(\frac{\hat{A}_{vv}(\Omega)}{\hat{A}_{vv}(\Omega_0)}\right). \tag{8}$$

The filter characteristic is approximated similarly to the frequency response by using an IIR peak (boost) filter [27] that employs a parallel allpass structure to limit the stopband's attenuation. The required design parameters include the boost factor (ratio between passband and stopband) and the 6 dB bandwidth of 8.4 Hz (Figure 4b). The sensor's performance in measuring magnetic fields (magnetic noise density $\hat{A}_{bb}(\Omega)$) is dependent on both frequency response and voltage noise density:

$$\hat{A}_{bb}(\Omega) = \frac{\hat{A}_{vv}(\Omega)}{|\hat{H}(e^{j\Omega})|}. \tag{9}$$

The resulting spectrum (c) presents a minimum in noise density of 4 pT at resonance with approximately linear slopes to both sides. Some form of magnetic noise density value (at a varying application-dependent frequency, e.g., [10]) is commonly used as an important performance metric for magnetometers. The curve is approximated by multiplying the inverse frequency response approximation and the voltage noise density approximation.

2.3. Signal Processing Structure

As previously characterized, the sensor offers a very limited 6 dB bandwidth of 12.5 Hz. Generally, it is beneficial for the SNR to adapt the bandwidth depending on the target application's requirements. Thus, an equalizing filter corresponding to the inverse approximated frequency response was applied, which resulted in a flat frequency response in the region of interest. $\hat{A}_{bb}(\Omega)$ (Figure 4c) is suitable for noise performance estimations, where a deviation from the sensor's 6 dB bandwidth is required. This is practically realized by some form of band limiting.

The sensor also picked up a multitude of undesired magnetic and electric fields as it was operated unshielded in the experimental setup. This effect was even worsened by the forceful amplification of low frequencies due to the equalizer. It might be countered by a band limiting to the operating range of $f_0 \pm 200$ Hz, which was subsequently applied using a 10th order Butterworth filter.

The final noise reduction step is a matched filter which corresponds to a bandwidth limitation and demodulation based on knowledge of the excitation signal $g(n)$. The length of the matched filter T_{mf} (N samples) is an assumption on how long the desired signal $d(n)$ is approximately constant. In the stationary case, one might choose a very long (e.g., multiple seconds) and therefore narrow matched filter to achieve a superior noise performance. In the 1D motion case (>0.1 m/s), the assumption of a constant $d(n)$ for such a filter is invalid. Consequently, the matched filter distorts the estimation result $\hat{d}(k)$ by suppressing high frequency components of $d(n)$.

In general, the trade-off between a high spatial accuracy for slow movements (such as the localization of a resting object) and a high temporal accuracy for fast movements (such as obtaining the high-frequency components of movement) has to be managed. A matched filter implementation with the (equivalent) correlator realization [28] might be beneficial here, as it enables the parallel use of a multitude of filters with different lengths N_j. Thereby, the process of multiplying (weighting) the enhanced signal $p(n)$ with the excitation signal $g(n)$ can be separated from the summation and is only required once. The multiplication requires a matching phase (group delay for non-sinusoidal signals). Processing is done at a lower common sample (update) rate f_{up} (period: T_{up}) of 100 Hz which is linked to the system sample rate f_s of 48 kHz by the conversion factor c. Figure 5a illustrates this process in principle:

$$\hat{d}_j(k) = \frac{1}{N_j}\sum_{i=0}^{N_j-1} p(k-i)\,g(k-i) \quad \text{with} \quad k = cn. \tag{10}$$

Our specific implementation for measurements (Figure 5a) covers two competing matched filters:

(1) $T_{mf,1}$ targets "fast" movements with a filter length of 10 ms, which corresponds to a two-sided bandwidth of 100 Hz.
(2) $T_{mf,2}$ targets "slow" movements with a filter length of 200 ms, which corresponds to a two-sided bandwidth of 5 Hz.

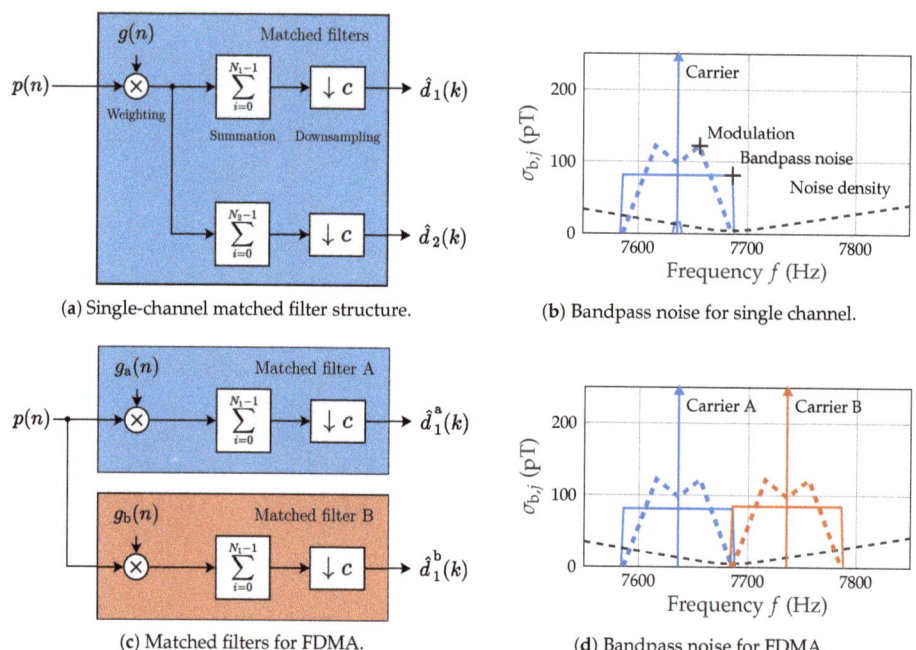

Figure 5. Matched filter structures and corresponding bandwidth considerations.

Integration of the magnetic noise density (ideal conditions, cf. Figure 4c) within the specified matched filter bandwidth illustrates the effect of the matched filter length (Figure 5b). The carrier is placed at 7636 Hz (approximately $f_0 - 50$ Hz). Modulation due to motion will require additional bandwidth apart from the carrier. Integration within an ideal bandpass is illustrated here as a rectangle with a width (integration bandwidth) and a height (square root of the integrated noise power $\sigma_{b,j}^2$). $T_{mf,1}$ yields an approximate theoretical noise of 80 pT, while $T_{mf,2}$ yields a significantly lower noise below 16 pT.

This matched filter implementation also allows for the separation of multiple simultaneously active carriers in a frequency division multiple access (FDMA) scheme (Figure 5c). The weighting signals $g_{a,b}(n)$ correspond to both active carriers. Subsequently, various matched filter lengths (e.g., N_1) might be applied in parallel. Figure 5d illustrates the placement of both carriers 100 Hz apart.

2.4. Simulation

In addition to the physical setup, we set up a simulation to provide a reference signal $\tilde{d}(k)$ for the estimation result $\hat{d}(k)$. As the processing strictly separates the excitation signal $g(n)$ from the desired magnetic signal $d(n)$, the simulation only includes a DC magnetic field based on a dipole approximation and an idealized 1D motion in this field:

$$\tilde{d}(k) = \hat{\vec{e}}_s \circ \underbrace{\hat{a}_0\, \hat{I}_{\mathrm{rms}}}_{\hat{b}_0} \cdot \frac{1}{\hat{\rho}^3} \cdot \vec{\Theta}(\hat{\vec{e}}_r) \quad \text{with} \quad \hat{\vec{r}}(k). \tag{11}$$

Figure 6 visualizes how the previously characterized parameters of the overall system are included in the simulation model. In comparison to the physical setup, there is no bandpass filter applied, so even fast simulated movements will not lead to a signal distortion.

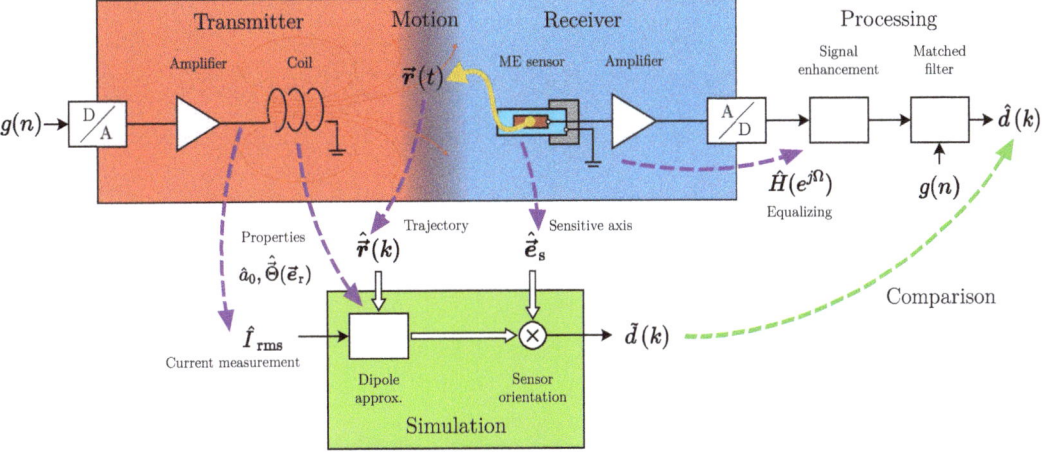

Figure 6. Overall signal chain including simulation.

3. Measurements and Results

3.1. Measurement Setup

Measurements were conducted in a simplified motion scenario (Figure 7) on a 1.2 m aluminum track. The sensor system is moving on a cart towed by stepper motors with its long axis $\hat{\vec{e}}_s$ in the direction of movement. The velocity $\dot{x}$ in the central part of the track (97 cm) is assumed to be constant. With a starting point x_0 and a y_0 displacement of 10 cm to each coil, the subsequent equation for the relative position is adopted for the field simulation:

$$\hat{\vec{r}}(k) = \begin{pmatrix} x(k) \\ y_0 \\ 0 \end{pmatrix} = \begin{pmatrix} \dot{x} \cdot kT_{\mathrm{up}} + x_0 \\ y_0 \\ 0 \end{pmatrix}. \tag{12}$$

The setup features two cylindrical coils for magnetic excitation fields powered by two composite amplifiers (based on [29]) at a current $\hat{I}_{\mathrm{rms}}$ of approximately 400 mA. Optical switches close to the beginning and the end of the track enable synchronization and average velocity measurements. The conversion of signals between the analog and the digital domain was done using a sound card (UFX+, RME). The basic validity of the simulated magnetic flux density along the track was verified using a fluxgate sensor (FLC 100, Stefan Mayer Instruments).

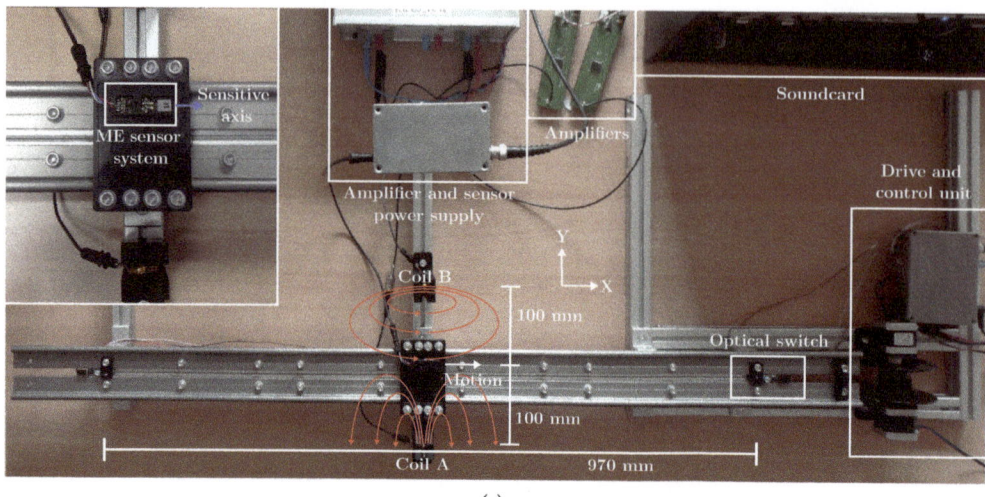

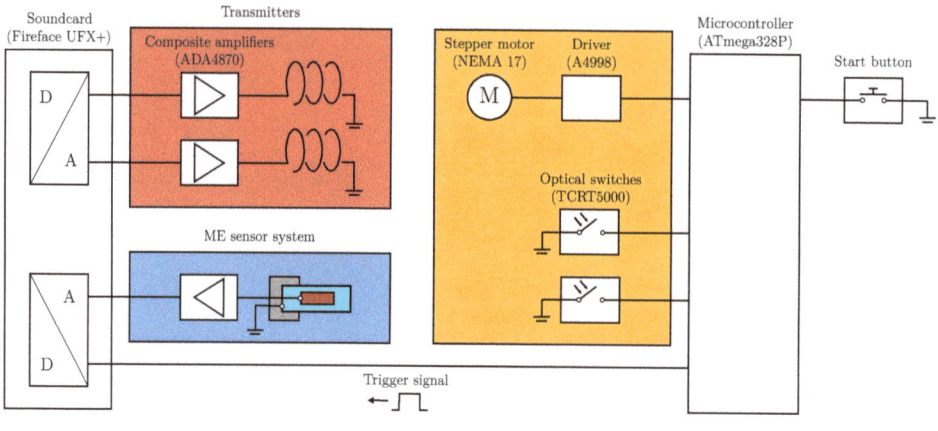

Figure 7. Experimental setup for magnetoelectric motion sensing. (**a**) Overview of the experimental setup. (**b**) Block diagram of key functional components.

3.2. Single Channel Measurements

As a first proof of concept, only coil A was used to generate a magnetic field, while the sensor was moving along the track multiple times at varying velocities. Based on the optical switches' trigger signals, each desired signal's time scale was rescaled to distance. As only the x position of the sensor is time-variant, we also refer to the measurement/simulation results as $\hat{d}(x)$ and $\tilde{d}(x)$, respectively. The previously introduced matched filter lengths N_1 and N_2 were applied on each signal to manage the trade-off between bandwidth and noise. Figure 8a shows accordance between the simulated signal and all three measured signals. However, the signals appear to be quite noisy. In Figure 8b, there is less noise but the signal measured at the fastest average velocity of 0.62 m/s is significantly flattened.

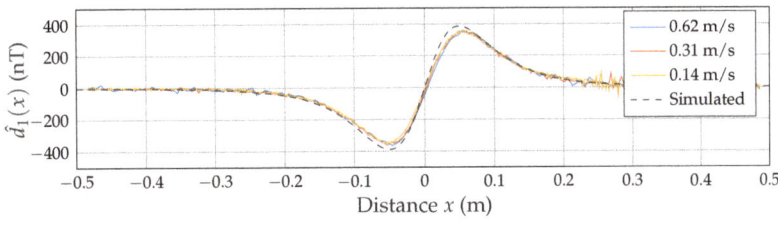

(a) Matched filter length of 10 ms.

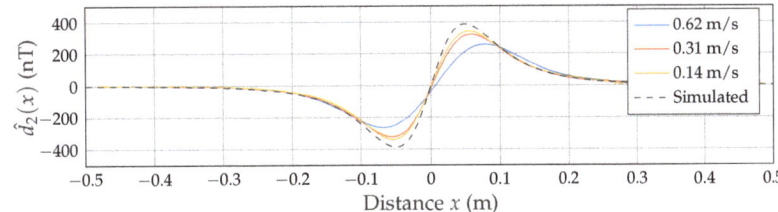

(b) Matched filter length of 200 ms.

Figure 8. Measured signals for varying speeds and matched filter lengths.

3.3. Noise Measurements

As we established the receiver noise as the main limitation of system performance in the previous chapter, we conducted noise measurements (without magnetic excitation) of resting and moving sensors on the test setup (cf. Figure 8). Three different scenarios were tested and processed using the previously introduced matched filter of length $T_{mf,1}$ and $T_{mf,2}$, respectively. The resting case (Figure 9a,b) is thereby compared with a slower movement (Figure 9c) at the higher MF length and with a faster movement (Figure 9d) at the shorter MF length. For each measurement, the corresponding noise standard deviation (e.g., $\sigma_{u,1}^a$) is provided.

While the resting case ($\sigma_{u,1}^a$ and $\sigma_{u,2}^b$) shows a very low noise close to the spectral integration (cf. Figure 5b), the moving cases yield a significant increase in noise (approximately 100 times) due to spikes ($\sigma_{u,1}^c$ and $\sigma_{u,2}^d$). The assumption of a stationary noise power ($\sigma_{u,1}^c$) during the motion is only valid for some sections of the time signal, as oscillation and multiple prominent peaks occur.

Based on the obtained noise values $\sigma_{u,1}^a$ to $\sigma_{u,2}^d$, multiple performance metrics regarding range and accuracy were calculated. Firstly, the noise floor was directly plotted in comparison to the positive half-wave of the simulated magnetic field (Figure 10a, cf. Figure 8). This leads directly to the SNR calculation as depicted in Figure 10b.

$$SNR_1^a(\vec{r})^{dB} = 20 \log_{10} \left(\frac{\tilde{d}(\vec{r})}{\sigma_{u,1}^a} \right) \quad (13)$$

While $\sigma_{u,1}^a$ and $\sigma_{u,2}^b$ show significant SNR levels even at at distance of 50 cm, the SNR reaches zero quickly for both other scenarios at 33 cm ($\sigma_{u,1}^c$) and 25 cm ($\sigma_{u,2}^d$), respectively. The noise can also be added on top of the simulated magnetic curve to obtain the uncertainty in the measurement of magnetic flux density for each position. This results in an upper and a lower boundary between which the measured value is to expected at the specified confidence of one standard deviation (Figure 10c). It is now possible to reverse search the corresponding spatial values $\hat{x}$ for the upper and the lower boundary at each position. This results in an upper and a lower estimated distance value. Error bars are included for relevant values of 1 and 5 cm (Figure 10d). While $\sigma_{u,2}^b$ ensures a high accuracy of 1 cm even at a distance of 42 cm, $\sigma_{u,1}^c$ reaches 1 cm at a distance of 15 cm and $\sigma_{u,2}^d$ never even reaches 1 cm.

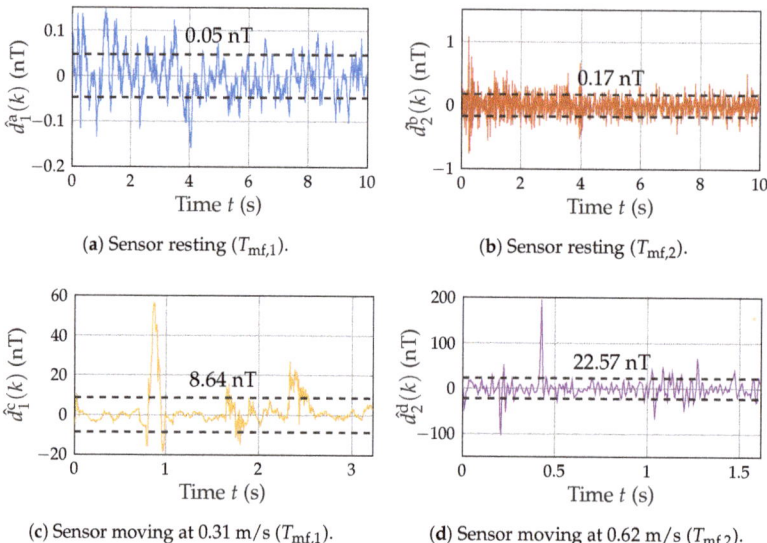

Figure 9. Noise measurements of moving and resting sensors. Each plot contains the noise at the matched filter output and the corresponding RMS value from $\sigma_{u,1}^a$ to $\sigma_{u,2}^d$.

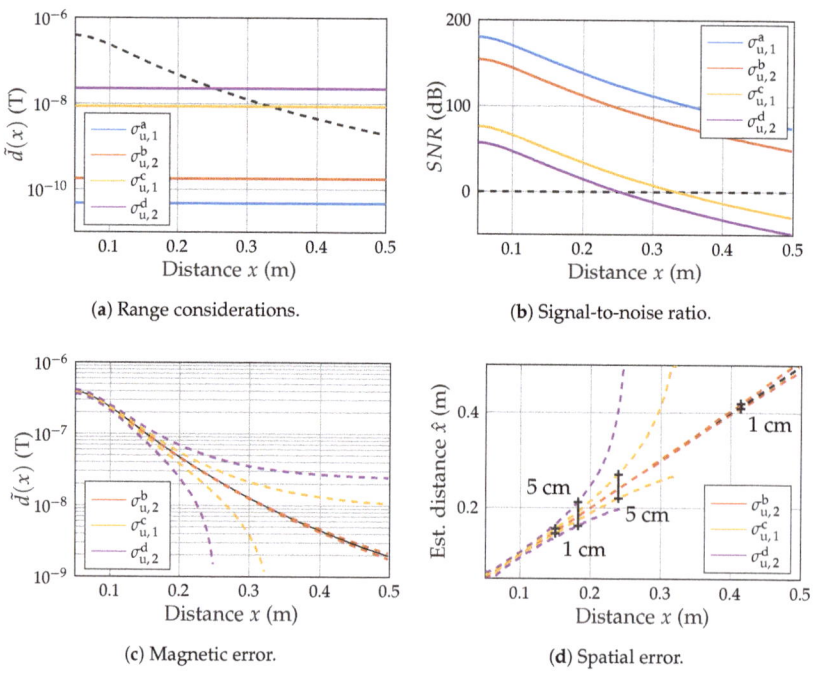

Figure 10. Performance metrics based on noise.

3.4. Frequency Division Multiple Access Measurement

The FDMA approach utilizes the second excitation coil to transmit another carrier signal in a separate frequency band. As coil B is rotated by 90° compared to coil A, the sensor measures the axial component of this magnetic field. Figure 11 shows the capability of the applied matched filter structure to smoothly separate both signals at the applied velocity of 0.31 m/s (with a matched filter length $T_{mf,1}$ of 5 ms) even with the high dynamic difference between both signals at a distance x close to zero.

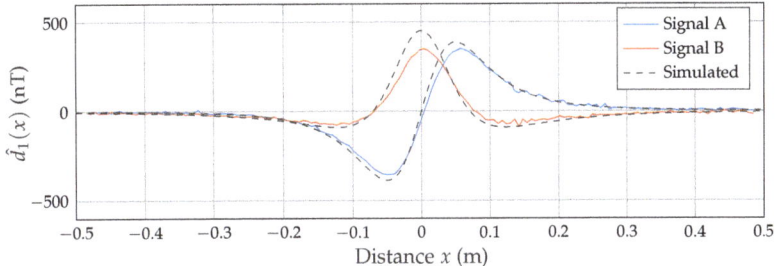

Figure 11. Estimated desired signals in an FDMA approach.

4. Conclusions and Outlook

In this article, we presented a concept for magnetic motion sensing using ME sensors and multiple artificial sources. We focused on signal-to-noise ratio as the key metric to determine the performance of this approach and showed the basic feasibility in an experimental setup.

The setup was targeted at the demonstration of sensor measurements during a defined movement. While the restriction of motion to a single axis is suitable for characterization, it does not cover the full range of a 6D tracking problem, where position and orientation of the sensor might change. The solution of such a problem might require supplemental sensors and sources.

The first proof-of-concept measurements using a single or dual excitation signal generally show accordance with the magnetic simulation. A visual comparison between both scenarios shows no significant influence of intercarrier interference. While the magnetic simulation for coil A matched quite well, there was a more significant deviation for coil B. We generally expected some uncertainty for both coils regarding the exact measurement of the current and the estimation of position, orientation, and geometric properties of the coil (e.g., mean radius). In particular, the current might vary between each coil as well as between physical setup and simulation. This is due to the low impedance (cf. Figure 2), which is consequently affected by parasitic resistances of connectors and cables as well as measurement inaccuracy.

In contrast to the localization in the quasi-stationary case, we considered bandwidth as a key factor in motion sensing. The proposed structure uses multiple matched filters in parallel, which define the bandwidth requirements. The selection of the instantaneous bandwidth manages the trade-off between a high spatial accuracy for slow movements and a sufficient temporal accuracy for fast movements. There might be potential in an adaptive selection based on velocity or position estimations.

The noise measurements during motion showed a significant increase in noise compared to the resting case, which also leads to a degradation in SNR. Some of the noise-increasing effects might be related to the experimental nature of the measurement setup. As the sensor cable is trailing the cart, it might cause mechanical oscillations, which propagate to the sensor. This is relevant for the peaks that regularly occurred during the measurements. It also affected the theoretical range of 25 cm for the highest bandwidth, which was below our initial expectations. These performance metrics (cf. Figure 10) were obtained for an approximately constant movement. When considering the bandwidth only, they should be comparable for any arbitrary motion with similar bandwidth characteristics in its corresponding magnetic (desired) signal. However, there might be other mechanical effects in practical measurements that contribute to the additional noise.

As receiver noise was identified to be the key factor for system performance, there are multiple potential improvements conceivable: a cantilever without a magnetostrictive layer (piezoelectric sensor [30]) might be employed to analyze the effects of mechanical noise due to motion and apply adaptive noise reduction schemes; electrical shielding might improve noise performance at the cost of sensitivity and a tilted sensitivity axis (as in [9]); additionally, an increased matched filter length is an option for slow and quasi-stationary movements.

However, we assessed magnetoelectric motion sensing as a preliminary development stage of a comprehensive human motion tracking system. This does not yet allow for a meaningful comparison with common motion tracking systems. Instead, the SNR is more of a ground floor, which can potentially be significantly improved by subsequent tracking and sensor fusion algorithms (e.g., extended Kalman-based) if a priori knowledge on the motion and the moving object is taken into account.

All of the applied scenarios were focused on simple movements that are not generally comparable to the complexity of human movements. As we aim for an application in movement analysis for neurodegenerative diseases like Parkinson's disease, an important next step is the application of a magnetoelectric motion sensing setup inside the clinical motor lab with our partners at the University Medical Center Schleswig-Holstein. State-of-the-art optical and IMU reference systems are available there to validate the magnetoelectric measurement results and evaluate the system's performance. As the signal processing is currently done offline, such measurements would also greatly benefit from a real-time implementation.

Author Contributions: Conceptualization, J.H. and E.E.; methodology, J.H.; software, J.H.; validation, J.H.; formal analysis, J.H.; investigation, J.H. and C.B.; data curation, J.H.; writing—original draft preparation, J.H. ; writing—review and editing, J.H., E.E., C.B. and G.S.; visualization, J.H.; supervision, E.E., C.B. and G.S.; project administration, W.M. and G.S.; funding acquisition, C.H., W.M. and G.S. All authors have read and agreed to the published version of the manuscript.

Funding: This work was supported by the German Research Foundation (Deutsche Forschungsgemeinschaft, DFG) through the project B9 of the Collaborative Research Centre CRC 1261 "Magnetoelectric Sensors: From Composite Materials to Biomagnetic Diagnostics".

Institutional Review Board Statement: Not applicable.

Informed Consent Statement: Not applicable.

Data Availability Statement: Not applicable.

Acknowledgments: The authors would like to thank Henrik Wolframm for providing the composite amplifiers and Christine Kirchhof and Phillip Durdaut for creating the sensor.

Conflicts of Interest: The authors declare no conflict of interest. The funders had no role in the design of the study; in the collection, analyses, or interpretation of data; in the writing of the manuscript, or in the decision to publish the results.

Abbreviations

The following abbreviations are used in this manuscript:

ME	Magnetoelectric
IMU	Inertial measurement unit
SNR	Signal-to-noise ratio
OPM	Optically pumped magnetometer
SQUID	Superconducting quantum interference device
DC	Direct current

AC	Alternating current
VLF	Very low frequency
TDMA	Time division multiple access
CDMA	Code division multiple access
FDMA	Frequency division multiple access
RMS	Root mean square
ASD	Amplitude spectral density
MF	Matched filter

References

1. Sabatini, A.M. Kalman-Filter-Based Orientation Determination Using Inertial/Magnetic Sensors: Observability Analysis and Performance Evaluation. *Sensors* **2011**, *11*, 9182–9206. [CrossRef]
2. Kortier, H.G.; Sluiter, V.I.; Roetenberg, D.; Veltink, P.H. Assessment of hand kinematics using inertial and magnetic sensors. *J. Neuroeng. Rehabil.* **2014**, *11*, 70. [CrossRef]
3. Mirelman, A.; Bernad-Elazari, H.; Thaler, A.; Giladi-Yacobi, E.; Gurevich, T.; Gana-Weisz, M.; Saunders-Pullman, R.; Raymond, D.; Doan, N.; Bressman, S.B.; et al. Arm swing as a potential new prodromal marker of Parkinson's disease. *Mov. Disord.* **2016**, *31*, 1527–1534. [CrossRef]
4. Espay, A.J.; Bonato, P.; Nahab, F.; Maetzler, W.; Dean, J.M.; Klucken, J.; Eskofier, B.M.; Merola, A.; Horak, F.; Lang, A.E.; et al. Technology in Parkinson disease: Challenges and Opportunities. *Mov. Disord.* **2016**, *31*, 1272–82. [CrossRef]
5. Thun-Hohenstein, C.; Klucken, J. Wearables as a Supportive Tool in the Care of Patients with Parkinson's Disease: A Paradigm Change. *Klin. Neurophysiol.* **2021**, *52*, 44–51. [CrossRef]
6. Roetenberg, D.; Slycke, P.J.; Veltink, P.H. Ambulatory Position and Orientation Tracking Fusing Magnetic and Inertial Sensing. *IEEE Trans. Biomed. Eng.* **2007**, *54*, 883–890. [CrossRef] [PubMed]
7. Pasku, V.; Angelis, A.D.; Angelis, G.D.; Arumugam, D.D.; Dionigi, M.; Carbone, P.; Moschitta, A.; Ricketts, D.S. Magnetic field-based positioning systems. *IEEE Commun. Surv. Tutor.* **2017**, *19*, 2003–2017. [CrossRef]
8. Bosch Sensortec GmbH. BMX055 IMU Specification. Available online: https://www.bosch-sensortec.com/products/motion-sensors/absolute-orientation-sensors/bmx055/ (accessed on 3 August 2021).
9. Spetzler, B.; Bald, C.; Durdaut, P.; Reermann, J.; Kirchhof, C.; Teplyuk, A.; Meyners, D.; Quandt, E.; Höft, M.; Schmidt, G.; et al. Exchange biased delta-E effect enables the detection of low frequency pT magnetic fields with simultaneous localization. *Sci. Rep.* **2021**, *11*, 5269. [CrossRef] [PubMed]
10. Stefan Mayer Instruments. FL1-100 Fluxgate Magnetometer Specification. Available online: https://stefan-mayer.com/de/produkte/magnetometer-und-sensoren/magnetfeldsensor-fl1-100.html (accessed on 22 November 2021).
11. QuSpin. QTFM Total Field Magnetometer Specification. Available online: https://quspin.com/qtfm/ (accessed on 3 August 2021).
12. Twinleaf. microSAM Total Field Magnetometer Specification. Available online: https://twinleaf.com/scalar/microSAM/ (accessed on 3 August 2021).
13. Wu, F.; Liang, Y.; Fu, Y.; Ji, X. A robust indoor positioning system based on encoded magnetic field and low-cost IMU. In Proceedings of the IEEE/ION Position, Location and Navigation Symposium (PLANS), Savannah, GA, USA, 11–14 April 2016; pp. 204–212. [CrossRef]
14. Hellmers, H.; Norrdine, A.; Blankenbach, J.; Eichhorn, A. An IMU/magnetometer-based Indoor positioning system using Kalman filtering. In Proceedings of the International Conference on Indoor Positioning and Indoor Navigation, Montbeliard, France, 28–31 October 2013. [CrossRef]
15. Polhemus. Viper Motion Tracking System. Available online: https://polhemus.com/viper (accessed on 3 August 2021).
16. Caruso, M.J.; Bratland, T.; Smith, C.H.; Schneider, R. A new perspective on magnetic field sensing. *Sensors* **1998**, *15*, 34–47.
17. Tumanski, S. *Handbook of Magnetic Measurements*; CRC Press Taylor and Francis: Boca Raton, FL, USA, 2011; pp. 6, 42, 45.
18. Omicron Labs. Vector Network Analyzer Bode 100 Rev. 1. 2021. Available online: https://www.omicron-lab.com/products/vector-network-analysis/bode-100/technical-data (accessed on 3 August 2021).
19. Bao, J.; Hu, C.; Lin, W.; Wang, W. On the magnetic field of a current coil and its localization. In Proceedings of the 2012 IEEE International Conference on Automation and Logistics, Zhengzhou, China, 15–17 August 2012; pp. 573–577. [CrossRef]
20. Rothwell, E.J.; Cloud, M.J. *Electromagnetics*; CRC Press: Boca Raton, FL, USA, 2001; pp. 158–160.
21. Paperno, E.; Plotkin, A. Cylindrical induction coil to accurately imitate the ideal magnetic dipole. *Sens. Actuators A Phys.* **2004**, *112*, 248–252. [CrossRef]
22. Gojdka, B.; Jahns, R.; Meurisch, K.; Greve, H.; Adelung, R.; Quandt, E.; Knöchel, R.; Faupel, F. Fully integrable magnetic field sensor based on delta-E effect. *Appl. Phys. Lett.* **2011**, *99*, 223502. [CrossRef]
23. Durdaut, P.; Penner, V.; Kirchhof, C.; Quandt, E.; Knöchel, R.; Höft, M. Noise of a JFET Charge Amplifier for Piezoelectric Sensors. *IEEE Sens. J.* **2017**, *17*, 7364–7371. [CrossRef]
24. Lage, E.; Woltering, F.; Quandt, E.; Meyners, D. Exchange biased magnetoelectric composites for vector field magnetometers. *J. Appl. Phys.* **2013**, *113*, 17C725. [CrossRef]

25. Jahns, R.; Knöchel, R.; Greve, H.; Woltermann, E.; Lage, E.; Quandt, E. Magnetoelectric sensors for biomagnetic measurements. In Proceedings of the 2011 IEEE International Symposium on Medical Measurements and Applications, Bari, Italy, 30–31 May 2011, pp. 107–110. [CrossRef]
26. Proakis, J.G.; Manolakis, D.G. *Digital Signal Processing: Principles, Algorithms and Applications*; Prentice Hall: Upper Saddle River, NJ, USA, 1996; pp. 340–343.
27. Zölzer, U. *Digital Audio Signal Processing*; Wiley: Hoboken, NJ, USA, 2008; p. 129.
28. Sklar, B. *Digital Communications: Fundamentals and Applications*; Number 5; Prentice Hall: Upper Saddle River, NJ, USA, 2007; pp. 122–126.
29. Analog Devices. ADA4870, High Speed, High Voltage, 1 A Output Drive Amplifier. 2021. Available online: https://www.analog.com/media/en/technical-documentation/data-sheets/ADA4870.pdf (accessed on 3 August 2021).
30. Reermann, J.; Bald, C.; Salzer, S.; Durdaut, P.; Piorra, A.; Meyners, D.; Quandt, E.; Höft, M.; Schmidt, G. Comparison of reference sensors for noise cancellation of magnetoelectric sensors. In Proceedings of the 2016 IEEE SENSORS, Orlando, FL, USA, 30 October–3 November 2016; pp. 1–3. [CrossRef]

Article

Phase Noise of SAW Delay Line Magnetic Field Sensors

Phillip Durdaut [1,*,†], Cai Müller [2,†], Anne Kittmann [3], Viktor Schell [3], Andreas Bahr [4,‡], Eckhard Quandt [3,‡], Reinhard Knöchel [1,‡], Michael Höft [1,‡] and Jeffrey McCord [2,‡]

1. Microwave Engineering, Institute of Electrical Engineering and Information Technology, Faculty of Engineering, Kiel University, Kaiserstr. 2, 24143 Kiel, Germany; rk@tf.uni-kiel.de (R.K.); mh@tf.uni-kiel.de (M.H.)
2. Nanoscale Magnetic Materials and Magnetic Domains, Institute for Materials Science, Faculty of Engineering, Kiel University, Kaiserstr. 2, 24143 Kiel, Germany; camu@tf.uni-kiel.de (C.M.); jmc@tf.uni-kiel.de (J.M.)
3. Inorganic Functional Materials, Institute for Materials Science, Faculty of Engineering, Kiel University, Kaiserstr. 2, 24143 Kiel, Germany; anki@tf.uni-kiel.de (A.K.); visc@tf.uni-kiel.de (V.S.); eq@tf.uni-kiel.de (E.Q.)
4. Sensor System Electronics, Institute of Electrical Engineering and Information Technology, Faculty of Engineering, Kiel University, Kaiserstr. 2, 24143 Kiel, Germany; ab@tf.uni-kiel.de
* Correspondence: pd@tf.uni-kiel.de; Tel.: +49-431-880-6167
† These authors contributed equally to this work.
‡ J. McCord, M. Höft, R. Knöchel, E. Quandt, and A. Bahr are with Kiel Nano, Surface and Interface Science (KiNSIS), Kiel University, Christian-Albrechts-Platz 4, 24118 Kiel, Germany.

Abstract: Surface acoustic wave (SAW) sensors for the detection of magnetic fields are currently being studied scientifically in many ways, especially since both their sensitivity as well as their detectivity could be significantly improved by the utilization of shear horizontal surface acoustic waves, i.e., Love waves, instead of Rayleigh waves. By now, low-frequency limits of detection (LOD) below $100\,\text{pT}/\sqrt{\text{Hz}}$ can be achieved. However, the LOD can only be further improved by gaining a deep understanding of the existing sensor-intrinsic noise sources and their impact on the sensor's overall performance. This paper reports on a comprehensive study of the inherent noise of SAW delay line magnetic field sensors. In addition to the noise, however, the sensitivity is of importance, since both quantities are equally important for the LOD. Following the necessary explanations of the electrical and magnetic sensor properties, a further focus is on the losses within the sensor, since these are closely linked to the noise. The considered parameters are in particular the ambient magnetic bias field and the input power of the sensor. Depending on the sensor's operating point, various noise mechanisms contribute to f^0 white phase noise, f^{-1} flicker phase noise, and f^{-2} random walk of phase. Flicker phase noise due to magnetic hysteresis losses, i.e. random fluctuations of the magnetization, is usually dominant under typical operating conditions. Noise characteristics are related to the overall magnetic and magnetic domain behavior. Both calculations and measurements show that the LOD cannot be further improved by increasing the sensitivity. Instead, the losses occurring in the magnetic material need to be decreased.

Keywords: Barkhausen noise; delay line sensor; Flicker noise; Kerr microscopy; magnetic domain networks; magnetic field sensor; magnetic noise; magnetoelastic delta-E effect; phase noise; surface acoustic wave

1. Introduction

Since the invention of the interdigital transducer (IDT) in 1965, surface acoustic waves (SAW) can be excited very efficiently on piezoelectric materials [1]. Due to their small size, low cost, and high reproducibility, SAW filters have taken on a key role in modern consumer and communication systems [2]. The same advantageous properties make SAW technology attractive for sensor applications [3,4].

Utilizing the inverse piezoelectric effect, a SAW is excited by applying an electrical voltage on an (input) IDT that is patterned on a piezoelectric material. The mechanical wave propagates perpendicular to the direction of the IDT in both directions on the surface of the piezoelectric substrate ([5], p. 139). For sensing applications the substrate's surface is frequently coated with an additional layer which reacts to changes of the physical quantity to measure and, in turn, alters the propagating wave in its amplitude and in its velocity. For the detection of externally affected wave properties such a device can be equipped with an additional (output) IDT, thus forming a so-called delay line structure. Due to the reciprocity of IDTs and via the direct piezoelectric effect the mechanical wave is then converted back into an electrical signal.

The operating principle of SAW delay line magnetic field sensors [6] is based on the magnetoelastic ΔE effect. It leads to changes of the Young's modulus E and the related shear modulus G, respectively, of an additional magnetostrictive layer as a function of the material's magnetization M, i.e., of an ambient magnetic flux density $B = \mu_0 H$ (μ_0 and H denote the vacuum permeability and the magnetic field strength). Due to the relation between the mechanical property G and the wave's propagation velocity v [7] the phase φ of a shear wave magnetic field sensor's output signal is a function of the magnetic flux density B with the phase sensitivity

$$S_{PM} = \frac{\partial \varphi}{\partial B} = \frac{\chi}{\mu_0} \cdot \frac{\partial G}{\partial M} \cdot \frac{\partial v}{\partial G} \cdot \frac{\partial \varphi}{\partial v} \tag{1}$$

where χ denotes the magnetic susceptibility $\chi = \partial M / \partial H$.

The first magnetoelastic SAW delay line devices were presented in the 1970s for the possible application as magnetically tunable phase shifters [8,9], e.g., in frequency-tunable oscillators [10]. The use of soft magnetic materials such as iron-boron (FeB) instead of hard magnetic nickel (Ni) leads to lower ambient magnetic flux densities B required for achieving a significant phase shift [11]. To further increase the effect or the maximum phase shift, respectively, the thickness of the magnetic layer was increased and various magnetic alloys were applied [12–19]. In 1992, Yokokawa et al. demonstrated that the magnetically induced phase shift can be significantly increased if shear horizontal surface acoustic waves, i.e., Love waves, instead of Rayleigh waves are excited [20].

The first magnetoelastic delay line magnetic field sensor capable of detecting changes of 1 µT was presented in 1987 [21]. It was not until 30 years later that magnetically coated delay lines were again operated as sensors [22] with an achieved detection limit of 140 nT [23]. The first sensor explicitly exploiting the high sensitivity of Love waves was presented in 2018 reaching a limit of detection (LOD) of $250\,\text{pT}/\sqrt{\text{Hz}}$ at a frequency of 10 Hz [24]. Besides utilizing higher Love modes in the Gigahertz regime [25] also resonant surface acoustic Love wave magnetic field sensors have been introduced in the last three years [26–28]. Meanwhile, Love wave delay line sensors reach limits of detection as low as $70\,\text{pT}/\sqrt{\text{Hz}}$ at a frequency of 10 Hz [29]. Thus, such sensors are already significantly more detective than state-of-the-art Hall effect sensors with typical limits of detection around $1\,\mu\text{T}/\sqrt{\text{Hz}}$ at a frequency of 10 Hz [30]. Currently, the LOD of SAW magnetic field sensors is most comparable with magnetoresistive sensors with values around $100\,\text{pT}/\sqrt{\text{Hz}}$ at 10 Hz [31,32]. However, giant magnetoimpedance [33] and fluxgate sensors [34], for example, still achieve significantly better low-frequency values around or even below $10\,\text{pT}/\sqrt{\text{Hz}}$.

Apart from the fact that even values on measured limits of detection are rarely given, no detailed results on the noise behavior of magnetoelastic SAW delay line sensors have been reported so far. In a recent study on the required readout electronics for the operation of such delay line sensors it was found that magnetostrictively coated SAW devices can exhibit significantly increased noise compared to bare devices without magnetic material [35]. In previous studies on the effective noise of other types of thin-film magnetic field sensors, direct links to the magnetic domain behavior have been determined [36]. Therefore, the LOD of SAW magnetic field sensors based on magnetic thin films can only be further improved by gaining a deep understanding of the existing sensor-intrinsic noise sources with an emphasis on the magnetic domain behavior and its impact on the sensor's overall performance.

This paper is organized as follows: Section 2 introduces the SAW magnetic field sensor under investigation and discusses its electrical and magnetic behavior. The comprehensive analysis of the intrinsic phase noise of SAW delay line magnetic field sensors is divided into two parts. First, in Section 3, phase noise occurring in magnetically saturated devices as well as in devices without any sensitive coating is discussed. Secondly, additional phase noise phenomena due to the magnetostrictive layer are presented and analyzed in Section 4. This article finishes with a summary of the findings in Section 5.

2. SAW Sensor

A delay line is formed using two split-finger IDT electrodes with 25 finger pairs, a periodicity, i.e., an acoustic wavelength, of $\lambda = 28\,\mu\text{m}$, and a finger width of $3.5\,\mu\text{m}$ with an IDT center-to-center length of $L = 4.64\,\text{mm}$. An SiO_2 layer with a thickness of $4.5\,\mu\text{m}$ deposited on top of the IDTs and the delay line acts as a guiding layer for the surface acoustic Love wave. A magnetostrictive material $(Fe_{90}Co_{10})_{78}Si_{12}B_{10}$ with a thickness of $z = 200\,\text{nm}$ and a length of $3.8\,\text{mm}$ is magnetron sputter-deposited on top of the guiding layer and between the IDTs. During deposition, for maximizing the sensor's magnetic sensitivity, a magnetic field is applied along the direction of the delay line to saturate the magnetic film and to introduce an easy axis of magnetization [29].

Further details about the fabrication can be found in [37]. The sensor mainly discussed in this paper has already been used in a previous study with a focus on the electrical readout systems which also contains a photography of the sensor [35].

2.1. Electrical Properties

For the electrical characterization, the two-port scattering parameters s_{ij} $(i, j \in \{1, 2\})$ of the sensor are measured with a calibrated vector network analyzer *E8361A* from *Agilent Technologies* at a signal power of $P_0 = 0\,\text{dBm}$. In order to counteract additional magnetic influences (will be discussed further below), the magnetostrictive layer is magnetically saturated with a permanent magnet ($B = B_\text{sat} \approx 10\,\text{mT}$) perpendicular to the wave propagation direction, i.e., along the magnetic hard axis. To also minimize mismatch losses due to reflections at the electrical-acoustical interfaces, an individual impedance matching to the system impedance of $Z_0 = 50\,\Omega$ was carried out using discrete inductors and capacitors prior to all measurements. In addition, to suppress significant signal-dropping in the transmission characteristics due to interference of electrical crosstalk, the delay line is connected symmetrically utilizing a balun (*ATB2012-50011* from *TDK*) at each port.

Values for the return loss of $\text{RL}(f_0) = -20\log_{10}(|s_{ii}(f_0)|)\,\text{dB} > 20\,\text{dB}$ ($i \in \{1,2\}$) are achieved at the sensor's synchronous frequency of $f_0 = 144.8\,\text{MHz}$ at each port (Figure 1a). The exactly measured values correspond with an overall mismatch loss ([38], pp. 64–65)

$$\text{ML}(f) = 10\log_{10}\left(\frac{1}{1-|s_{11}(f)|^2} \cdot \frac{1}{1-|s_{22}(f)|^2}\right)\text{dB} \qquad (2)$$

as low as $\text{ML}(f_0) = 0.04\,\text{dB}$ which is negligible for the sensor under investigation. Thus, the measured insertion loss (Figure 1b) at f_0 with a typical value for Love wave delay lines [39] of $\text{IL}(f_0) = -20\log_{10}(|s_{21}(f_0)|)\,\text{dB} = 20\,\text{dB}$ is virtually solely determined by the SAW device itself.

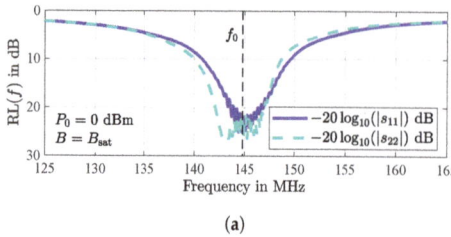 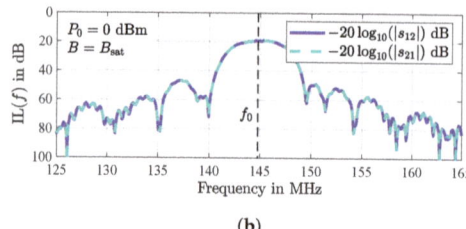

(a)   (b)

Figure 1. Measured two-port scattering parameters of the SAW sensor yielding a return loss (RL) better than 20 dB at each port (**a**) and an insertion loss (IL) with a value of 20 dB (**b**), both at the synchronous frequency of $f_0 = 144.8$ MHz. During the measurements, performed for an input power of $P_0 = 0$ dBm, the sensor has been magnetically saturated at $B = B_{sat} \approx 10$ mT in order to avoid additional magnetic influences (which will be discussed below).

2.2. Magnetic Properties

For the magnetic characterization, magnetooptical Kerr effect (MOKE) magnetometry and magnetic domain observations were applied using a home-build large view MOKE setup [40,41] with telecentric optics. The magnetization loop measured along and perpendicular to the device dimensions shown in Figure 2 displays a well-defined soft magnetic uniaxial anisotropy behavior. With the saturation polarization $B_s = 1.5$ T [42] a uniaxial anisotropy constant of $K_u \approx 960$ J/m^3 is obtained. This corresponds to a relative permeability of $\mu_{r,ha} \approx 950$ along the magnetic hard axis. The easy axis maximum permeability, governed by magnetic domain wall motion, is significantly higher and in the order of $\mu_{rmax,ea} \approx 10{,}000$.

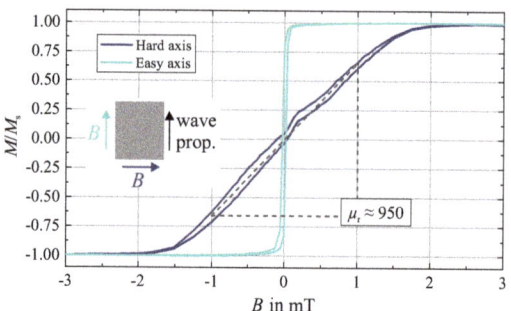

Figure 2. Magnetization loops along (easy axis) and perpendicular (hard axis) to the direction of wave propagation of the SAW device measured by magnetooptical magnetometry. The relative permeability perpendicular to the direction of wave propagation is around $\mu_r \approx 950$. The magnetic film geometry and the measurement directions are sketched.

Relevant insight into the actual magnetization behavior is obtained from the magnetic domain behavior. A comparison of the magnetic domain behavior for ascending and descending loop branches perpendicular to the direction of wave propagation is shown in Figure 3. Two important points become obvious in the domain arrangement. First, the magnetic domain behavior is asymmetric. Different magnetic domain characteristics are found for the ascending and descending loop for a given magnetic field. Coming from magnetic saturation spike domains develop at the edges and magnetization rotation takes place in the center of the magnetic layer structure. The sign of initial magnetization rotation does not depend on the sign of applied saturation field B, it is counterclockwise (ccw). After remanence, the spike domains grow and further on penetrate the whole sample, forming large magnetic domains. The switched domains then rotate clockwise (cw) with further increase of field magnitude. Secondly, the spike domains as well as the central domain walls are slightly tilted with respect to the dimension of the device. Both findings indicate a slightly tilted magnetic anisotropy axis. Consequently, the magnetization firstly rotates towards the preferential anisotropy axis, counterclockwise for ascending and descending external field variations. Due to the resulting symmetry breaking, after reversing the magnetic field, the magnetization reversal from one to the now favored axis of anisotropy takes place by domain wall motion.

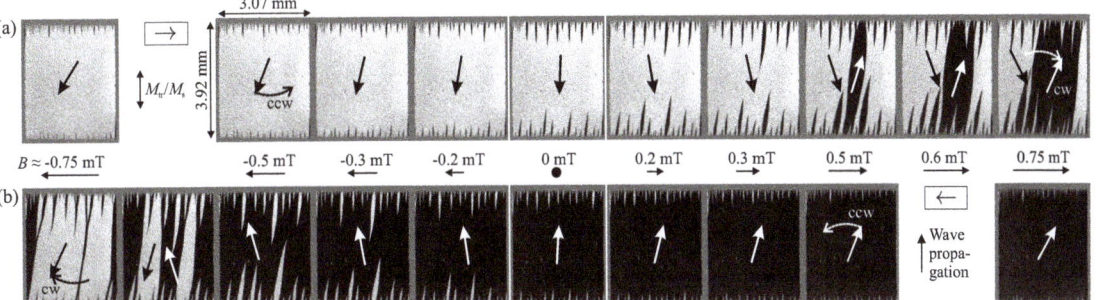

Figure 3. Magnetic domain evolution for increasing (**a**, →) and decreasing (**b**, ←) direction of external field *B*. Magnetic field values are indicated. The corresponding basic alignment of magnetization inside the magnetic film is sketched. The magnetooptical sensitivity is transverse to the applied magnetic field direction.

To obtain a measure for the asymmetric magnetic domain and domain wall behavior, a simple Sobel filter implemented in the image processor ImageJ [43] is used for estimating the relative magnetic domain wall length with variation of *B*, the results of which are displayed in Figure 4. The development of domain walls displays a strong hysteresis with a relatively monotonous increase and then decrease in domain wall density. The domain wall density peaks around $B \approx \mp 0.5$ mT for the reversed magnetic field application. Recapping, the magnetization process is asymmetric and reversing in characteristics with reversing magnetic field history, which is directly reflected in the magnetic domain wall density variation with field.

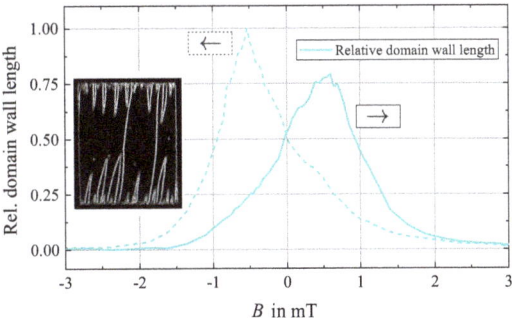

Figure 4. Relative magnetic domain wall evolution for increasing (→) and decreasing (←) direction of external field *B* obtained via edge detection from the magnetooptical micrographs. The intensity from the domain wall contrast obtained by the edge detection operation is interpreted as a value related to the magnetic domain wall length. An example on an edge detection filtered intensity analyzed image is shown.

2.3. Electrically Induced Changes of the Magnetization Behavior

A quantitative measure of the magnetic domain switching behavior with varying electrical input power P_0 is obtained from an analysis of the magnetization behavior by MOKE transverse magnetometry [44] with the MO sensitivity aligned perpendicular to the magnetic field excitation. Exemplary transverse loop data on the switching behavior is displayed in Figure 5a.

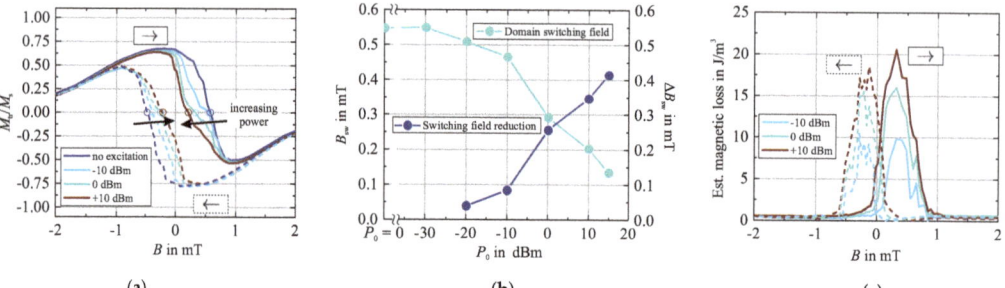

Figure 5. Exemplary transverse sensitivity magnetization loops for different values of electrical input power P_0 (**a**) and corresponding magnetic domain switching fields B_{sw} and magnetic switching field reduction ΔB_{sw} with P_0 (**b**) derived from the transverse loops. The magnetic energy transfer (**c**) is estimated taking into account (**a**) and the easy axis magnetization characteristics (Figure 2). A static easy axis hysteresis loss of $P_{dw} \approx 60\,\text{J/m}^3$ is estimated from the easy axis loop. The varying electrical input power P_0 was set at or close to the synchronous frequency $f_0 = 144.8\,\text{MHz}$ of the device.

No change in the regular magnetization loops with an increase of input power P_0 was found. Starting from negative values of B, the transverse magnetization component increases corresponding to a dominating ccw rotation of magnetization for the given MOKE settings (dark contrast in Figure 3, $M_{tr}/M_s > 0$). After remanence, the transverse magnetization component decreases due to the growth of reversed magnetized magnetic domains. The domain switching field B_{sw}, with the same fraction of upward and downward magnetized domains, we define at the field with $M_{tr}/M_s = 0$ (note that these values are not equal to the coercive fields). The positions of B_{sw} are indicated in Figure 5a. With further reversing B ($M_{tr}/M_s < 0$) the sense of magnetization rotation inverts to cw rotation, again confirming the slightly tilted magnetic anisotropy axis. The switching process is accompanied by irregular stepwise change in the transverse magnetization, corresponding to domain wall or Barkhausen jumps. With the application of an electrical input power of P_0 the general magnetization behavior remains unchanged. Yet, the domain switching field decreases with increasing P_0. The overall decrease of B_{sw} with P_0 and the directly related reduction of the switching field ΔB_{sw} is displayed in Figure 5b. Even for small levels of input power ($P_0 = -20\,\text{dBm}$) a measurable influence on the magnetic domain walls depinning fields is visible in the data. This effect on the magnetic domains, respectively magnetic domain walls we interpret as an energy transfer from the surface acoustic waves to the magnetic domain walls.

An estimation of the energy transfer with B is not straightforward but comparing the difference in the transverse magnetization loops relative to the zero-input state should give a rough approximation of the energy transferred to the magnetic domain walls. The difference $\Delta M_{tr}/M_s$ is then compared to the hysteretic energy loss of the easy axis magnetization loop ($\oint H\, dB$) where the overall magnetization response from negative to positive saturation is $2 \cdot M_{tr}/M_s$. This process is as well characterized by magnetic domain wall motion. Assuming an idealized square easy axis loop the hysteric loss P_{dw} is then defined by

$$P_{dw} = \frac{1}{2}\Delta M_{tr}/M_s \cdot \oint H_{ea}\, dB_{ea}. \qquad (3)$$

The corresponding field dependency is shown in Figure 5c. For the easy axis magnetization behavior the relevant energy densities are much smaller than the uniaxial anisotropy density. The estimated energy transfer peaks at the domain stability field, respectively, the domain switching field, as well as with the maximum in domain wall length (compare to Figure 2). Yet, the regime of relevant energy transfer is reduced and more asymmetric, respectively, enhanced in the domain switching regime. The linked relation of the magnetization response on the electrical properties is discussed next.

2.4. Magnetically Induced Changes of the Electrical Properties

Characteristic magnetic influences on the SAW delay line sensor's electrical behavior are extracted from a series of measurements of the sensor's two-port scattering parameters as already described above but additionally for various ambient static magnetic flux densities B. The results are depicted in Figure 6.

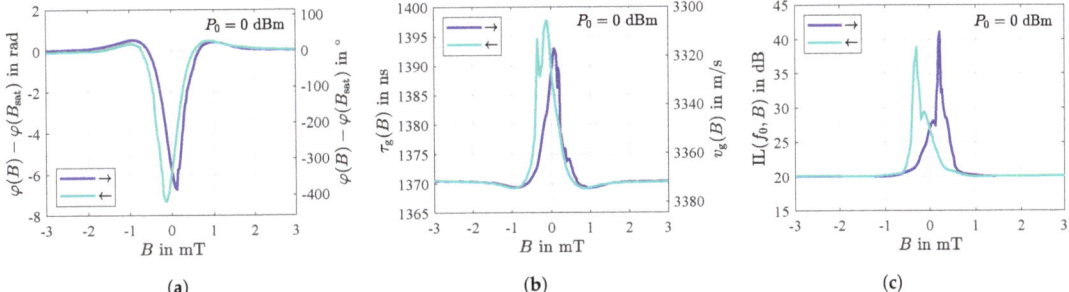

Figure 6. Phase response (**a**), group delay and group velocity (**b**), and insertion loss (**c**) of the SAW delay line sensor obtained from a series of measurements of the two-port scattering parameters for various static magnetic flux densities B. The results of all three sub-figures are based on data from the same series of measurements which was performed for an input power of the sensor of $P_0 = 0$ dBm and at or around, respectively, its synchronous frequency of $f_0 = 144.8$ MHz.

The static magnetic fields are generated by means of a programmable current source B2962A from *Keysight* and a solenoid [45]. The sensor and the surrounding solenoid are placed inside an ultra high magnetic field shielding mu-metal cylinder ZG1 from *Aaronia AG* in order to avoid significant static offsets by earth's magnetic field. The magnetic flux density is swept from negative to positive values and also backwards after magnetically saturating the sensor at $B_{\text{sat}} = \mp 10$ mT before each magnetic field sweep.

The static phase response $\varphi(B) = \arg(s_{21}(f_0, B))$ shown in Figure 6a exhibits a significant dependence on the magnetic flux density B. Compared to the value $\varphi(B_{\text{sat}})$ in magnetic saturation, the phase changes by up to about 7 rad ($\approx 400°$). Since the phase changes are significant especially in the ranges around $B \approx \pm 0.2$ mT, these regions are of particular interest for a later sensor operation (discussed below in Section 2.5). As a consequence of a slightly tilted magnetic anisotropy axis in the magnetic layer (Section 2.2) the phase responses are asymmetric and hysteretic such that the minimum values are reached just below or slightly above $B = 0$, respectively [29].

The phase response can be expressed by the group delay $\tau_g(B) = -\partial \varphi(f, B)/(2\pi\, \partial f)$ or, analogously, by the group velocity $v_g(B) = L/\tau_g(B)$ as depicted in Figure 6b where the derivative of the phase response $\varphi(f)$ was calculated in its linear regime (compare Figure 2a in [37]) around the sensor's synchronous frequency ($f_0 \pm 2$ MHz). With a value in the range of about 3340 m/s, the latter lies well between the theoretical bulk shear velocities $v_{\text{sh}} = \sqrt{G/\rho}$ of quartz (4309 m/s) and the magnetostrictive FeCoSiB (2737 m/s) with G and ρ representing the shear modulus and the specific mass, respectively, of the individual material [37].

In addition to the phase changes $\varphi(B)$, an analysis of the magnetic field dependent insertion loss $\text{IL}(f_0, B)$ reveals another significant dependence. As shown in Figure 6c, the insertion loss increases from the fundamental value of 20 dB in magnetic saturation to values of up to 39 dB and 41 dB, respectively, depending on the direction of the previously performed magnetic saturation. In fact, these high values occur in those ranges where the phase also changes significantly with the external magnetic field (compare Figure 6a). However, obviously, maximum losses only occur on one side with regard to $B = 0$, namely after zero crossing. In contrast, each phase response has two steep slopes. This loss is related to the corresponding magnetic domain behavior as discussed in Sections 2.2 and 2.4. The regime of increased electrical losses coincides with the occurrence of a multi-domain state and the shown energy transfer into the magnetic film. This leads to the dependence on the ambient magnetic field, due to the obvious hysteretic effects. These results clearly indicate an additional loss mechanism in the magnetic layer. Due to the general relation between losses and fluctuations the losses are of particular interest and are therefore characterized and discussed in detail further below.

2.5. Sensor Operation

When operating a SAW delay line sensor, the output signal is typically compared with its input signal to extract the desired information about the measurement signal. Although, in contrast to such open-loop systems, closed-loop or self-oscillating systems, respectively, are also common, an open-loop analysis of the sensor can be performed without any loss of generality [46].

Assuming an ideal oscillator signal, i.e., a sinusoidal signal without any fluctuations in amplitude and phase

$$P_{\text{in}}(t) = P_0\sqrt{2} \cdot \cos(2\pi f_0 t) \tag{4}$$

to excite the sensor at its synchronous frequency f_0 with an input power of P_0, the sensor's output signal can be described by

$$P_{\text{out}}(t) = P_0\sqrt{2}\,|s_{21}(f_0, B)|^2 \cdot \cos(2\pi f_0 t + \varphi(B) + \Delta\varphi(t)). \tag{5}$$

The term $\Delta\varphi(t)$ describes random phase fluctuations due to the sensor itself which are analyzed in detail in Sections 3 and 4. In real sensor operation, a magnetic measurement signal $B_x(t)$ should generally be detected with high sensitivity. Therefore B is to be interpreted as the sum of $B_x(t)$ and an ambient static magnetic bias flux density B_{bias} which is generally applied for maximizing the sensitivity S_{PM}. Thus, when neglecting any further changes and fluctuations of the signal's amplitude, Equation (5) can be written as

$$P_{\text{out}}(t) = P_0\sqrt{2}\,|s_{21}(f_0, B_{\text{bias}})|^2 \cdot \cos(2\pi f_0 t + S_{\text{PM}}(B_{\text{bias}})B_x(t) + \Delta\varphi(t)). \tag{6}$$

According to Equation (1), the phase sensitivity S_{PM} can principally be obtained by the derivative of the phase response $\varphi(B)$ (Figure 6a). However, it was found that this procedure leads to partially non-reproducible and incorrect results. In fact, a more precise distinction must be made. The slope of the phase response which corresponds with low insertion losses (Figure 6c) is typically unproblematic with regard to a numerical derivation. However, the slope that corresponds with significant insertion losses is often impaired by small phase jumps due to sudden and irreversible magnetic domain wall behavior (Section 2.3) which, in turn, will get even more pronounced when the derivative is calculated, thus, erroneously resulting in apparently high sensitivities. To overcome this issue, dynamic phase measurements for the determination of S_{PM} can be performed that are explained in Section 4.

2.6. Noise

The frequency dependent noise floor of a magnetic field sensor system is usually given by an amplitude spectral density in units of $\text{T}/\sqrt{\text{Hz}}$, often also referred to as equivalent magnetic noise floor, detectivity, or limit of detection (LOD)

$$\text{LOD}(f, B_{\text{bias}}, P_0) = \frac{\sqrt{S_\varphi}}{S_{\text{PM}}}. \tag{7}$$

It not only depends on the frequency f but also on the magnetic bias flux density B_{bias} and the sensor's input power P_0. Although the phase sensitivity decreases above a certain cutoff frequency depending on the sensor's geometry and its delay time [47], it is constant for frequencies below 10 kHz for the sensors under investigation. The term S_φ describes the one-sided power spectral density (in units of rad^2/Hz) of the sensor's random phase fluctuations $\Delta\varphi(t)$ ([48], p. 22). Its logarithmic representation $10\log_{10}(S_\varphi(f))$ is given in units of dB rad^2/Hz. For historical reasons, the two-sided phase noise density spectrum $\mathscr{L}(f)$ defined as $\mathscr{L}(f) = 1/2\, S_\varphi(f)$ and usually given in units of dBc/Hz (dB below the carrier) is often used [49].

A useful model for describing the frequency dependence of a power spectral density of random phase fluctuations is the polynomial law

$$S_\varphi(f) = \sum_{i=-n}^{0} b_i f^i \tag{8}$$

with usually $n \leq 4$. The exponents $i = 0$ and $i = -1$ refer to white phase noise and $1/f$ flicker phase noise, respectively, which are usually the main noise processes in two-port components ([48], p. 23) like amplifiers [50]. However, under certain circumstances, magnetostrictively coated SAW delay line devices can also exhibit random walk of phase ($i = -2$).

In the following, it will be shown that a total of five different types of phase noise phenomena, namely

(1) f^0 white phase noise and
(2) f^{-1} flicker phase noise

due to the SAW device itself and

(3) f^0 white phase noise,
(4) f^{-1} flicker phase noise, and
(5) f^{-2} random walk of phase

due to the additional magnetic material are observed depending on the sensor's operating point.

3. Phase Noise in Magnetic Saturation

In this section the phase noise of SAW delay line elements both without any magnetostrictive coating as well as delay lines of which the sensitive layer is magnetically saturated by means of a permanent magnet field ($B = B_{\text{sat}} \approx 10$ mT) is investigated. All phase noise measurements were performed at room temperature ($T = 290$ K) utilizing an FSWP phase noise analyzer from *Rohde & Schwarz* while the SAW device itself is located inside an electrically, magnetically, and acoustically shielded measurement environment.

3.1. White Phase Noise

Energy equipartition in classical thermodynamics states that the thermal energy is $1/2\, k_B T$ per degree of freedom with $k_B \approx 1.38 \times 10^{-23}$ J/K representing the Boltzmann constant ([51], pp. 264–266). For signals in the frequency range well below 6 THz (at room temperature) an overall noise energy of

$$N = k_B T = 4 \times 10^{-21}\,\text{J} \,\widehat{\approx}\, -174\,\frac{\text{dBm}}{\text{Hz}} \tag{9}$$

is equally partitioned into the two degrees of freedom, i.e., amplitude and phase ([48], p. 42). Thus, for a sensor's output signal with a power of $P_0|s_{21}|^2$ (Equation (6)), the white thermal phase noise is a factor of

$$\mathscr{L}^{-1} = \frac{2P_0|s_{21}|^2}{N} \tag{10}$$

below the carrier. This leads to a one-sided white phase noise power density ([48], p. 42) of

$$b_0 = \frac{N}{P_0 |s_{21}|^2} = \frac{\text{IL } N}{P_0} = \frac{F \, N}{P_0} \qquad (11)$$

which is often expressed by the device's noise figure $F = \text{IL} = |s_{21}|^{-2}$ (linear representations of F and IL) and linearly decreases with higher input power levels P_0. Therefore, thermal phase noise, i.e., white phase noise, is referred to as *additive (phase) noise* ([48], p. 35).

The measurement results shown in Figure 7 were acquired with an additional amplifier ZFL-1000LN+ from *Mini-Circuits* with a previously determined noise figure of $F_{\text{AMP}} = 1.875 \cong 2.73\,\text{dB}$ to amplify the sensor's output signal. For such a chain of two devices, the overall white phase noise power density at the amplifier's output can be determined by the adapted *Friis formula* ([48], p. 48)

$$b_0^{\text{chain}} = \left(F + \frac{F_{\text{AMP}} - 1}{|s_{21}|^2} \right) \frac{N}{P_0}. \qquad (12)$$

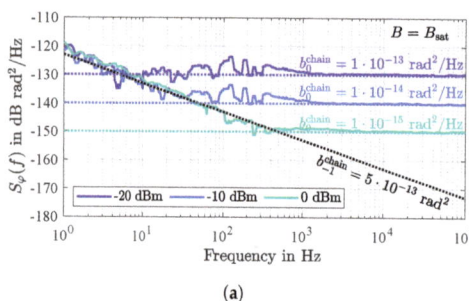 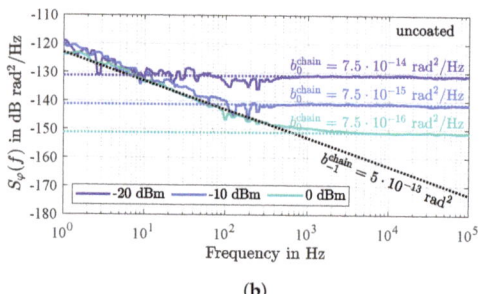

(a) (b)

Figure 7. Measured power spectral densities $S_\varphi(f)$ of the random phase fluctuations $\Delta\varphi(t)$ of the magnetostrictively coated sensor in magnetic saturation (a) and of an uncoated reference delay line (b). Both devices are measured for various input power levels P_0 and with an additional preamplifier. In agreement with Equation (11), the additive white phase noise decreases with P_0. Equal values for the parametric $1/f$ flicker phase noise are observed because both devices are located on the same chip.

For the SAW sensor under investigation previously introduced in Section 2, Figure 7a shows measured power spectral densities of random phase fluctuations for various input power levels P_0. As expected according to Equations (11) and (12), the white phase noise decreases by 10 dB each time the input power level is increased by 10 dB. Due to utilization of the additional amplifier, the measured coefficients b_0^{chain} contain additional phase noise of the amplifier. Calculating the sensor's noise figure based on Equation (12) yields a value of $F = 21.3\,\text{dB}$ which, within the measurement accuracy, agrees with the insertion loss of the sensor itself (20 dB) and additional losses of the connecting coaxial cables (previously determined to 1.2 dB). Thus, with regard to white phase noise, magnetostrictively coated SAW delay line sensors behave exactly as described by the existing noise theory. The white phase noise can be reduced by a higher input power level but increases with the insertion loss, regardless of the physical causes for the losses.

For comparison, a second series of measurements with the same measurement setup was performed for an uncoated reference delay line on the same chip as the previously measured magnetostrictively coated delay line (a photography of this chip can be found in [35]). The measurement results in Figure 7b show that the measured white phase noise is about 1.25 dB lower than for the sensitive delay line (Figure 7a) because the insertion loss of the uncoated device is lower by about the same amount. The reason for the slightly higher losses of the magnetically coated element are probably small defects in one of its two interdigital transducers (microscopic images of this imperfect transducer can be found in ([52], p. 362).

Losses due to eddy-currents always occur as soon as the coating material is electrically conductive. In [24] it was shown that the insertion losses significantly increase when a delay line is coated with thicker magnetic layers. If the thickness z of the magnetic layer is less than one skin depth δ (for the sensor under investigation $\delta \approx 1.4\,\mu m$ ([53], p. 19), the power loss due to eddy-currents can be calculated by

$$P_{\text{eddy}} = \frac{(2\pi f_0)^2 \hat{B}_0^2 V z^2}{24\rho} \qquad (13)$$

where V is the volume of the magnetic layer, ρ is the magnetic material's resistivity, and $\hat{B}_0$ is the amplitude of the magnetic flux density in the magnetic layer [54]. With $\hat{B}_0^2$ being proportional to the sensor's input power P_0, Equation (13) can be written as

$$P_{\text{eddy}} = \frac{(2\pi f_0)^2 \gamma P_0 V z^2}{24\rho} \qquad (14)$$

with $\gamma = \hat{B}_0^2/P_0$. With this definition, the sensor's additional insertion loss due to eddy-currents yield (linear representation)

$$\text{IL}_{\text{eddy}} = \frac{P_0}{P_0 - P_{\text{eddy}}} = \left(1 - \frac{(2\pi f_0)^2 \gamma V z^2}{24\rho}\right)^{-1}. \qquad (15)$$

Analytically, γ is not trivial to determine. However, based on time-resolved MOKE microscopy, the normalized amplitude of the magnetization $\hat{M}_0/M_s$ due to the surface acoustic wave and via the inverse magnetostrictive effect (Villari effect [55]) could be determined to values $\hat{M}_0/M_s < 0.1$ for sensors with a magnetic layer thickness of $z = 200\,nm$ at an input power of $P_0 = 10\,mW \cong 10\,dBm$. With $\hat{M}_0/M_s = \hat{B}_0/B_s$ the coefficient γ can also be expressed as

$$\gamma = \frac{\hat{B}_0^2}{P_0} = \frac{\left(\frac{\hat{M}_0}{M_s} B_s\right)^2}{P_0} \qquad (16)$$

yielding a value of $\gamma < 2.25\,T^2/W$ when assuming a saturation flux density of $B_s = 1.5\,T$ [42] for the utilized $(Fe_{90}Co_{10})_{78}Si_{12}B_{10}$ alloy. With a volume of $V = 3.07\,mm \cdot 3.92\,mm \cdot z$ ($z = 200\,nm$) and a resistivity of $\rho = 1.1\,\mu\Omega m$ [56] for the amorphous FeCoSiB alloy, the calculated power loss and the insertion loss due to eddy-currents yield $P_{\text{eddy}} < 68\,\mu W$ (at $P_0 = 10\,mW$) and $\text{IL}_{\text{eddy}} < 1.0068 \cong 0.03\,dB$, respectively, which are neglectable values for such sensors. However, white phase noise due to eddy-current losses is not generally neglectable. For layer thicknesses of $z = 650\,nm$ the insertion loss yields $\text{IL}_{\text{eddy}} = 1\,dB$ and further increases significantly for thicker layers, e.g., to $\text{IL}_{\text{eddy}} = 8.2\,dB$ for $z = 1\,\mu m$.

3.2. Flicker Phase Noise

Unlike frequency-independent (white) noise, the noise power of other noise phenomena is often confined at low frequencies. Although the power spectral densities describing these phenomena can have various spectral shapes, the most prominent example is the $1/f$ flicker noise which, with regard to the frequency, decreases with $10\,dB/\text{decade}$. Hence, $1/f$ noise is primarily disturbing in low-frequency applications. However, as soon as an additional carrier signal with a comparatively high amplitude is present, the noise also becomes noticeable around the carrier frequency, thus impairing the spectral components of a modulating signal ([48], p. 35). Besides a nonlinear mechanism, temporal fluctuations of the system properties can also cause the up-conversion of low-frequency noise ([48], pp. 44–45). An important characteristic of such *parametric amplitude and phase noise* is the independence from the carrier power.

For frequencies below the corner frequency

$$f_c = \frac{b_{-1}}{b_0} \tag{17}$$

white phase noise b_0 becomes neglectable and the overall power spectral density $S_\varphi(f)$ is dominated by $1/f$ flicker phase noise described by the coefficient b_{-1}.

In a chain of several components, e.g., a delay line followed by an amplifier, the white phase noise of each component adds up according to Equation (12). On the contrary, the $1/f$ flicker phase noise at the output of such a chain

$$b_{-1}^{\text{chain}} = b_{-1} + b_{-1}^{\text{AMP}} \tag{18}$$

is directly given by the sum of the individual $1/f$ flicker phase noise coefficients [48] (p. 49) (here b_{-1} and b_{-1}^{AMP} represent the $1/f$ flicker phase noise components of the SAW device and of an additional amplifier).

In advance to the noise measurements of which the results are shown in Figure 7, the flicker phase noise coefficient of the utilized preamplifier ZFL-1000LN+ from *Mini-Circuits* was determined to $b_{-1}^{\text{AMP}} = 6 \times 10^{-14}$ rad^2. Thus, with measured flicker phase noise coefficients of $b_{-1}^{\text{chain}} = 5 \times 10^{-13}$ rad^2 the SAW devices contribute a flicker phase noise of $b_{-1} = 4.4 \times 10^{-13}$ rad^2. Interestingly, both SAW devices, i.e., the magnetostrictively coated sensor as well as the uncoated reference delay line, show exactly the same flicker phase noise indicating that the additional magnetic layer does not contribute any further dominant flicker phase noise, at least if the sensitive layer is magnetically saturated. As characteristic for parametric noise, the flicker phase noise does not change with the input power.

Figure 8 shows the output power of both devices as a function of the input power P_0, each revealing strict linearity. Thus, a nonlinear mechanism for the up-conversion of the $1/f$ flicker noise can be excluded. Instead, a quasi-linear parametric mechanism ([48], p. 45) due to fluctuating transmission characteristics of the delay line leads to noticeable noise around the carrier frequency. Various effects can cause these fluctuations whereas, so far, only a dominant flicker phase noise contribution of the magnetically saturated magnetostrictive layer can be excluded.

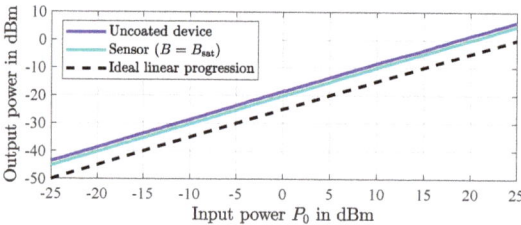

Figure 8. Measured output power as a function of input power P_0 of the two investigated SAW delay lines, i.e., the magnetostrictively uncoated device and the sensor in magnetic saturation. A strict linearity is revealed for both devices. The measurements were performed with a calibrated setup consisting of a signal generator *SMBV100A* and a signal and spectrum analyzer *FSV*, both from *Rohde & Schwarz*.

Previous studies identified IDT metalizations [57–59] and the piezoelectric substrate [58,59] as the major sources of flicker noise in SAW devices. Mobile impurities or defects in the substrate cause fluctuations in the local acoustic wave velocity [59], thus leading to random phase fluctuations. In addition, due to a very strong sensitivity to surface conditions, the surface acoustic wave velocity is modulated by gas molecules adsorbed onto the surface [58,59]. For example, as early as 1979, it was reported that the flicker noise depends on the cleanliness of the surface [60,61].

Obviously, the elements of the sensor that are most critical in terms of fluctuations are those that are most involved in the acoustic wave generation, propagation, and reconversion. Therefore, for the special case of surface acoustic Love wave devices, the additional guiding layer (here SiO_2 with a thickness of 4.5 µm) is also expected to contribute to the overall flicker phase noise. Figure 9 shows the measured phase noise density spectra of several delay lines of basically the same design but from different wafers that are not only based on quartz but also on $LiTaO_3$ substrates. Although the actual partial component responsible for the flicker phase noise is not apparent from this, the significant variance indicates differences in the purity of the materials. Apart from few studies on phase noise in SAW components mentioned above, most of which date back 30 to 40 years, surface acoustic Love wave elements in particular are still rarely investigated offering opportunities for future studies. However, as discussed in the following Section 4, in the special case of magnetostrictively coated SAW devices, additional phase noise of magnetic origin occurs to which the phase noise of bare devices is generally yet neglectable.

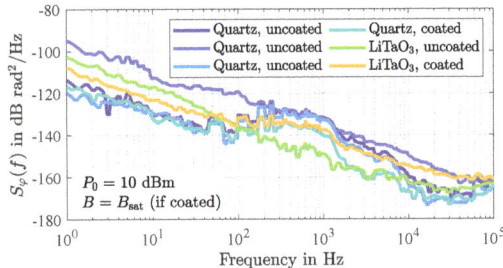

Figure 9. Measured power spectral densities of random phase fluctuations of several delay lines of basically the same design but from different wafers. The significant variance between the measured noise floors indicates differences in the purity of the materials that cause fluctuations in the local acoustic wave velocity. The labels *coated* and *uncoated* refer to the presence of a magnetostrictive layer. The measurements were performed at an electrical input power level of 10 dBm and without any additional amplifier.

4. Phase Noise in Magnetic Operating Points

In this section, the phase noise, the phase sensitivity, and the insertion loss of the previously introduced magnetostrictively coated SAW delay line magnetic field sensor is analyzed for various practically relevant magnetic operating points.

4.1. Measurement Setup

The automatized measurement system depicted in Figure 10 is designed to enable the measurement of the sensor's phase sensitivity S_{PM}, the phase noise S_φ, and the insertion loss IL as a function of both the sensor's input power P_0 as well as the ambient magnetic bias flux density B_{bias}.

As before, the SAW sensor itself is located in a magnetically (ZG1 from *Aaronia AG*), electrically and acoustically shielded measuring chamber and is surrounded by two solenoids. These coils are used to generate the static magnetic bias flux density B_{bias} by means of an in-house built and battery-based low-noise direct current source and for the generation of the dynamic flux density $B_x(t)$ utilizing a commercial precision current source (*Keithley 6221*).

The internal generator of the phase noise analyzer (*FSWP* from *Rohde & Schwarz*) is used to excite the sensor at its synchronous frequency $f_0 = 144.8$ MHz. Because the output power of this integrated generator cannot be finely adjusted, a programmable step attenuator (*RSC* from *Rohde & Schwarz*) is utilized between the generator and the sensor which allows to alter the sensor's input power P_0 in a wide range. In a separate analysis it was ensured that the flicker phase noise of the step attenuator can be neglected compared to the flicker phase noise of the SAW sensor under investigation. In fact, due to its design based on mechanical switches, the step attenuator's flicker phase noise is even below the inherent flicker phase noise of the *FSWP* phase noise analyzer when configured to 100 cross-correlations.

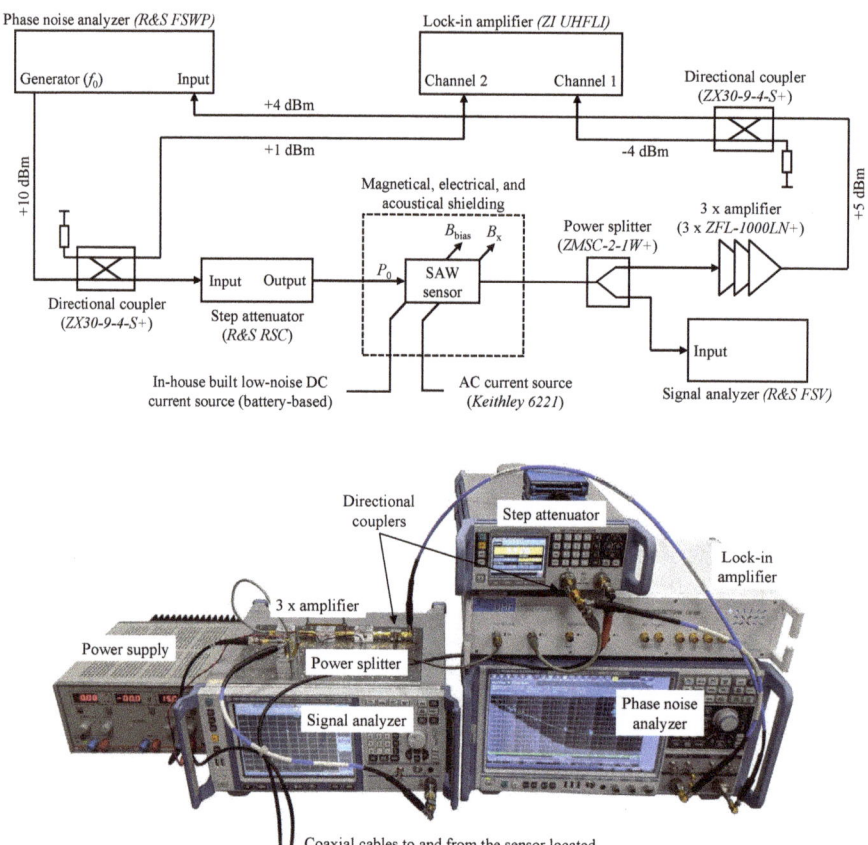

Figure 10. Block diagram (**top**) and photography (**bottom**) of the utilized system for the automatized measurement of the sensor's phase sensitivity S_{PM}, the phase noise S_φ, and the insertion loss IL as a function of both the sensor's input power P_0 as well as the magnetic bias flux density B_{bias}.

After amplifying the sensor's output signal utilizing three amplifiers connected in series (3 × *ZFL-1000LN+* from *Mini-Circuits*) with an overall gain of approx. 75 dB the signal is fed back to the input of the phase noise analyzer. The high gain is necessary in cases of low input power P_0 or high insertion loss IL, respectively, because the *FSWP* phase noise analyzer is not equipped with an internal preamplifier. On the other hand, at least one of these amplifiers is operated in compression if the sensor's input power P_0 is relatively high or the sensor's insertion loss is low. This leads to an increased noise figure F_{AMP} of the respective amplifier [50]. However, the flicker phase noise b_{-1}^{AMP} of these amplifiers does not increase when operated in compression, yielding an overall flicker phase noise of the amplifier chain of $b_{-1}^{\text{chain}} = 3 \times b_{-1}^{\text{AMP}} = 1.8 \times 10^{-13}$ rad^2.

In order to allow a determination of the sensor's insertion loss, the sensor signal is analyzed by an additional and carefully calibrated signal analyzer (*FSV* from *Rohde & Schwarz*) after this signal is divided into two branches by means of a conventional 3 dB power splitter (*ZMSC-2-1W+* from *Mini-Circuits*). Utilizing two 9 dB directional couplers (*ZX30-9-4-S+* from *Mini-Circuits*) the amplified sensor output signal and the generator signal (phase reference) are fed into a lock-in amplifier (*UHFLI* from *Zurich Instruments*). It is operated as a phase demodulator whose output signal is used for the determination of the phase sensitivity S_PM (Equation (1)) by evaluating the amplitude spectrum of the demodulated phase signal $\varphi(t)$ for calibrated amplitudes $\hat{B}_\text{x} = 1\,\text{µT}$ of the dynamic flux density $B_\text{x}(t) = \hat{B}_\text{x} \cos(2\pi f_\text{x} t)$ at a frequency of $f_\text{x} = 10\,\text{Hz}$. In addition, synchronously to noise measurements with the phase noise analyzer (while $\hat{B}_\text{x} = 0$), the phase demodulator, i.e., the lock-in amplifier, is used to record the random phase fluctuations $\Delta\varphi(t)$ (Equation (6)) as a time-domain signal.

The additional flicker phase noise of the passive components, i.e., the directional couplers and the power splitter, is usually as low as $b_{-1}^\text{passive} < 1 \times 10^{-17}\,\text{rad}^2$ and is thus negligible [62].

4.2. White Phase Noise

As mentioned in Section 3.1, a signal's overall thermal noise floor of $N = k_\text{B} T$ corresponds with an *additive* white phase noise quantified by Equation (11). White phase noise b_0 decreases with higher signal power, i.e., for sensors with low insertion losses IL and high input powers levels P_0. Only the insertion losses are relevant here, regardless of the physical causes leading to the losses.

For the sensor under investigation, typical values for the white phase noise b_0 are depicted in Figure 11a as a function of the ambient bias magnetic flux density B_bias and for various input power levels P_0. For the sensor virtually being magnetically saturated (at $B_\text{bias} = \pm 1\,\text{mT}$) the white phase noise simply decreases by the same amount P_0 is increased. The same trend is also observed for small magnetic flux densities around $B_\text{bias} = 0$. However, in this region additional magnetically induced insertion losses occur (compare Figures 6c and 13b that lead to increased white phase noise. This is consistant with the nucleation and presence of magnetic domain walls with the variation of B_bias discussed above.

According to Equation (7) and using Equations (11) and (9), the limit of detection above the corner frequency f_c (Equation (17)), i.e., in the white noise regime

$$\text{LOD}(f) \overset{f \geq f_\text{c}}{=} \frac{\sqrt{b_0}}{S_\text{PM}} = \frac{1}{S_\text{PM}} \sqrt{\frac{\text{IL}\, k_\text{B} T}{P_0}}, \tag{19}$$

directly scales with the phase sensitivity S_PM and further improves with lower insertion losses (linear representation of IL), lower temperatures, and higher input power levels. Based on Equation (19), values for the LOD in the white noise regime are displayed in Figure 11b where the underlying phase sensitivity S_PM will be discussed further below (Figure 13c). In principle, for input power levels above 0 dBm, white noise detectivities below $1\,\text{pT}/\sqrt{\text{Hz}}$ can be reached. However, please note that such values are only reachable if the white noise corner frequency f_c (Equation (17)) is below the cutoff frequency of the phase sensitivity [47].

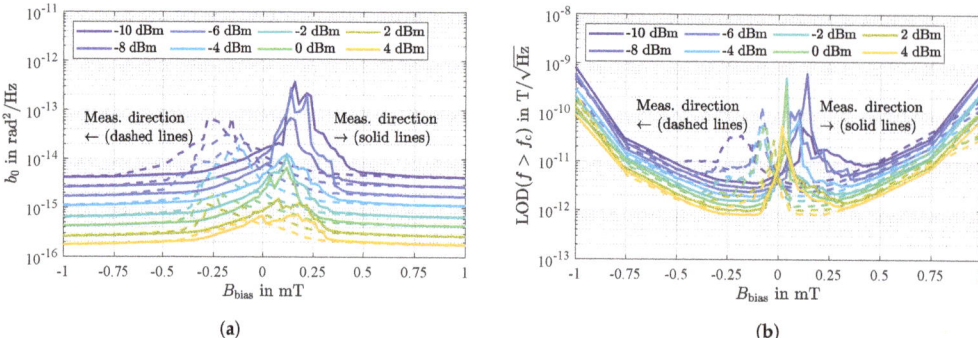

Figure 11. (a) Measured white phase noise b_0 as a function of the ambient bias magnetic flux density B_{bias} and for various input power levels P_0. The measurements were performed at the sensor's synchronous frequency of $f_0 = 144.8$ MHz and reveal the general trend of decreasing white phase noise for increasing input power levels. For magnetic flux densities around $B = 0$ additional magnetically induced insertion losses occur that lead to increased white phase noise. (b) Calculated limit of detection in the white noise regime according to Equation (19). The underlying phase sensitivity is shown in Figure 13c.

4.3. Flicker Phase Noise

In Section 3.2 it was shown that a magnetically uncoated SAW delay line device contributes a flicker phase noise as low as $b_{-1} = 4.4 \times 10^{-13}$ rad^2. The same value is reached for the magnetically coated device when operated in magnetic saturation. However, sensors coated with a magnetostrictive layer that are operated outside magnetic saturation are impaired by additional low-frequency phase noise that depends on the magnetic bias flux density B_{bias}.

As shown in Figure 12 for the sensor being operated exemplary at an input power of $P_0 = -10$ dBm, this phase noise decreases proportionally to $1/f$ so that it can also be referred to as flicker phase noise. It is also noticeable that this additional flicker phase noise increases with the ambient bias magnetic flux density B_{bias} up to a certain value (here 0.14 mT) and then decreases again. Noticeably the points of maximum flicker phase noise switch lower values of B_{bias} for higher input power levels P_0, indicating again a connection to magnetic domain wall occurrence. Extracting the flicker phase noise coefficients b_{-1} as a function of B_{bias} results in a characteristic as shown in Figure 13a. Noticeably, maximum flicker phase noise coincides with the highest magnetically induced insertion losses (Figure 13b) and decreases when the insertion losses IL decrease, i.e., for B_{bias} approaching magnetic saturation and for higher input power levels P_0. In comparison to the previously determined flicker phase noise in magnetic saturation ($b_{-1} = 4.4 \times 10^{-13}$ rad^2, Figure 7a) highest insertion losses correspond with an increase in flicker phase noise power by more than a factor of 40,000 or 46 dB, respectively. The regime of highest noise coincides with the regime of high magnetic energy transfer into magnetic domain walls (Figure 5), indicating a connection to magnetic domain wall processes. In investigations on magnetoresistive sensors an identical behavior could be observed in the past [63,64]. These sensors also show the largest noise for operating points of maximum sensitivity which was attributed to random fluctuations of the magnetization due to magnetic domain wall movements and rotations [64,65].

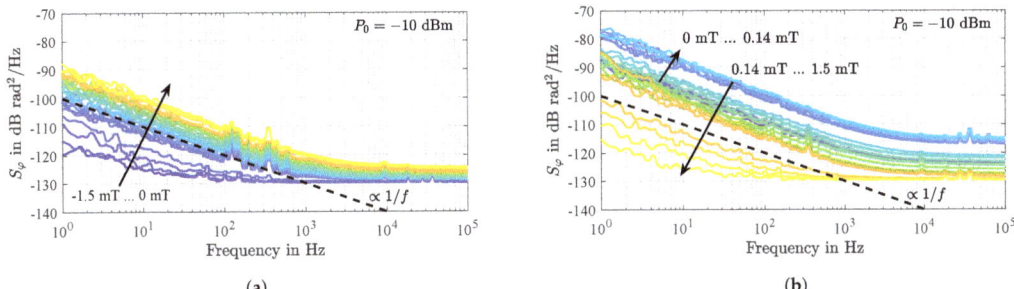

Figure 12. Measured power spectral densities of random phase fluctuations of the SAW delay line magnetic field sensor for increasing magnetic bias flux densities B_{bias} from -1.5 mT to 0 mT (**a**) and from 0 mT to 1.5 mT (**b**). Obviously, the phase noise significantly depends on the ambient magnetic flux density B_{bias}. As for magnetically uncoated devices as well as for magnetically saturated sensors (Figure 7), in the low-frequency regime, a clear $1/f$ frequency dependence of the additionally and magnetically induced phase noise is revealed.

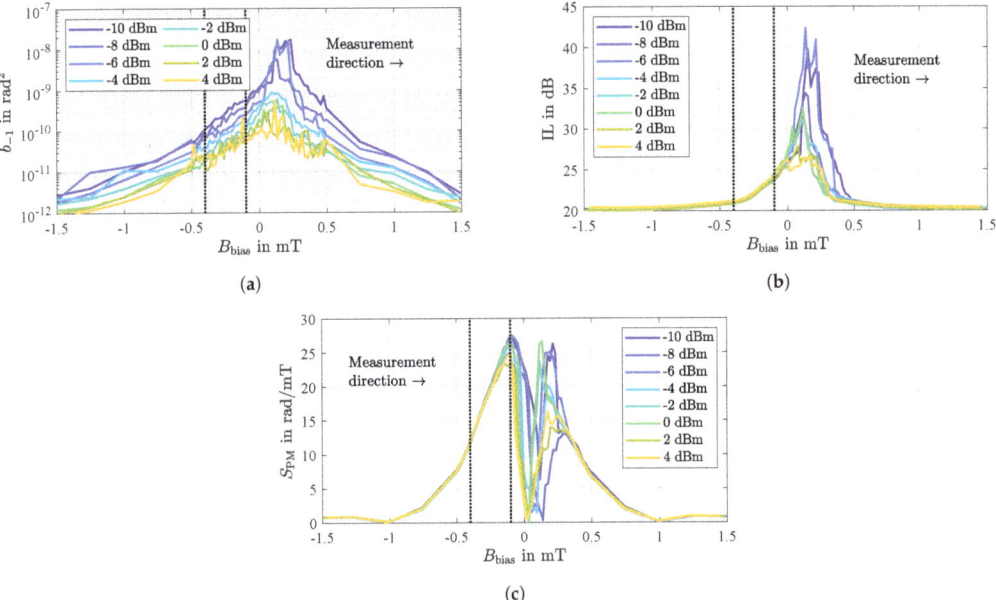

Figure 13. Measured flicker phase noise (**a**), insertion losses (**b**), and phase sensitivity (**c**) of the SAW magnetic field delay line sensor as a function of the ambient magnetic bias flux density B_{bias} and for various input power levels P_0. The sensor is operated preferably in a certain range (marked by dotted lines, here approximately between -0.4 mT and -0.1 mT) where the insertion losses and thus the flicker phase noise are comparatively low but the sensitivity is still high.

Apparently, when also considering the phase sensitivity S_{PM} (Figure 13c), there is a certain magnetically stable range (marked by dotted lines and for this sensor approximately between -0.4 mT and -0.1 mT) in which the sensor is to be operated preferably, i.e., where the insertion losses and thus the flicker phase noise are comparatively low but the sensitivity is still relatively high. This region coincides with the low domain wall density regime (Figure 3) with low magnetic losses discussed in detail in Section 2.4. Note that these measurements were performed for increasing magnetic bias flux densities after saturation in negative direction. An inverted measurement started at positive magnetic saturation virtually yields identical results only with reversed signs (compare e.g., Figure 6c), again coinciding with the bias field asymmetry of magnetic domain behavior and density (Figure 4).

Due to the significant relation to the additional magnetic insertion losses it is obvious to describe the magnetically induced phase noise using the fluctuation-dissipation theorem. Based on that theorem, a power spectral density of random fluctuations of the magnetization

$$S_M(f) = \frac{4 k_B T}{2\pi f\, V} \frac{\mu''_{r,\text{eff}}}{\mu_0} \tag{20}$$

with the physical dimension $(A/m)^2/Hz$ can be derived [66,67]. It can be referred to as flicker magnetization noise since the power spectral density $S_M(f)$ decreases with $1/f$. This expression is typically given as a function of the imaginary part μ''_r of the magnetic material's complex permeability $\mu_r = \mu'_r - j\mu''_r$. In general, however, μ''_r is also used to account for other losses, in particular eddy-current losses, which in turn do not correspond with flicker noise but with frequency-independent white noise [68,69]. Therefore, an effective complex permeability $\mu_{r,\text{eff}} = \mu'_r - j\mu''_{r,\text{eff}}$ is introduced to cover only for magnetic hysteresis losses corresponding with $1/f$ flicker noise. Furthermore, μ_0 and V denote the vacuum permeability and the volume of the magnetic material. Note that f denotes the Fourier frequency which is not equal to the delay line sensor's synchronous frequency f_0.

With the magnetic susceptibility $\chi = \partial M/\partial H = \mu'_r - 1$ as the relationship between the magnetic field strength H and the magnetization M, the relation between phase changes and magnetization changes

$$\frac{\partial \varphi}{\partial M} = \frac{\partial \varphi}{\partial H} \cdot \frac{\partial H}{\partial M} = \frac{\partial \varphi}{\partial \mu_0 H} \cdot \frac{\mu_0}{\chi} = S_{\text{PM}} \cdot \frac{\mu_0}{\mu'_r - 1} \tag{21}$$

can be expressed as a function of the phase sensitivity S_{PM} (Equation (1)). Thus, the flicker phase noise coefficient yields

$$b_{-1} = f \cdot \left(\frac{\partial \varphi}{\partial M}\right)^2 \cdot S_M(f) = S_{\text{PM}}^2 \cdot \frac{4 k_B T}{2\pi\, V} \frac{\mu_0 \mu''_{r,\text{eff}}}{(\mu'_r)^2} \tag{22}$$

when assuming that $\mu'_r \gg 1$, which is true for our soft magnetic material (Figure 2). Note that for the discussion here, the domain wall susceptibility or the easy axis magnetic field behavior might be the relevant figure of merit. Equivalently, for the low-frequency flicker noise regime below the corner frequency f_c (Equation (17)) the power spectral density of random phase fluctuations is given by

$$S_\varphi(f) \stackrel{f \le f_c}{=} \frac{b_{-1}}{f} = S_{\text{PM}}^2 \cdot \frac{4 k_B T}{2\pi f\, V} \frac{\mu_0 \mu''_{r,\text{eff}}}{(\mu'_r)^2}. \tag{23}$$

According to Equation (7), the limit of detection in the flicker noise regime then yields

$$\text{LOD}(f) \stackrel{f \le f_c}{=} \frac{\sqrt{b_{-1} f^{-1}}}{S_{\text{PM}}} = \sqrt{\frac{2 k_B T}{V\, \pi f} \frac{\mu_0 \mu''_{r,\text{eff}}}{(\mu'_r)^2}} \approx \underbrace{\frac{36.5\,\text{nT}}{\sqrt{f}} \cdot \sqrt{\frac{\mu''_{r,\text{eff}}}{(\mu'_r)^2}}}_{\text{for } V = 2.41 \times 10^{-12}\,\text{m}^3} \tag{24}$$

which, on the contrary to the LOD in the white noise regime (Equation (19)), no longer depends on the phase sensitivity S_{PM}. In fact, the LOD in the flicker noise regime is mainly determined by the complex-valued permeability of the magnetic material. Recently, we have confirmed this result in two studies. An investigation on SAW delay lines with magnetic layers of different thicknesses has shown that, although the sensitivity increases significantly with thicker layers, the LOD in the flicker noise regime remains constant due to increasing magnetic losses [24]. A comparison between the operation of such a sensor in the fundamental and the first higher Love wave mode also showed that, although both sensitivities significantly differ, similar limits of detection in the flicker noise regime resulted [70]. Another recently published investigation on ferrite flux concentrators utilized with diamond magnetometers [69] also comes to the same conclusion that the relative loss factor $\mu''_{r,\text{eff}}/\mu'^{2}_r$ must be limited in order to minimize hysteresis noise.

For the sensor under investigation operated at an ambient magnetic bias flux density of $B_{\text{bias}} = -0.25$ mT, detectivities as depicted in Figure 14 were measured for various input power levels P_0. In accordance with Equation (24), all measured equivalent magnetic noise spectra improve with $1/\sqrt{f}$ confirming that magnetic hysteresis losses, i.e., random fluctuations of the magnetization, dominate under these operating conditions. However, because the flicker phase noise depends on P_0 (Figure 13a), the LOD also improves with increasing input power levels up to a value of about 70 pT/$\sqrt{\text{Hz}}$ at an exemplary frequency of 10 Hz for optimum input power levels between 0 dBm and 4 dBm. Thus, the magnetic losses represented by $\mu''_{r,\text{eff}}$ depend on the input power level P_0.

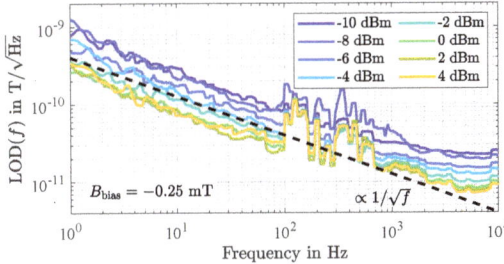

Figure 14. Measured equivalent magnetic noise floors for various input power levels P_0 at an ambient magnetic bias flux density of $B_{\text{bias}} = -0.25$ mT (after a negative magnetic saturation). The $1/\sqrt{f}$ dependency confirms Equation (24), i.e., magnetic hysteresis losses dominate under these operating conditions.

The previously discussed measurement results (Figure 13) also revealed a significant dependence of the flicker phase noise and the phase sensitivity on the ambient magnetic bias flux density. Nevertheless, as shown in Figure 15a, the LOD remains virtually constant over a comparatively large range with respect to B_{bias} between -0.4 mT and -0.1 mT (marked by dotted lines), thus confirming the independence of the phase sensitivity S_{PM}. In contrast, the dependence on the input power is significant, indicating again a dependence of the magnetic properties on P_0.

Based on the measurement results and Equation (22) the imaginary part of the magnetic material's effective complex permeability

$$\mu''_{r,\text{eff}} = \frac{b_{-1} V \pi}{2 k_B T \mu_0} \cdot \left(\frac{\mu'_r}{S_{\text{PM}}} \right)^2 \tag{25}$$

can be determined. Depending on the input power of the sensor, $\mu''_{r,\text{eff}}$ is in the range between about 500 ($P_0 = -10$ dBm) and 50 ($P_0 = 4$ dBm) corresponding with magnetic loss factors $\tan \delta = \mu''_{r,\text{eff}}/\mu'_r$ ([71], p. 33) of about 0.6 and 0.06 (Figure 15b). Because the ferromagnetic resonance frequency of FeCoSiB thin films is typically above 1 GHz rather low losses in the frequency range around 150 MHz would have been expected [72,73] assuming simple Landau-Lifshitz-Gilbert (LLG) resonance behavior [74]. Yet, for similar amorphous magnetic films [75] and FeCoSiB films of similar thicknesses [76], domain wall resonance effects in the lower 100 MHz regime have been reported and the losses were directly connected to magnetic domain wall resonances. Eddy-current effects should not play a role in the magnetic domain wall losses [77], only internal damping is of relevance. If one considers the magnetic quality factor $Q = 1/\tan \delta$ with values of up to about 25 or the relative magnetic loss factor $\tan \delta / \mu'_r$ with values slightly below 10^{-4} (each for

$P_0 = 4$ dBm) comparable values can be found in literature [66,78]. In fact, a similar value for the relative magnetic loss factor of 1.6×10^{-4} has recently been found for a resonant magnetic field sensor with a magnetostrictive thin-film of the same alloy utilized here [79].

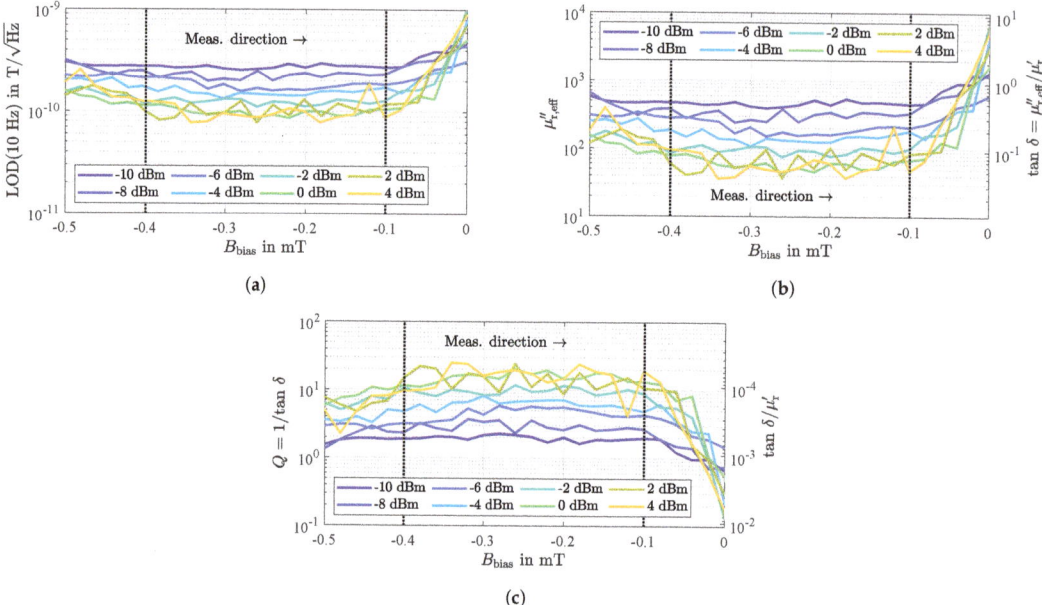

Figure 15. Measured limit of detection at an exemplary frequency of 10 Hz (**a**) and determined magnetic key figures (**b**,**c**) on the basis of Equation (25). On one hand, the results confirm the detectivity's independence of the phase sensitivity and constant magnetic properties in a wide range of magnetic bias flux densities (dotted lines). On the other hand, a significant dependence of the magnetic properties, and thus the sensor performance, on the input power level is revealed. The underlying measurements were performed after a previously performed magnetic saturation in negative direction.

4.4. Random Walk of Phase

So far, the sensor's phase noise was primarily considered as a function of the frequency and the ambient magnetic bias flux density. However, measurements at selected power levels up to 4 dBm have already shown a significant influence of the sensor's electrical input power on the phase noise performance.

The results of a series of measurements as a function of the sensor's input power P_0 at a constant ambient magnetic bias flux density of $B_{\text{bias}} = 0$ (after magnetically saturating the sensor in negative direction) are depicted in Figure 16. As observed before, the insertion losses decrease with higher P_0 by about 1.8 dB in the considered range from −30 dBm to 8 dBm (Figure 16a). Decreasing P_0 again virtually results in the same values. Furthermore, two additional measurements performed (gray) confirm the repeatability of this experiment. The only differences are marginally shifted curves due to different magnetic domain configurations after magnetically saturating the sensor. Although the losses decrease only moderately with higher P_0, a significant reduction of the phase noise (here exemplary at a frequency of 10 Hz) by a factor of about 640 is observed over a wide range (Figure 16b). For all power levels approximately below 4 dBm the power spectral densities of random phase fluctuations progress strictly proportional to $1/f$ as shown exemplary in Figure 16c for P_0 increasing from −12.5 dBm to −7.5 dBm, referred to as *region A*. These results again confirm the previously discussed dominance of $1/f$ flicker phase noise due to random fluctuations of the magnetization directly related to magnetic hysteresis losses.

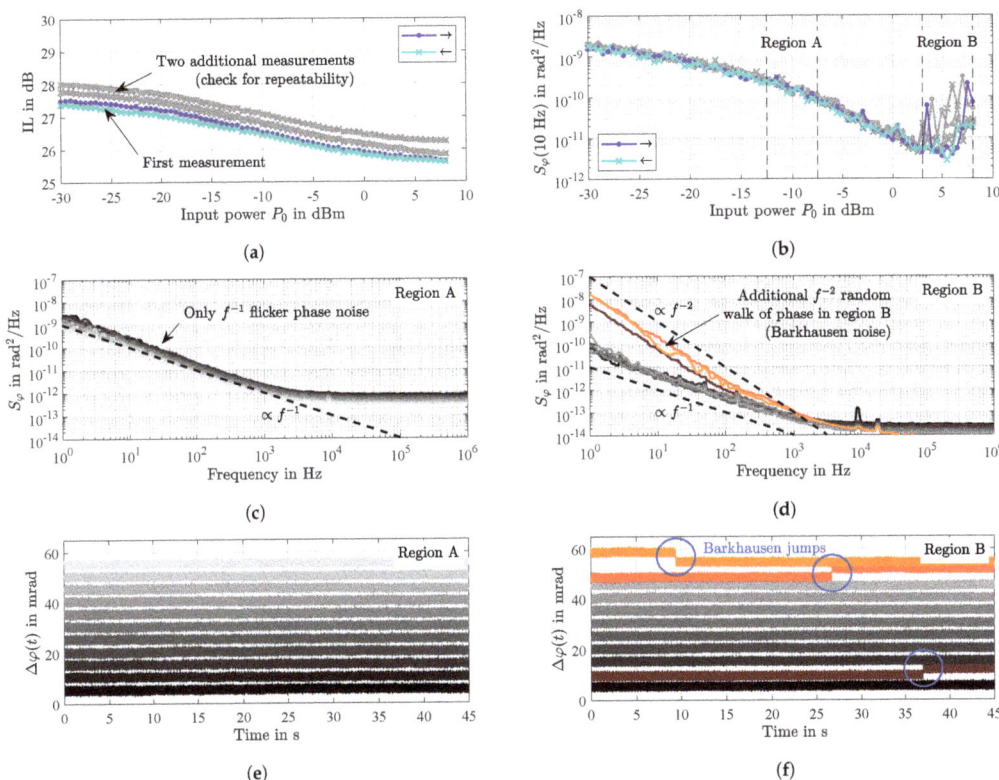

Figure 16. Measured loss and phase noise characteristics as a function of the sensor's electrical input power P_0 at a constant ambient magnetic bias flux density $B_{bias} = 0$ (after negative saturation). A direct relation between (hysteresis) losses (**a**) and phase noise at an exemplary frequency of 10 Hz (**b**) is revealed for input power levels P_0 approx. below 4 dBm. In this regime, i.e., in *region A*, the power spectral densities of random phase fluctuations progress proportionally to $1/f$ (**c**). At higher power levels, i.e., in *region B*, random walk of phase ($1/f^2$) occurs (**d**) that is caused by *Barkhausen jumps* (**f**) that do not occur at lower power levels (**e**). Please note the artificial phase offsets for clearer illustration in (**e**,**f**).

However, if P_0 is further increased, the phase noise in all three series of measurements increases again, partly significantly (Figure 16b), although the losses continue to decrease slightly or stagnate at a constant level (Figure 16a) indicating the occurrence of an additional mechanism. This regime is referred to as *region B* and occurs approximately above 3 dBm for the sensor under investigation. A consideration of the associated power spectral densities of random phase fluctuations (Figure 16d) reveals that this increased phase noise corresponds with slopes of $1/f^2$ (highlighted by reddish colors), referred to as random walk of phase ([48], p. 23). In contrast to region A (Figure 16e), the corresponding time signals in region B show jumps, also highlighted by reddish colors (Figure 16e). From literature it is well-known that stochastic changes of the size of magnetic domains correspond with $1/f^2$ slopes in the associated power spectral densities ([80], p. 281). Therefore, the random walk of phase is caused by so-called *Barkhausen jumps*. In the following we prove that the magnetic fluctuations are related to hopping of magnetic domain walls.

The assumption of low-frequency domain wall switching events is proven by in-situ magnetic domain observations. In Figure 5 we have shown that the magnetic switching process is altered with the electrical input power. The sporadic reorientation of magnetic domains without an alteration of the magnetic field is demonstrated in Figure 17. In addition to the magnetic domain states in Figure 17a–d the difference in the magnetic domain states over time is displayed in Figure 17e–h. The magnetic domains reorient across several seconds even for the given small input power. Magnetic domain walls move to a more stable state with time, where the probability of occurrence of magnetic domain reorientation increases with the electrical input power. Therefore, for each domain wall jump, the magnetization component M_{tr}/M_s increases. For the shown example, the overall process is mostly limited to two domain walls. The reorientation process in the negative bias field is a direct consequence of the slight misorientation of the magnetic anisotropy axis in the system. It should further be noted that the probability of domain switching events will also depend on the reverse magnetic bias field value, increasing drastically approaching the domain reorientation field discussed in Section 2.3. Electrically induced changes of the magnetic structure with zero field are also visible from the transverse magnetization curve data (Figure 5a), where a reduction of transverse remanence B_r becomes already visible with $P_0 = -10$ dBm.

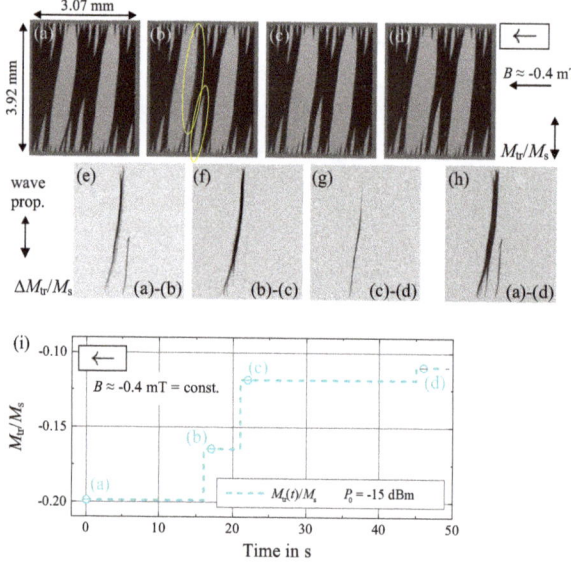

Figure 17. Magnetic domain observations with a constant magnetic bias field of $B \approx -0.4$ mT. (**a–d**) Magnetic domain structure over time with an input power of -15 dBm. (**e–h**) Differential domain images displaying the alteration in the magnetic domain states over time. (**i**) Change of the transversal magnetization component M_{tr}/M_s with time. The positions of high domain activity are indicated in (**b**).

5. Conclusions

In this paper, the noise behavior of SAW delay line magnetic field sensors coated with a thin-film of magnetostrictive material is investigated by means of extensive measurements, the results of which were used to describe the noise analytically. Such sensors utilize the magnetoelastic ΔE effect that leads to a magnetically induced alteration of the surface acoustic wave's propagation velocity. Electroacoustic transducers at the sensor's input and output port are utilized to generate the SAW and to provide an electrical signal whose phase contains the information about the magnetic field strength. Due to various sensor-intrinsic phenomena the output signal is impaired by phase noise that limits the detectivity.

Besides a discussion of the sensor's electrical properties around the synchronous frequency of 144.8 MHz, insights into its magnetization behavior are given based on the magnetic domain behavior obtained by means of magnetooptical Kerr effect microscopy and magnetometry. An asymmetric domain behavior is revealed in which, coming from magnetic saturation, spike domains develop at the edges. After remanence, the spike domains grow and further on penetrate the whole sample, forming large magnetic domains that are directly linked with a magnetic energy transfer from a generated surface acoustic wave into the magnetic layer. These asymmetric and bias field dependent losses are also reflected in the electrical transmission properties of the sensor.

With regard to the spectral shape, it is revealed that SAW delay line magnetic field sensors exhibit three different types of phase noise, each with various causes.

Fundamental f^0 thermal phase noise, i.e., white phase noise, is directly linked to the sensor's insertion loss, regardless of the physical causes for the loss. Typically, the insertion loss results from the static losses of the delay line structure and from the above mentioned dynamic hysteresis losses in the magnetic layer. In contrast, eddy-current losses are negligible in this frequency range and for such thin magnetic layers. White noise is additive noise that decreases with increasing signal power. For an optimal LOD in the white noise regime a high input power should be chosen at a magnetic operating point where magnetic losses are as low as possible while maintaining high phase sensitivity. In principle, the LOD in the white noise region can be reduced arbitrarily by increasing the input power. However, since the corner frequency of the white phase noise must remain below the cutoff frequency of the phase sensitivity, achievable values for the LOD in the white noise regime are typically in the range around $1\,\mathrm{pT}/\sqrt{\mathrm{Hz}}$.

Every SAW delay line device exhibits fundamental f^{-1} flicker phase noise due to the delay line structure itself, originating e.g., from mobile impurities or defects in the substrate and the guiding layer that cause fluctuations in the local acoustic wave velocity. This quasi-linear parametric mechanism is characterized by the fact that f^{-1} flicker phase noise is generally independent of the sensor's input power. It was found that magnetostrictively coated delay lines operated in magnetic saturation exhibit exactly the same f^{-1} flicker phase noise as bare devices, i.e., delay lines without any additional magnetostrictive layer. Outside magnetic saturation, however, the f^{-1} flicker phase noise significantly increases depending on the ambient magnetic bias field by more than 40 dB where maximum flicker phase noise coincides with highest magnetically induced insertion losses, i.e., with the occurrence of magnetic domain walls. Therefore, such sensors are preferably operated in magnetic operating points with low magnetic losses, i.e., in the low domain wall density regime. In this regime, and in agreement with calculations based on the fluctuation-dissipation theorem, measurements confirmed the independence of the LOD from the phase sensitivity. In contrast to the white noise regime, the LOD in the flicker noise regime cannot be improved by increasing the phase sensitivity. Instead, the magnetic losses must be limited in order to minimize hysteresis loss. Although flicker phase noise is inherently independent of the sensor's input power, a significant dependence of flicker phase noise on the input power was found. This is due to the fact that the magnetic losses are power dependent, i.e., the magnetic losses decrease with higher input power levels. For optimal power levels in the range between 0 dBm and 4 dBm and at an exemplary frequency of 10 Hz, an LOD of $70\,\mathrm{pT}/\sqrt{\mathrm{Hz}}$ could be achieved that corresponds with a relative magnetic loss factor of about 10^{-4}.

If the electrical input power of the sensor is increased further, the phase noise power spectral density no longer shows a strict f^{-1} slope. Instead, dominant f^{-2} random walk of phase noise occurs. It was found that this random walk of phase is directly linked to sporadic reorientations of magnetic domains without an alteration of the magnetic field, i.e., Barkhausen jumps. Therefore, for best performance of such sensors, the electrical input power should generally be chosen as high as possible, but below the power range in which domain network reorientation processes occur.

Author Contributions: Conceptualization: P.D., C.M., A.K., E.Q., R.K., M.H. and J.M.; Data curation: P.D., C.M., A.K. and J.M.; Formal analysis: P.D., C.M. and J.M.; Funding acquisition: E.Q., R.K., M.H. and J.M.; Investigation: P.D., C.M., A.K., V.S. and J.M.; Methodology: P.D., C.M. and J.M.; Project administration: E.Q., R.K., M.H. and J.M.; Resources: A.K., V.S., E.Q., R.K., M.H. and J.M.; Supervision: E.Q., R.K., M.H. and J.M.; Validation: P.D., C.M., A.K., V.S., A.B., E.Q., R.K., M.H. and J.M.; Visualization: P.D., C.M. and J.M.; Writing—original draft: P.D. and J.M.; Writing—review &

editing: P.D., C.M., A.K., V.S., A.B., E.Q., R.K., M.H. and J.M. All authors have read and agreed to the published version of the manuscript.

Funding: This work was funded by the German Research Foundation (Deutsche Forschungsgemeinschaft, DFG) through the Collaborative Research Centre CRC 1261 *Magnetoelectric Sensors: From Composite Materials to Biomagnetic Diagnostics*.

Institutional Review Board Statement: Not applicable.

Informed Consent Statement: Not applicable.

Data Availability Statement: Not applicable.

Acknowledgments: The authors gratefully thank Enrico Rubiola for fruitful discussions and advice on phase noise theory and measurements.

Conflicts of Interest: The authors declare no conflict of interest.

References

1. White, R.M.; Voltmer, F.W. Direct Piezoelectric Coupling to Surface Elastic Waves. *Appl. Phys. Lett.* **1965**, *7*, 314–316. [CrossRef]
2. Ruppel, C.; Dill, R.; Fischerauer, A.; Fischerauer, G.; Gawlik, W.; Machui, J.; Müller, F.; Reindl, L.; Ruile, W.; Scholl, G.; et al. SAW Devices for Consumer Communication Applications. *IEEE Trans. Ultrason. Ferroelectr. Freq. Control* **1993**, *40*, 438–452. [CrossRef]
3. White, R.M. Surface Acoustic Wave Sensors. In Proceedings of the 1985 IEEE Ultrasonics Symposium, San Francisco, CA, USA, 16–18 October 1985; pp. 490–494. [CrossRef]
4. Liu, B.; Chen, X.; Cai, H.; Mohammad Ali, M.; Tian, X.; Tao, L.; Yang, Y.; Ren, T. Surface acoustic wave devices for sensor applications. *J. Semicond.* **2016**, *37*, 021001. [CrossRef]
5. Fischerauer, G.; Mauder, A.; Müller, R., Acoustic Wave Devices. In *Sensors Set*; Wiley-VCH: Weinheim, Germany, 2008; pp. 135–180. [CrossRef]
6. Yang, Y.; Mishra, H.; Han, T.; Hage-Ali, S.; Hehn, M.; Elmazria, O. Sensing Mechanism of Surface Acoustic Wave Magnetic Field Sensors Based on Ferromagnetic Films. *IEEE Trans. Ultrason. Ferroelectr. Freq. Control* **2021**, *68*, 2566–2575. [CrossRef] [PubMed]
7. Smole, P.; Ruile, W.; Korden, C.; Ludwig, A.; Quandt, E.; Krassnitzer, S.; Pongratz, P. Magnetically tunable SAW-resonator. In Proceedings of the 2003 IEEE International Frequency Control Symposium and PDA Exhibition Jointly with the 17th European Frequency and Time Forum, Tampa, FL, USA, 4–8 May 2003; Number 2, pp. 903–906. [CrossRef]
8. Ganguly, A.; Davis, K.; Webb, D.; Vittoria, C.; Forester, D. Magnetically tuned surface-acoustic-wave phase shifter. *Electron. Lett.* **1975**, *11*, 610. [CrossRef]
9. Forester, D.W.; Vittoria, C.; Webb, D.C.; Davis, K.L. Variable delay lines using magnetostrictive metallic glass film overlays. *J. Appl. Phys.* **1978**, *49*, 1794–1796. [CrossRef]
10. Robbins, W.P.; Simpson, E.U. Surface Acoustic Wave Properties of RF Sputtered Nickel Films on Lithium Niobate. In Proceedings of the 1978 Ultrasonics Symposium, Cherry Hill, NJ, USA, 25–27 September 1978; pp. 658–661. [CrossRef]
11. Webb, D.; Forester, D.; Ganguly, A.; Vittoria, C. Applications of amorphous magnetic-layers in surface-acoustic-wave devices. *IEEE Trans. Magn.* **1979**, *15*, 1410–1415. [CrossRef]
12. Yamaguchi, M.; Hashimoto, K.; Kogo, H.; Naoe, M. Variable SAW delay line using amorphous $TbFe_2$ film. *IEEE Trans. Magn.* **1980**, *16*, 916–918. [CrossRef]
13. Hashimoto, K.; Yamaguchi, M.; Kogo, H.; Naoe, M. Magnetostrictive properties of sputtered Co-Cr film on surface acoustic wave. *IEEE Trans. Magn.* **1981**, *17*, 3181–3183. [CrossRef]
14. Altan, B.; Robbins, W. Tunable ZnO-Si SAW Devices Using Magnetostrictive Thin Films. In Proceedings of the 1981 IEEE Ultrasonics Symposium, Chicago, IL, USA, 14–16 October 1981; pp. 311–314. [CrossRef]
15. Hietala, A.; Robbins, W. Properties of Sputtered Tb-Fe for Tunable SAW Device Applications. In Proceedings of the 1982 Ultrasonics Symposium, San Diego, CA, USA, 27–29 October 1982; pp. 354–358. [CrossRef]
16. Koeninger, V.; Matsumura, Y.; Uchida, H.; Uchida, H. Surface acoustic waves on thin films of giant magnetostrictive alloys. *J. Alloy. Compd.* **1994**, *211–212*, 581–584. [CrossRef]
17. Li, W.; Dhagat, P.; Jander, A. Surface Acoustic Wave Magnetic Sensor using Galfenol Thin Film. *IEEE Trans. Magn.* **2012**, *48*, 4100–4102. [CrossRef]
18. Zhou, H.; Talbi, A.; Tiercelin, N.; Bou Matar, O. Multilayer magnetostrictive structure based surface acoustic wave devices. *Appl. Phys. Lett.* **2014**, *104*, 114101. [CrossRef]
19. Elhosni, M.; Petit-Watelot, S.; Hehn, M.; Hage-Ali, S.; Aissa, K.A.; Lacour, D.; Talbi, A.; Elmazria, O. Experimental Study of Multilayer Piezo-magnetic SAW Delay Line for Magnetic Sensor. *Procedia Eng.* **2015**, *120*, 870–873. [CrossRef]
20. Yokokawa, N.; Tanaka, S.; Fujii, T.; Inoue, M. Love-type surface-acoustic waves propagating in amorphous iron-boron films with multilayer structure. *J. Appl. Phys.* **1992**, *72*, 360–366. [CrossRef]
21. Hanna, S. Magnetic Field Sensors Based on SAW Propagation in Magnetic Films. *IEEE Trans. Ultrason. Ferroelectr. Freq. Control* **1987**, *34*, 191–194. [CrossRef] [PubMed]

22. Mazzamurro, A.; Talbi, A.; Dusch, Y.; Elmazria, O.; Pernod, P.; Matar, O.B.; Tiercelin, N. Highly Sensitive Surface Acoustic Wave Magnetic Field Sensor Using Multilayered TbCo2/FeCo Thin Film. *Proceedings* **2018**, *2*, 902. [CrossRef]
23. Wang, W.; Jia, Y.; Xue, X.; Liang, Y. Magnetostrictive effect in micro-dotted FeCo film coated surface acoustic wave devices. *Smart Mater. Struct.* **2018**, *27*, 105040. [CrossRef]
24. Kittmann, A.; Müller, C.; Durdaut, P.; Thormählen, L.; Schell, V.; Niekiel, F.; Lofink, F.; Meyners, D.; Knöchel, R.; Höft, M.; et al. Sensitivity and Noise Analysis of SAW Magnetic Field Sensors with varied Magnetostrictive Layer Thicknesses. *Sens. Actuators A Phys.* **2020**, *311*, 1–8. [CrossRef]
25. Mazzamurro, A.; Dusch, Y.; Pernod, P.; Bou Matar, O.; Addad, A.; Talbi, A.; Tiercelin, N. Giant Magnetoelastic Coupling in a Love Acoustic Waveguide Based on TbCo$_2$/FeCo Nanostructured Film on ST-Cut Quartz. *Phys. Rev. Appl.* **2020**, *13*, 044001. [CrossRef]
26. Liu, X.; Tong, B.; Ou-Yang, J.; Yang, X.; Chen, S.; Zhang, Y.; Zhu, B. Self-biased vector magnetic sensor based on a Love-type surface acoustic wave resonator. *Appl. Phys. Lett.* **2018**, *113*, 082402. [CrossRef]
27. Mishra, H.; Streque, J.; Hehn, M.; Mengue, P.; M'Jahed, H.; Lacour, D.; Dumesnil, K.; Petit-Watelot, S.; Zhgoon, S.; Polewczyk, V.; et al. Temperature compensated magnetic field sensor based on love waves. *Smart Mater. Struct.* **2020**, *29*, 045036. [CrossRef]
28. Yang, Y.; Mishra, H.; Mengue, P.; Hage-Ali, S.; Petit-Watelot, S.; Lacour, D.; Hehn, M.; MrJahed, H.; Han, T.; Elmazria, O. Enhanced Performance Love Wave Magnetic Field Sensors with Temperature Compensation. *IEEE Sens. J.* **2020**, *20*, 11292–11301. [CrossRef]
29. Schell, V.; Müller, C.; Durdaut, P.; Kittmann, A.; Thormählen, L.; Lofink, F.; Meyners, D.; Höft, M.; McCord, J.; Quandt, E. Magnetic anisotropy controlled FeCoSiB thin films for surface acoustic wave magnetic field sensors. *Appl. Phys. Lett.* **2020**, *116*, 073503. [CrossRef]
30. Dauber, J.; Sagade, A.A.; Oellers, M.; Watanabe, K.; Taniguchi, T.; Neumaier, D.; Stampfer, C. Ultra-sensitive Hall sensors based on graphene encapsulated in hexagonal boron nitride. *Appl. Phys. Lett.* **2015**, *106*, 193501. [CrossRef]
31. Zimmermann, E.; Verweerd, A.; Glaas, W.; Tillmann, A.; Kemna, A. An AMR sensor-based measurement system for magnetoelectrical resistivity tomography. *IEEE Sens. J.* **2005**, *5*, 233–241. [CrossRef]
32. Stutzke, N.A.; Russek, S.E.; Pappas, D.P.; Tondra, M. Low-frequency noise measurements on commercial magnetoresistive magnetic field sensors. *J. Appl. Phys.* **2005**, *97*, 10Q107. [CrossRef]
33. Dufay, B.; Saez, S.; Dolabdjian, C.P.; Yelon, A.; Menard, D. Characterization of an optimized off-diagonal GMI-based magnetometer. *IEEE Sens. J.* **2013**, *13*, 379–388. [CrossRef]
34. Ripka, P. Advances in fluxgate sensors. *Sens. Actuators A Phys.* **2003**, *106*, 8–14. [CrossRef]
35. Durdaut, P.; Kittmann, A.; Rubiola, E.; Friedt, J.M.; Quandt, E.; Knöchel, R.; Höft, M. Noise Analysis and Comparison of Phase- and Frequency-Detecting Readout Systems: Application to SAW Delay Line Magnetic Field Sensor. *IEEE Sens. J.* **2019**, *19*, 8000–8008. [CrossRef]
36. Urs, N.O.; Golubeva, E.; Röbisch, V.; Toxvaerd, S.; Deldar, S.; Knöchel, R.; Höft, M.; Quandt, E.; Meyners, D.; McCord, J. Direct Link between Specific Magnetic Domain Activities and Magnetic Noise in Modulated Magnetoelectric Sensors. *Phys. Rev. Appl.* **2020**, *13*, 024018. [CrossRef]
37. Kittmann, A.; Durdaut, P.; Zabel, S.; Reermann, J.; Schmalz, J.; Spetzler, B.; Meyners, D.; Sun, N.X.; McCord, J.; Gerken, M.; et al. Wide Band Low Noise Love Wave Magnetic Field Sensor System. *Sci. Rep.* **2018**, *8*, 278–287. [CrossRef]
38. Besser, L.; Gilmore, R. *Practical RF Circuit Design for Modern Wireless Systems*; Artech House: Boston, MA, USA, 2003.
39. Du, J.; Harding, G.; Ogilvy, J.; Dencher, P.; Lake, M. A study of Love-wave acoustic sensors. *Sens. Actuators A Phys.* **1996**, *56*, 211–219. [CrossRef]
40. McCord, J. Progress in magnetic domain observation by advanced magneto-optical microscopy. *J. Phys. D Appl. Phys.* **2015**, *48*, 333001. [CrossRef]
41. Urs, N.O.; Mozooni, B.; Mazalski, P.; Kustov, M.; Hayes, P.; Deldar, S.; Quandt, E.; McCord, J. Advanced magneto-optical microscopy: Imaging from picoseconds to centimeters - imaging spin waves and temperature distributions (invited). *AIP Adv.* **2016**, *6*, 055605. [CrossRef]
42. Glasmachers, S.; Frommberger, M.; McCord, J.; Quandt, E. Influence of strain on the high-frequency magnetic properties of FeCoBSi thin films. *Phys. Status Solidi A* **2004**, *201*, 3319–3324. [CrossRef]
43. Schneider, C.A.; Rasband, W.S.; Eliceiri, K.W. NIH Image to ImageJ: 25 years of image analysis. *Nat. Methods* **2012**, *9*, 671–675. [CrossRef]
44. McCord, J.; Schäfer, R.; Mattheis, R.; Barholz, K.U. Kerr observations of asymmetric magnetization reversal processes in CoFe/IrMn bilayer systems. *J. Appl. Phys.* **2003**, *93*, 5491–5497. [CrossRef]
45. Durdaut, P.; Wolframm, H.; Höft, M. Low-Frequency Magnetic Noise in Statically-Driven Solenoid for Biasing Magnetic Field Sensors. *arXiv* **2020**, arXiv:2006.08515.
46. Durdaut, P.; Höft, M.; Friedt, J.M.; Rubiola, E. Equivalence of Open-Loop and Closed-Loop Operation of SAW Resonators and Delay Lines. *Sensors* **2019**, *19*, 185. [CrossRef] [PubMed]
47. Labrenz, J.; Bahr, A.; Durdaut, P.; Höft, M.; Kittmann, A.; Schell, V.; Quandt, E. Frequency Response of SAW Delay Line Magnetic Field/Current Sensor. *IEEE Sens. Lett.* **2019**, *3*, 1–4. [CrossRef]
48. Rubiola, E. *Phase Noise and Frequency Stability in Oscillators*; Cambridge University Press: Cambridge, UK, 2009.
49. IEEE Standard Definitions of Physical Quantities for Fundamental Frequency and Time Metrology—Random Instabilities. *IEEE Stand.* **2009**, *1139*, 1999. [CrossRef]

50. Boudot, R.; Rubiola, E. Phase Noise in RF and Microwave Amplifiers. *IEEE Trans. Ultrason. Ferroelectr. Freq. Control* **2012**, *59*, 2613–2624. [CrossRef] [PubMed]
51. Sears, F.W.; Salinger, G.L. *Thermodynamics, Kinetic Theory, and Statistical Thermodynamics*, 3rd ed.; Addison-Wesley Publishing Company: Reading, MA, USA, 1975.
52. Durdaut, P. *Ausleseverfahren und Rauschmodellierung für magnetoelektrische und magnetoelastische Sensorsysteme*; Shaker: Aachen, Germany, 2020.
53. Pozar, D.M. *Microwave Engineering*, 4th ed.; Wiley: Hoboken, NJ, USA, 2012.
54. Yao, D.; Sullivan, C.R. Effect of Capacitance on Eddy-Current loss in Multi-Layer Magnetic Films for MHz Magnetic Components. In Proceedings of the 2009 IEEE Energy Conversion Congress and Exposition, San Jose, CA, USA, 20–24 September 2009; pp. 1025–1031. [CrossRef]
55. Villari, E. Ueber die Aenderungen des magnetischen Moments, welche der Zug und das Hindurchleiten eines galvanischen Stroms in einem Stabe von Stahl oder Eisen hervorbringen. *Ann. Der Phys. Und Chem.* **1865**, *202*, 87–122. [CrossRef]
56. Kohmoto, O. Electrical Resistivity of Amorphous Co-Rich FeCo-SiB Alloys. *Phys. Status Solidi A* **1984**, *85*, K155–K157. [CrossRef]
57. Jungerman, R.; Baer, R.; Bray, R. Delay Dependence of Phase Noise in SAW Filters. In Proceedings of the 1985 Ultrasonics Symposium, San Francisco, CA, USA, 16–18 October 1985; pp. 258–261. [CrossRef]
58. Parker, T.; Andres, D.; Greer, J.; Montress, G. 1/f noise in etched groove surface acoustic wave (SAW) resonators. *IEEE Trans. Ultrason. Ferroelectr. Freq. Control* **1994**, *41*, 853–862. [CrossRef] [PubMed]
59. Enguang, D.. Surface-related phase noise in SAW resonators. *IEEE Trans. Ultrason. Ferroelectr. Freq. Control* **2002**, *49*, 649–655. [CrossRef] [PubMed]
60. Parker, T. 1/f Phase Noise in Quartz Delay Lines and Resonators. In Proceedings of the 1979 Ultrasonics Symposium, New Orleans, LA, USA, 26–28 September 1979; pp. 878–881. [CrossRef]
61. Parker, T. 1/f phase noise in quartz s.a.w. devices. *Electron. Lett.* **1979**, *15*, 296. [CrossRef]
62. Rubiola, E.; Giordano, V. Advanced interferometric phase and amplitude noise measurements. *Rev. Sci. Instrum.* **2002**, *73*, 2445–2457. [CrossRef]
63. Hardner, H.T.; Weissman, M.B.; Salamon, M.B.; Parkin, S.S.P. Fluctuation-dissipation relation for giant magnetoresistive 1/f noise. *Phys. Rev. B* **1993**, *48*, 16156–16159. [CrossRef]
64. van de Veerdonk, R.J.M.; Beliën, P.J.L.; Schep, K.M.; Kools, J.C.S.; de Nooijer, M.C.; Gijs, M.A.M.; Coehoorn, R.; de Jonge, W.J.M. 1/f noise in anisotropic and giant magnetoresistive elements. *J. Appl. Phys.* **1997**, *82*, 6152–6164. [CrossRef]
65. Ingvarsson, S.; Xiao, G.; Parkin, S.S.P.; Gallagher, W.J.; Grinstein, G.; Koch, R.H. Low-Frequency Magnetic Noise in Micron-Scale Magnetic Tunnel Junctions. *Phys. Rev. Lett.* **2000**, *85*, 3289–3292. [CrossRef]
66. Durin, G.; Falferi, P.; Cerdonio, M.; Prodi, G.A.; Vitale, S. Low temperature properties of soft magnetic materials: Magnetic viscosity and 1/f thermal noise. *J. Appl. Phys.* **1993**, *73*, 5363–5365. [CrossRef]
67. Briaire, J. 1/f Noise in Permalloy. Ph.D. Thesis, Eindhoven University of Technology, Eindhoven, The Netherlands, 2000; [CrossRef]
68. Lee, S.K.; Romalis, M.V. Calculation of magnetic field noise from high-permeability magnetic shields and conducting objects with simple geometry. *J. Appl. Phys.* **2008**, *103*, 084904. [CrossRef]
69. Fescenko, I.; Jarmola, A.; Savukov, I.; Kehayias, P.; Smits, J.; Damron, J.; Ristoff, N.; Mosavian, N.; Acosta, V.M. Diamond magnetometer enhanced by ferrite flux concentrators. *Phys. Rev. Res.* **2020**, *2*, 023394. [CrossRef]
70. Schmalz, J.; Kittmann, A.; Durdaut, P.; Spetzler, B.; Faupel, F.; Höft, M.; Quandt, E.; Gerken, M. Multi-Mode Love-Wave SAW Magnetic-Field Sensors. *Sensors* **2020**, *20*, 3421. [CrossRef]
71. Boll, R. *Soft Magnetic Materials: Fundamentals, Alloys, Properties, Products, Applications, the Vacuumschmelze Handbook*; Vacuum-schmelze GmbH Hanau, Siemens AG Munich, Heyden & Sons Ltd.: London, UK, 1979.
72. Ludwig, A.; Tewes, M.; Glasmachers, S.; Löhndorf, M.; Quandt, E. High-frequency magnetoelastic materials for remote-interrogated stress sensors. *J. Magn. Magn. Mater.* **2002**, *242–245*, 1126–1131. [CrossRef]
73. Frommberger, M.; McCord, J.; Quandt, E. High-Frequency Properties of FeCoSiB Thin Films With Crossed Anisotropy. *IEEE Trans. Magn.* **2004**, *40*, 2703–2705. [CrossRef]
74. Gilbert, T. A phenomenological theory of damping in ferromagnetic materials. *IEEE Trans. Magn.* **2004**, *40*, 3443–3449. [CrossRef]
75. Queitsch, U.; McCord, J.; Neudert, A.; Schäfer, R.u.; Schultz, L.; Rott, K.; Brückl, H. Domain wall induced modes of high-frequency response in ferromagnetic elements. *J. Appl. Phys.* **2006**, *100*, 093911. [CrossRef]
76. Mozooni, B.; von Hofe, T.; McCord, J. Picosecond wide-field magneto-optical imaging of magnetization dynamics of amorphous film elements. *Phys. Rev. B* **2014**, *90*, 054410. [CrossRef]
77. Aharoni, A.; Jakubovics, J. Factors affecting domain wall mobility in thin ferromagnetic metal films. *Philos. Mag. B* **1982**, *46*, 253–272. [CrossRef]
78. Frommberger, M.; Ludwig, A.; Zanke, C.; Sehrbrock, A.; Quandt, E. High frequency magnetic properties of FeCoBSi/SiO$_2$ and (FeCo/CoB)/SiO$_2$ multilayer thin films. In Proceedings of the 2003 IEEE International Magnetics Conference, Rome, Italy, 27 July–1 August 2003; p. HD-09. [CrossRef]

79. Durdaut, P.; Rubiola, E.; Friedt, J.M.; Müller, C.; Spetzler, B.; Kirchhof, C.; Meyners, D.; Quandt, E.; Faupel, F.; McCord, J.; et al. Fundamental Noise Limits and Sensitivity of Piezoelectrically Driven Magnetoelastic Cantilevers. *J. Microelectromech. Syst.* **2020**, *29*, 1347–1361. [CrossRef]
80. Bertotti, G. *Hysteresis in Magnetism: For Physicists, Materials Scientists and Engineers*; Academic Press: San Diego, CA, USA, 1998.

Article

Sputter Deposited Magnetostrictive Layers for SAW Magnetic Field Sensors

Lars Thormählen [1,*], Dennis Seidler [1], Viktor Schell [1], Frans Munnik [2], Jeffrey McCord [1] and Dirk Meyners [1]

[1] Institute for Materials Science, Kiel University, Kaiserstraße 2, 24143 Kiel, Germany; dese@tf.uni-kiel.de (D.S.); visc@tf.uni-kiel.de (V.S.); jmc@tf.uni-kiel.de (J.M.); dm@tf.uni-kiel.de (D.M.)
[2] Institute of Ion Beam Physics and Materials Research, Helmholtz-Zentrum Dresden-Rossendorf (HZDR), 01328 Dresden, Germany; f.munnik@hzdr.de
* Correspondence: lath@tf.uni-kiel.de

Abstract: For the best possible limit of detection of any thin film-based magnetic field sensor, the functional magnetic film properties are an essential parameter. For sensors based on magnetostrictive layers, the chemical composition, morphology and intrinsic stresses of the layer have to be controlled during film deposition to further control magnetic influences such as crystallographic effects, pinning effects and stress anisotropies. For the application in magnetic surface acoustic wave sensors, the magnetostrictive layers are deposited on rotated piezoelectric single crystal substrates. The thermomechanical properties of quartz can lead to undesirable layer stresses and associated magnetic anisotropies if the temperature increases during deposition. With this in mind, we compare amorphous, magnetostrictive FeCoSiB films prepared by RF and DC magnetron sputter deposition. The chemical, structural and magnetic properties determined by elastic recoil detection, X-ray diffraction, and magneto-optical magnetometry and magnetic domain analysis are correlated with the resulting surface acoustic wave sensor properties such as phase noise level and limit of detection. To confirm the material properties, SAW sensors with magnetostrictive layers deposited with RF and DC deposition have been prepared and characterized, showing comparable detection limits below 200 pT/Hz$^{1/2}$ at 10 Hz. The main benefit of the DC deposition is achieving higher deposition rates while maintaining similar low substrate temperatures.

Keywords: magnetron sputter deposition; FeCoSiB; ERDA; XRD; film stress; magnetic field sensor; magnetic properties; magnetic domains; SAW

1. Introduction

Surface acoustic wave (SAW) devices constitute a multifunctional sensor concept [1]. The use of different piezoelectric substrates and crystallographic orientations allows the excitation of various types of surface acoustic waves, like seismic waves such as Rayleigh waves or shear horizontal waves called Love waves. Depending on the different surface waveforms, sensors can be used for different measurement applications such as biological molecule detection [2,3] and temperature [4]. A large area of application for SAW sensors is a magnetic field sensor [4–8] and a tabular comparison can be found in the Appendix A (see Table A1). The sensors are based on various substrates and magnetic sensitive layers. Typical substrates are St-cut Quartz in different cutting directions, LiNbO$_3$ and LiTaO$_3$ [2–12]. In that context, a wide range of ferromagnetic materials are used for the magnetic sensitive layer, e.g., FeCo, FeGa, Fe$_2$Tb and (Fe$_{90}$Co$_{10}$)$_{78}$Si$_{12}$B$_{10}$ (FeCoSiB) [10–12], which are deposited on the SAW sensor as full films, multilayers or patterned structures [12–14]. The variety of magnetic layers display different magnetic behavior, which is strongly correlated to the magnetic anisotropies of the magnetic films. Controlling and adapting these for the respective layer system and utilizing them is one of the major challenges in the development of these magnetic systems for SAW devices. The shown research is based on ST-cut quartz SAW devices with a full film FeCoSiB layer based on previous studies [5,8,9] with a focus on the deposition process of the magnetic sensitive layer with regard to material composition, magnetic anisotropy, film stress, amorphicity and overall sensor performance.

In general, the working principle of SAW sensors is the generation of elastic waves based on the inverse piezoelectric effect by applying a high frequency voltage to interdigital transducers (IDT) deposited on a piezoelectric substrate.

As mentioned before, the Love wave sensors utilize horizontal shear waves. The waves propagate along a delay line while the wave energy is concentrated at the sensor's surface by a guiding layer. To allow external influences to affect the sensor, it can be equipped with an additional functional layer on the delay line. For the purpose of magnetic field sensing, this functional layer consists of a magnetostrictive material. External magnetic fields cause the magnetostrictive material to change the effective shear and young's modulus, affecting the velocity of transmitted Love waves in the guiding layer. Using opposite output IDTs, the voltage generated in the piezoelectric material by the transmitted and influenced shear wave is recorded and the phase change between input and output signal serves as a measure of the applied magnetic field. In this case, a magnetostrictive layer was deposited from a magnetron sputter target with the composition $(Fe_{90}Co_{10})_{78}Si_{12}B_{10}$ (FeCoSiB) on top of the delay line. The general characteristics of the used sensor concept are shown in ref Kittmann et al. [5].

In previous studies, it was shown that the preparation of a magnetostrictive layer presents a particular challenge in terms of control of magnetic anisotropy and magnetic domain structure [8]. Although methods exist to subsequently adjust the magnetic anisotropy of the film by means of heat treatments, these cannot be directly transferred to quartz substrates. Unlike silicon (100)-based magnetic field sensors, quartz substrates for SAW applications exhibit anisotropic thermal expansion, resulting in uniaxial film stress in the magnetostrictive layer with thermal load. As a result, these (uniaxial) film stresses contribute directly to the effective magnetic anisotropy in these films [15]. To minimize the effects of film stress, changes due to anisotropic thermal expansion, deposition processes have to be adapted to quartz to reduce the substrate temperature during deposition. On the other hand, film stress generated by the deposition can be beneficially utilized to influence the magnetic properties of the magnetostrictive film. By introducing an additional stress anisotropy to the induced magnetic anisotropy, it is possible to imprint an improved alignment of the magnetic anisotropy and, thus, the magnetic domain structure in soft magnetic material [16].

One of the most commonly used magnetron sputtering methods is based on radio frequency (RF) sputter operation which has the advantage that thicker magnetic target materials with strong magnetron fields can be used. For this reason, it is possible to use the powder-sintered material FeCoSiB, as the manufacturing process and machining of the target material requires a base thickness of several millimeters. However, the method has a high energy input on the substrate and the resulting deposition rates are low. To compensate for exhibited substrate heating, complex and expensive methods for sample cooling must be applied. As an alternative, cooling phases with plasma interruption were added to the deposition process [8,9].

This publication investigates the application of direct current (DC) deposition for magnetostrictive layer preparation on quartz substrates with the aim of reducing the energy input into the substrate and increasing the deposition rate. To enable DC operation, a magnetically adapted FeCoSiB target is used. An important aspect is the chemical composition of the resulting thin films. The soft-magnetic behavior of FeCoSiB depends on the concentration of silicon and boron atoms, which serve as glass formers between iron and cobalt [17]. The use of these elements makes the sputter-deposited layers amorphous and by this eliminates magneto-crystalline anisotropy effects on the magnetic properties. It must therefore be ensured that despite the change of deposition mode, the composition of the films is unaffected so that the film structure remains amorphous. These amorphous magnetic films show good soft-magnetic properties [18] together with a high magnetostrictive response, as desired for use in SAW-based magnetic field sensors. It will be shown that a changeover to DC deposition enables significantly higher deposition rates without a significant alteration in structural and magnetic properties, as well as in SAW sensor performance. For this purpose, the chemical, structural and magnetic properties of DC and RF with different sputter pressure-deposited FeCoSiB films are investigated. The corresponding SAW sensor performance with RF and DC deposited magnetic films is compared in terms of sensitivity and noise.

2. Materials and Methods

The required magnetostrictive FeCoSiB layers for the experiments are deposited in a VonArdenne CS730S sputtering system. The system has nine planar magnetron sputter sources and allows the deposition of materials as adhesion or passivation layers without breaking the vacuum. FeCoSiB depositions are performed from a multi-element target with 99.95% purity and at constant power for RF and DC of 200 W with argon gas flow of 40 sccm. The chamber pressure is varied for the depositions in the range of 1.0×10^{-3} mbar to 1.0×10^{-2} mbar with a step size of 1.0×10^{-3} mbar. The depositions cannot be performed in one cycle. Deposition time limits are necessary to keep the energy input into the sample and the resulting sample temperature low to minimize thermal expansion-induced film stress. Depending on the chamber pressures, deposition times are adjusted to deposit 5 nm FeCoSiB by RF mode and 20 nm FeCoSiB by DC mode. As a result, for the preparation of in total 200 nm thick FeCoSiB layers, 40 cycles and 10 cycles are run for RF and DC, respectively. To ensure that the maximum temperature of the substrate does not exceed 30 °C, cooling breaks of 300 s (RF) and 180 s (DC) are introduced between deposition cycles. As an additional measure, 2 mm thick AlO_2 spacers are placed below the samples to thermally decouple the SAW substrate from the aluminum sample carrier.

For the fabrication of FeCoSiB layers on SAW substrates with a thickness of 200 nm, the Ar pressure is chosen to achieve a compressive film stress of 200 MPa. Thus, for an Ar pressure of 2.5×10^{-3} mbar, the rate for RF is 0.16 nm/s with a cycle sequence of 30 s deposition time and 300 s deposition pause. For DC, on the other hand, with a chamber pressure of 2.0×10^{-3} mbar, the rate is 0.4 nm/s with a cycle sequence of 50 s deposition/180 s cooling. As a result, the total processing time is reduced from approx. 4.5 h for RF to 1.5 h for DC for the aimed FeCoSiB thickness. A uniaxial magnetic anisotropy is induced in the films parallel to the SAW propagation direction by applying a static permanent magnetic field of $\mu_0 H_{dep}$ = 70 mT during film deposition (see Figure 1). This field is generated by two parallel aligned $Nd_2Fe_{14}B$ permanent magnets with dimensions of 25 mm × 6 mm × 2 mm. The magnets are connected by an iron yoke to increase the strength and homogeneity of the generated magnetic field. The magnet is used for all depositions.

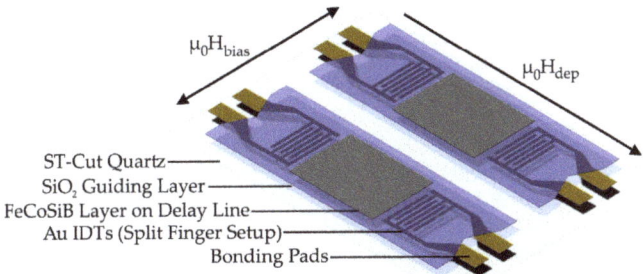

Figure 1. Schematic drawing of a SAW chip in use with two SAW delay line sensors, each with an independent guide layer. For magnetic anisotropy imprinting a magnetic field $\mu_0 H_{dep}$ is used during deposition of the magnetic layer FeCoSiB. The sensors are operated with $\mu_0 H_{bias}$ at the magnetic operating point during sensor operation.

The deposited FeCoSiB films are chemically characterized by elastic recoil detection analysis (ERDA) [19], which enables the detection of light elements in a heavy element matrix. For the chemical analysis, 10 mm × 10 mm silicon substrates (100) with native oxide are used. These substrates are coated with a stack of 5 nm Cr/400 nm FeCoSiB/10 nm Ta. Here, chromium serves as an adhesion promoter. Tantalum is used for surface passivation to prevent early corrosion of the FeCoSiB. The ERDA measurements are performed with a Cl^{7+} ion beam and two different energies of 43 MeV and 35 MeV. The angle of incidence on the film is 75° with respect to the normal sample. The irradiated film area has a size of 2 mm × 2 mm. The scattered Cl ions and recoiled target atoms are detected at an angle of 30° with a Bragg ionization chamber (BIC), which allows the energy and the atomic number to be determined. The last is necessary to identify the particle. To measure the total energy, the particles must be stopped completely in the BIC, which is filled with isobutane at a pressure of 125 mbar. This pressure is optimal for the measurement of recoil atoms of C to P and scattered Cl ions produced by a beam energy of 43 MeV. Because the recoil B atoms are light and have a long-range, a separate measurement with a beam energy of 35 MeV has been performed, which is sufficiently low for the B atoms to be completely stopped in the detector. The output of the BIC detector is a 2D-histogram where the number of particles is displayed as a function of energy and Bragg peak signal [20]. In this 2D plot, the elements can be identified in branches, detaching from the main branch of scattered Cl ions. The analysis depth for the recoil atoms is based on where the recoil branch detaches from the main branch and it is highest for atoms with low atomic number (Z). For Si, the analysis depth is relatively low as can be observed in the depth profiles (Section 3.1). H recoil atoms have been detected with a separate solid-state detector at a scattering angle of 40°. This detector is preceded by a 25 μm Kapton foil to stop scattered ions and heavy recoil ions. The depth resolution of this system is reduced because of energy loss straggling in the foil. A separate detector is needed because the range of the H atoms is too large to stop in the BIC detector. The processing of the measured spectra and calculation of elemental depth profiles is performed using NDF v9.3g [21].

As described, B and Si serve as glass formers and prevent the formation of crystalline iron-cobalt phases. Provided that enough of the glass formers is incorporated, an amorphous film structure is expected to form during sputter deposition. To investigate the structure by X-ray diffraction (XRD) method, samples are prepared from silicon (100) substrates with native surface oxide. The sample size is 15 mm × 15 mm to avoid beam overspill. The substrates are coated by 10 nm Ta/200 nm FeCoSiB. These layers are patterned by lithography and lift-off process to create a circular FeCoSiB layer with an area of 150 mm^2. No passivation layer is used for the XRD measurements, which are executed timely after layer deposition. To examine for the X-ray amorphous state, a Rigaku Smartlab 9 kW XRD tool is operated in grazing incidence (GIXRD) geometry. Generating Cu Kα_1 and Kα_2 radiation, the source is run at a power setting of 200 mA and 45 kV. The resulting diffraction patterns are recorded with a 2D detector.

Another important aspect when considering the magnetic performance on the SAW sensors is the film stress of the magnetostrictive layer. For the determination of the respective film stress, cantilevers made of "UPILEX-S®" in a size of 25 mm × 2.5 mm × 50 μm are used. The cantilevers are coated with a 10 nm layer of Ta before magnetic coating and reference measurement, to improve adhesion and reflection properties. The bending of the cantilevers before and after deposition of 200 nm FeCoSiB at varying chamber pressures is recorded and evaluated using a Keyence VK100 laser confocal microscope. The Stoney equation [22] is used to calculate the resulting film stresses. In order to investigate the influence of film stress on the imprinting of magnetic anisotropy by the external magnetic field, additional magnetic analysis samples based on 10 mm × 10 mm silicon (100) substrates with native oxide are prepared for magnetic property investigations. These samples are coated with the layer system 10 nm Ta/200 nm FeCoSiB/10 nm Ta. In this case, the Ta layers act as adhesion promoter and passivation layer, respectively. The deposition of

FeCoSiB for the RF and DC samples is performed at the deposition pressures between 2×10^{-3} mbar and 1×10^{-2} mbar.

The SAW sensors used for this work are based on the above-mentioned ST-Cut quartz and have two SAW sensors with independent delay line on a sensor chip of size 14 mm × 8.9 mm. The single sensors have two pairs of split finger IDTs with a periodicity of 28 µm and Figure 2 shows the scattering parameters (S12 and S21) for a sensor used in this publication. The center frequency is at 142.5 MHz. All sensors investigated show similar scattering parameters in the range between 141 MHz and 146 MHz. The variance of the center frequency is due to thickness differences in the SiO_2 wave guiding layer.

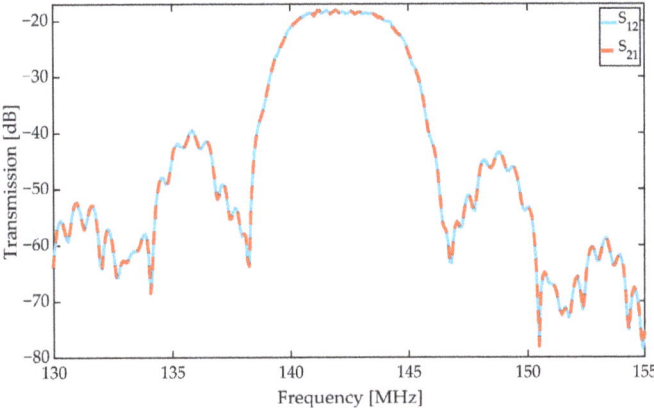

Figure 2. Scattering parameters (S12 and S21) of an SAW sensor used in this publication measured in magnetic saturation. The center frequency of the Love wave device is at 142.5 MHz.

The design and fabrication of the SAW sensors has been described in detail in publications [5,8]. To evaluate the sensor properties, the SAW sensors are mounted on printed circuit boards (PCB) and the impedance of input and output is matched to 50 ohm. The measurements are performed using a Zurich Instruments UHFLI lock-in amplifier, a vector network analyzer and a Rohde and Schwarz FSWP phase noise analyzer in a shielded chamber [23,24]. Scattering parameters, magnetic sensitivity and phase noise are measured. A more detailed description of the measurement procedure can be found in Schell et al. [8]. The cross-sensitivity to temperature of SAW sensors is well known and can be mitigated by using a second SAW device without magnetostrictive coating for compensation [25,26]. In the experiments presented here with a focus on the influence of the magnetic layer, the temperature in the lab environment was kept constant and a compensation method was not applied. An additional approach to reduce these influences on ST-Cut quartz resonators is shown in Mishra et al. [27].

For a better understanding of the magnetic layer behavior of the fabricated SAW magnetic field sensors, large view magneto-optical Kerr effect (MOKE) microscopy investigations are performed [28]. The large view microscope has the ability to image the entire delay line and capture the magnetization loops and magnetic domain images, also on the device level. For this purpose, the above-mentioned magnetic analysis samples, as well as the magnetic SAW sensors, were considered. Magnetic domain images are taken in the demagnetized state and around the devices' working points to analyze the orientation of the magnetic domains and their overall domain structure. To investigate the magnetic behavior in more detail, longitudinal and transversal magnetization loops of the device samples were measured with the application of magnetic field parallel and perpendicular to the imprinting field $\mu_0 H_{dep}$ in the MOKE microscope. The magnetic hysteresis losses are extracted from the magnetization loops obtained for films with different conditions of film depositions.

3. Results

3.1. Chemical and Structural Analysis

The ERDA depth profiles of DC (a) and RF (b) deposited FeCoSiB layers in Figure 3 show the elemental distributions in at.% as a function of profile depth in at/cm^2. Based on the estimated densities of the measured elements, an overall measurement depth of 420 nm (equal to 3526 at/cm^2) can be assumed. In Figure 3, the depth profiles recorded at accelerator energies 43 MeV and 35 MeV for a depth of 120 nm are merged for the measured elements iron (Fe), cobalt (Co), silicon (Si), oxygen (O), carbon (C), boron (B) and hydrogen (H).

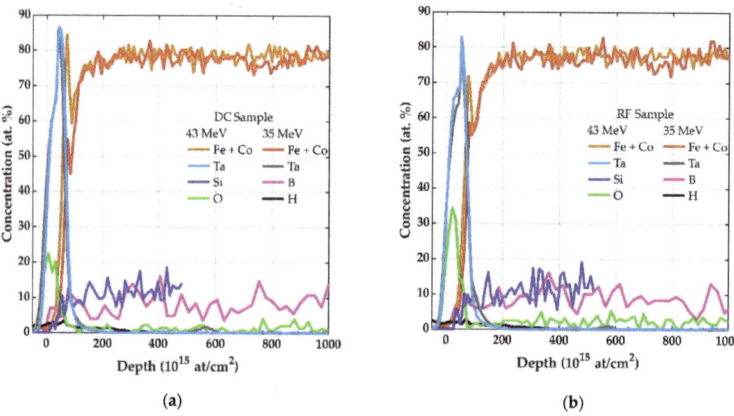

Figure 3. (**a,b**) ERDA depth profile of FeCoSiB deposited with DC and RF. The samples consist of a 5 nm Cr/400 nm FeCoSiB/10 nm Ta layer. The tantalum layer serves as a protection layer to prevent oxidation. The displayed measurement depth range corresponds to approximately 120 nm. As indicated, two different accelerating voltages 43 MeV and 35 MeV were used to obtain optimal results for all elements.

The concentration data are derived from the measured ERDA and Cl scattering spectra and have been calculated using NDF v9.3g. A special case arises for the elements Fe and Co. Due to their similar masses, Fe and Co signals cannot be differentiated from each other and are therefore summed. The uncertainty of the determined elemental concentration of the performed measurement is in the 1 at % range for the elements (Fe+Co), Si and B. Due to the low amounts of oxygen, carbon and hydrogen (close to the detection limit) the relative error of O content is 10% and 20–30% for C and H. Furthermore, the depth profiles are convoluted by the system resolution, physical effects such as straggling and isotope distribution for B, and also by the statistical fluctuations of the spectra [19].

From the elemental depth profiles, it can be concluded that the 10 nm thick Ta top layer has sufficient oxidation resistance and can prevent deep penetration of O. The concentration of O drops sharply with penetration depth, falling within the Ta layer to a low amount of about 2 at.%, which remains constant till the end of the depth profile. Apparently, O is incorporated into the film during deposition. Possible oxygen sources include chamber leakage, sputter gas impurities, and the use of powder-sintered target material. The last one is assumed to be the most likely source due to the manufacturing method of powder-sintered targets. Furthermore, it can be seen in Figure 3 that in both deposition cases the element distribution after the Ta top layer is constant. This provides information that no measurable compositional variation is formed by the pause times between the deposition cycles for DC and RF and the FeCoSiB grows as a continuous film. The origin of the elements C and H are assumed to be due to a residual gas content in the deposition chamber. Furthermore, manufacturing and processing of the FeCoSiB target for the magnetic adaption could cause the incorporation of the detected C impurities.

However, it should be noted that the configuration of the ERDA setup with BIC was chosen to enable the detection of the target elements, whereas the Ar detection requires a configurational change. To improve the mass separation between Ar atoms and Cl ions, ERDA can be used with a time-of-flight analysis system (ToF). Additional ToF-ERDA experiments reveal an upper limit of 0.5 at.% for the Ar concentration in the FeCoSiB layers. The elemental concentrations averaged across the measured FeCoSiB cross-sections are summarized in Table 1.

Table 1. Averaged elemental concentrations calculated from the FeCoSiB depth profile for an RF and DC sample deposited from a target material with a nominal composition of $(Fe_{90}Co_{10})_{78}Si_{12}B_{10}$.

Sample	Fe+Co (at.%)	Si (at.%)	O (at.%)	C (at.%)	B (at.%)	H (at.%)
RF	77.4	11.4	2.32	0.44	8.40	0.075
DC	78.2	12.0	1.63	0.52	7.63	0.056

Fe and Co together reach a proportion of about 78 at.%, which equals the content in the sputter target with the nominal composition of $(Fe_{90}Co_{10})_{78}Si_{12}B_{10}$. Si also reaches the target value of 12 at.%. B, on the other hand, deviates from the desired concentration of 10 at.% by more than 1 at.%. Within the mentioned measurement error, the ERDA measurements reveal the same elemental concentrations for both deposition methods RF and DC.

The structural analysis was performed by XRD as described above. Figure 4 shows the measured X-ray intensities of the two FeCoSiB films deposited by RF (DC) mode at a pressure of 2.5×10^{-3} mbar (2.0×10^{-3} mbar). The XRD spectra show two broad and smaller diffractions at 2 Theta equal to 34° and 39°, which originate from the buried Ta layer. The Ta adhesion promoter is still measurable due to the GIXRD penetration depth of more than 200 nm. Due to the low intensity, it is not possible to determine any precise information about the Ta structure. We assume a nanocrystalline or an amorphous structure. An amorphous structure was reported for thin tantalum layers prepared by magnetron sputter deposition [29]. The third reflex, and largest, extends around 44°. In this 2 Theta range, diffraction peaks can emerge from Fe at 44.77°, Co at 47.69° and FeCo at 44.97°. It is assumed that the main source of the signal is the high amount of Fe in the film. Broad and low-intensity distributions are measured in this 2 Theta range for possible Fe and Co signals. No further peaks are exhibited. Based on the results of the GIXRD, it can be assumed the amorphous FeCoSiB quality is the same, regardless of RF or DC deposition mode.

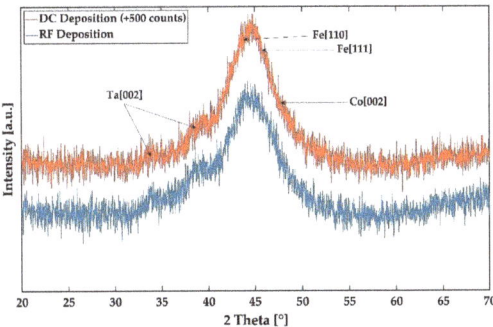

Figure 4. XRD result of Ta/FeCoSiB thin films deposited by RF and DC mode at 2.5×10^{-3} mbar and 2.0×10^{-3} mbar, respectively. For both deposition processes, the power was set to 200 W at an Ar flow of 40 sccm. Grazing incidence XRD was used with 2 Theta ranging from 20° to 70° with steps of 0.25°. For a better representation, an offset for the XRD result for the thin film deposited by DC mode was used. The adhesion promoter tantalum and a signal combination of the amorphous FeCoSiB are noticeable.

3.2. FeCoSiB Film Stress and Magnetic Behavior

After it has been demonstrated that both deposition methods DC and RF lead to the same layer composition and structure, the control of film stress and imprinted magnetic anisotropy become important parameters for the following magnetic samples and SAW sensors. The film stresses determined for this purpose are shown in Figure 5 as a function of Ar deposition pressure for RF and DC sputter deposition conditions.

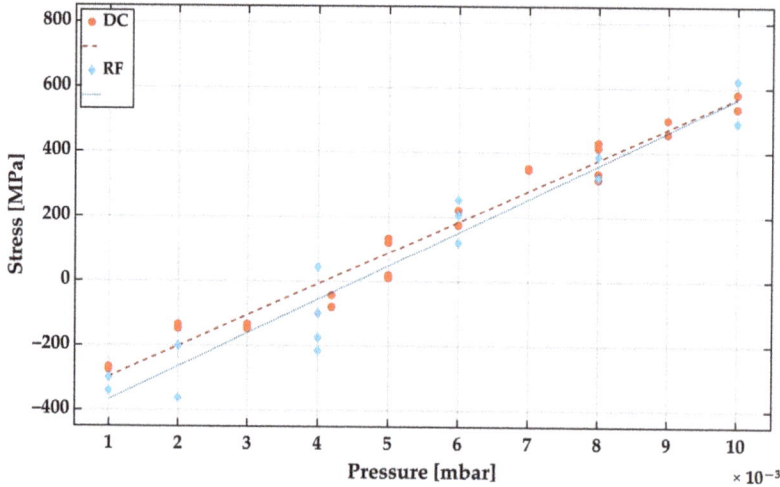

Figure 5. Determined film stress measurements for depositions performed with DC (red) and RF (blue) with different Ar deposition pressures. The resulting FeCoSiB film thickness was set to 200 nm and the initial and resulting bending is measured with a laser confocal microscope. The film stresses are calculated by the Stoney equation from the curvature of the cantilevers before and after film deposition. (The data are fitted with a simple linear regression).

It is clearly visible that both deposition modes show a similar film stress to deposition pressure relation. The overall course of the stress vs. Ar deposition pressure curves can be explained by the pressure-dependent mean free path of the film formers. At lower pressure the sputtered species retain their kinetic energy and a more compact film growth [30]. According to the assumed linear regression, the transition from compressive to tensile stress appears at around 4.0×10^{-3} mbar and 4.5×10^{-3} mbar for DC and RF deposition, respectively. For both deposition modes, the Ar deposition pressure can be adjusted to obtain FeCoSiB layers with low-stress magnitudes.

To relate the changes in film stress to the magnetic properties of the film, 10 mm × 10 mm sized samples with a 200 nm thick FeCoSiB layer were analyzed utilizing MOKE magnetometry and microscopy. For the measurement of the magnetization loops the region of interest was positioned in the center of the sample with a size of 5 mm × 5 mm, which approximately corresponds to the SAW sensor magnetic film size and position. Concentrating on the center of the films minimizes edge effects in the measurements of the magnetization reversal and thereby reflects the actual magnetic film properties.

Magnetization loops obtained field from RF and DC deposited samples at deposition pressures of 2.0×10^{-3} mbar and 8.0×10^{-3} mbar are shown in Figure 6. The corresponding film stress is $\geq$200 MPa compressive and $\geq$300 MPa tensile for 2.0×10^{-3} mbar and 8.0×10^{-3} mbar, respectively. The magnetic field was applied from -3 mT to $+3$ mT parallel and perpendicular to the magnetic deposition field $\mu_0 H_{dep}$.

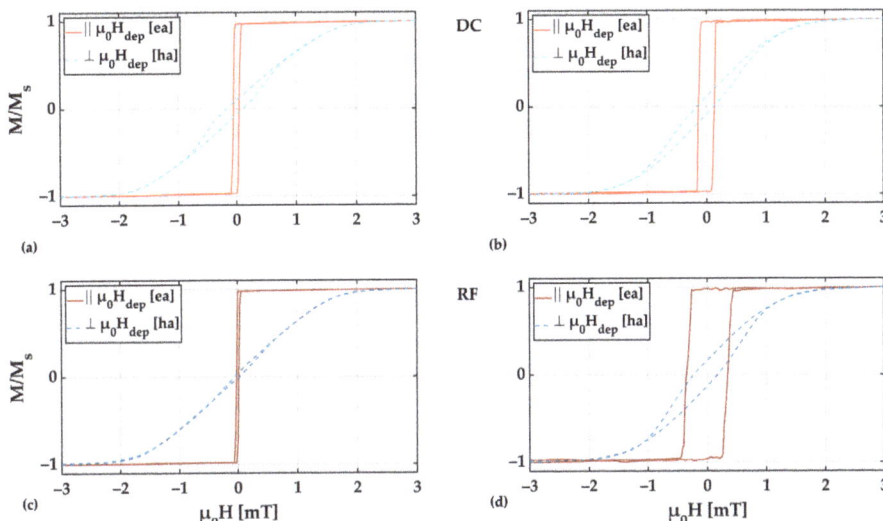

Figure 6. Magnetization curves for 200 nm thick FeCoSiB films deposited with DC (**a,b**) and RF (**c,d**) at an Ar deposition pressure of 2.0×10^{-3} mbar (**a,c**) and 8.0×10^{-3} mbar (**b,d**) measured along and perpendicular to $\mu_0 H_{dep}$, respectively along the presumed easy axis (ea) and hard axis (ha) of magnetization. The measured area of 5 mm × 5 mm was set to the sample center to avoid edge effects. The measurement field was applied parallel (red) and perpendicular (blue) to $\mu_0 H_{dep}$.

From the magnetization loops, clear uniaxial anisotropy characteristics can be derived with the easy axis (ea) of magnetization aligned along the direction of $\mu_0 H_{dep}$. In all cases, the coercivity of the magnetic films increases with tensile film stress.

In Figure 7, corresponding magnetic domain images obtained at zero field from the same RF and DC deposited samples are shown. Small tilts (up to 5°) of the induced anisotropy axis are visible as a result from slight field inhomogeneities in the deposition field setup ($\mu_0 H_{dep}$) and/or deviations in the position in the sputtering chamber.

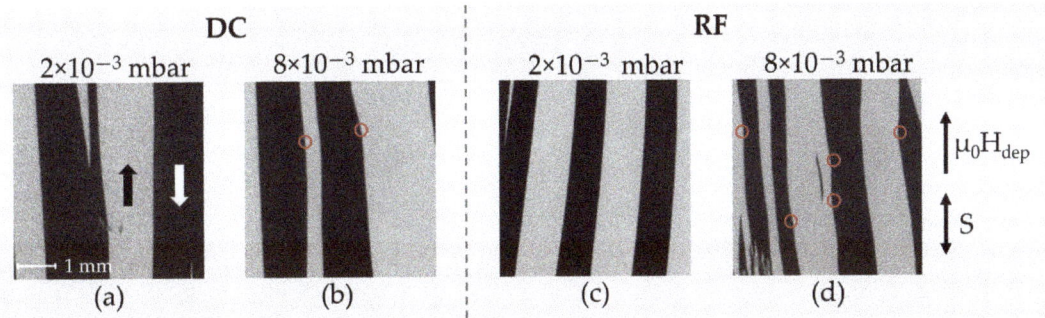

Figure 7. MOKE microscopy images of demagnetized magnetic samples deposited with 10 nm Ta/200 nm FeCoSiB/10 nm Ta with an applied magnetic deposition field $\mu_0 H_{dep}$ during deposition, at Ar deposition pressures of 2.0×10^{-3} mbar (**a,c**) and 8.0×10^{-3} mbar (**b,d**) for DC (**a,b**) and RF (**c,d**) sputter-deposited samples. The magneto-optical sensitivity S is aligned parallel to the $\mu_0 H_{dep}$. The FeCoSiB films with low deposition pressure and high compressive film stresses (**a,c**) show a slight misalignment of the magnetic domain walls by up to 5° as compared to the direction of $\mu_0 H_{dep}$. At high deposition pressures and high tensile stresses (**b,d**), magnetic domain wall pinning centers (red circles) appear in the magnetic layer.

For both deposition pressures the formation of large and well-aligned magnetic domains becomes visible. In contrast to the 2.0×10^{-3} mbar samples, the 8.0×10^{-3} mbar samples show areas of local domain wall pinning (red circles in Figure 7), which are particularly evident for the case of RF deposition. This is an indication of magnetically active defects, which are related to magnetic property inhomogeneities. The domain wall pinning becomes also visible during the magnetization reversal along the magnetic easy axis (not shown). It is directly reflected in the variations seen in the magnetization loops (Figure 6). Considering the high tensile stresses, the presumed magnetic property variations are assumed to be due to local magneto–elastic interactions, which directly influence the magnetic behavior [31]. The exact nature of the defects cannot be concluded from our measurements.

Overall, comparing the measured magnetization loops at low and high deposition pressure, a significant broadening is visible for higher pressures, which correlates directly with the observed changes in magnetic domain characteristics and substantiates the domain wall pinning by local stress fields due to an increase in stress magnitude.

An accurate parameter to quantify the change in the magnetization loops is the hysteresis loss. The hysteresis loss is especially important as it is strongly connected to the exhibited magnetic noise in magnetoelectric composite magnetic field sensors [32–34]. Data for the whole parameter range are displayed in Figure 8a for DC and Figure 8b for RF deposition for Ar deposition pressures ranging from 2.0×10^{-3} mbar to 8.0×10^{-3} mbar. A saturation magnetization of M_s = 1193 kA/m [35] was assumed for the extraction from the MOKE loops.

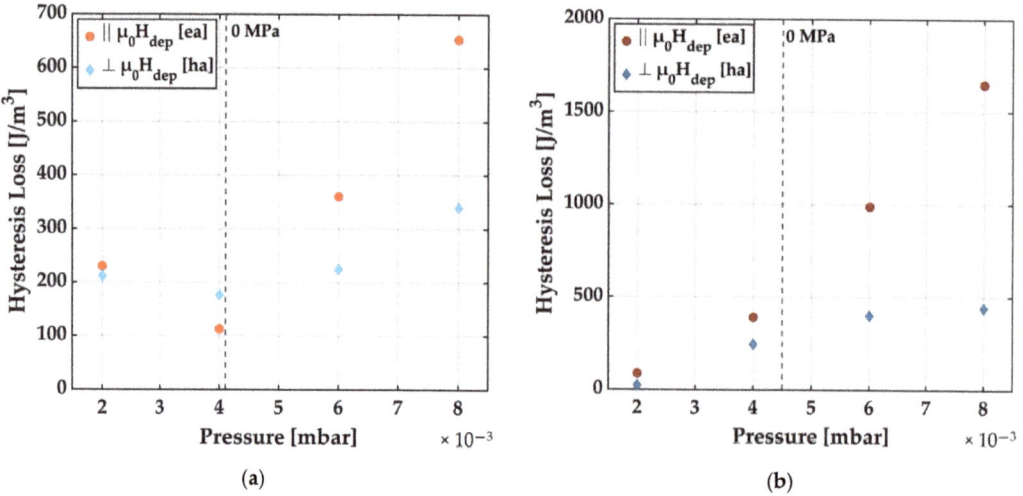

Figure 8. Hysteresis losses for DC (**a**) and RF (**b**) deposited FeCoSiB films as a function of Ar deposition pressure. The values in (**a**,**b**) were obtained from hysteresis curves of magnetic test samples with a measured area of 5 mm × 5 mm located in the center of the samples. To determine the hysteretic losses, the enclosed area of the hysteresis curves was determined according to the orientation parallel and perpendicular to $\mu_0 H_{dep}$. The approximate positions of zero stresses are indicated.

For both deposition methods, a trend to higher hysteresis losses with higher Ar deposition pressure is visible. Although we assume that the FeCoSiB films deposited with RF and DC have the same chemical composition (ERDA) and structure (XRD), we see a clear trend in the change of magnitude of magnetic losses. From the magnetic domain studies, we relate this to inhomogeneities in the magnetic films that, in connection with the magnetostrictive material, hinder the magnetization process. Magnetic domain wall pinning and a reduction of small field permeability arise from the magnetoelastic interactions. It is well known that CoFe-based amorphous soft-magnetic alloys display a minimum in coercivity with zero magnetostriction [36] due to the exhibited coupling between stresses and magnetic properties.

3.3. SAW Sensor Performance

As shown in the previous chapter, a uniaxial anisotropy cannot be induced precisely with increasing sputtering pressure. Therefore, the FeCoSiB films on the SAW devices are deposited at sputtering pressures, which correspond to film stresses of −200 MPa (compressive stress), i.e., 2.0×10^{-3} mbar for DC deposition and 2.5×10^{-3} mbar for RF deposition (see Figure 5).

An analogous magnetic analysis as shown before, including magnetization loop measurements and magnetic domain analysis, was performed directly on the SAW devices. The magnetic analysis was further restricted to the active region of the SAW device. Magnetization loops with the magnetic field along and perpendicular to the direction of wave propagation are shown in Figure 9. The magnetic field direction perpendicular to the deposition field ($\perp \mu_0 H_{dep}$) corresponds to the applied bias field in SAW sensor operation defining the working point. To get a clearer view on the overall magnetization reversal behavior also the transverse magnetization loops [32,37] are analyzed.

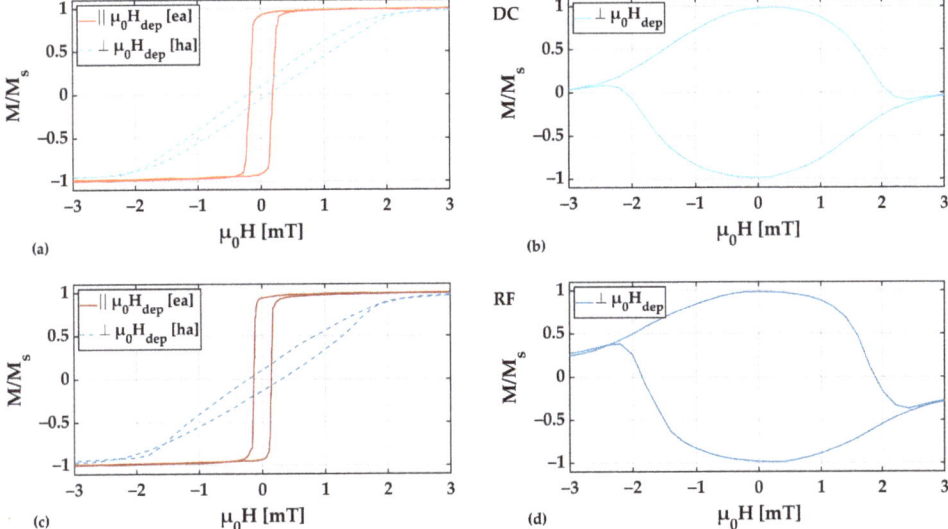

Figure 9. Longitudinal and transversal magnetization curves for a 200 nm thick FeCoSiB film deposited with DC sputter deposition at an Ar pressure of 2.0×10^{-3} mbar (**a,b**) and with RF sputter deposition at an Ar pressure of 2.5×10^{-3} mbar (**c,d**) from SAW devices. Longitudinal curves were measured along (red) and perpendicular (blue) to $\mu_0 H_{dep}$, the latter being perpendicular to the SAW propagation direction. The measured area was set in accordance with the active SAW device area.

Corresponding magnetic domain images at magnetic bias fields $\mu_0 H_{bias}$ of −0.35 mT and +0.35 mT coming from negative and positive saturation, respectively, are displayed in Figure 10.

The magnetic hysteresis effects are more pronounced as in the case of the extended thin films, which we attribute to magnetic domain effects at the sample edges. Spike domains extending in the active region of the sensor become visible in Figure 10e, which hinder the magnetization reversal and which lead to an increase in coercivity for the magnetic field directions along and perpendicular to $\mu_0 H_{dep}$. For the magnetic field direction perpendicular to $\mu_0 H_{dep}$ we obtain a coercivity field of $\mu_0 H_c \approx 0.15$ mT for RF and $\mu_0 H_c \approx 0.3$ mT in the case of DC sputter deposition. From the transverse magnetization loops (Figure 9b,d), a mostly coherent rotation of magnetization is obtained for the horizontally aligned magnetic bias fields. Despite the rotational process, unity of magnetization is not reached due to the spike domain formation at the top and bottom edge. The lower amplitude in the general magnetic and magnetic domain behavior of the magnetostrictive phase of the SAW sensors is comparable. The hysteretic magnetization reversal behavior has a direct influence on the SAW sensor characteristics.

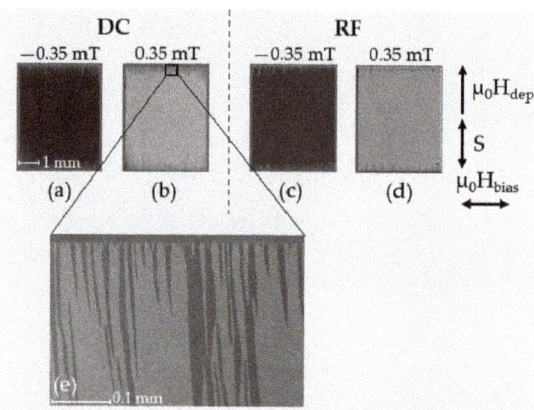

Figure 10. MOKE images from DC (**a**,**b**,**e**) and RF (**c**,**d**) sputter deposited SAW structures at magnetic bias fields $\mu_0 H_{bias}$ of -0.35 mT and $+0.35$ mT aligned perpendicular to $\mu_0 H_{dep}$ (as indicated). The magneto-optical sensitivity (S) is aligned parallel to $\mu_0 H_{dep}$. Nearly single magnetic domain behavior occurs. Yet, spike domains (**e**) [38] at the bottom and top edges ($\perp$ to $\mu_0 H_{dep}$)) form. Shown in (**e**) is a higher resolution (20×) version of image (**a**) in the area of the upper edge and shows the spike domains.

Figure 11a shows the phase change of SAW sensors with DC and RF deposited FeCoSiB films under the intrinsic compressive stress as a function of an applied DC bias field $\mu_0 H_{Bias}$. The field was changed from -10 mT to $+10$ mT and vice versa with a step size of 25 µT in the most relevant field range from -1 mT to $+1$ mT.

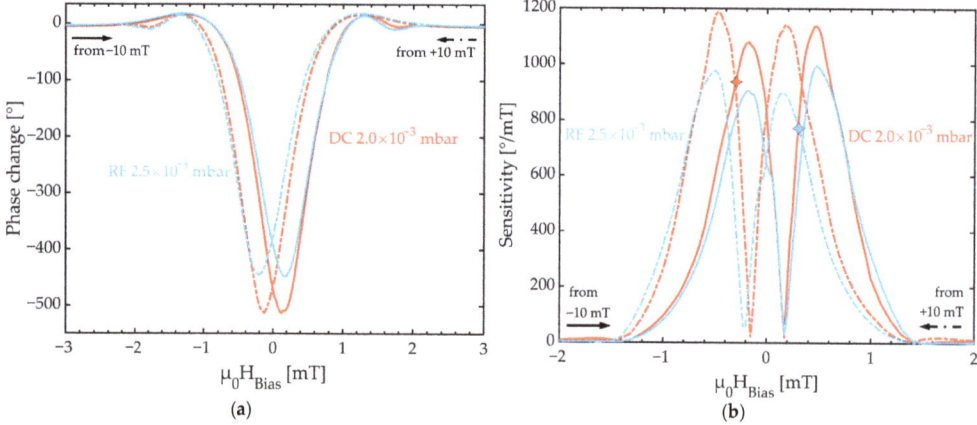

Figure 11. (**a**) Phase change of SAW sensors as a function of applied DC bias field for DC (red) and RF (blue) deposited FeCoSiB films with low compressive film stress. (**b**) Measured sensitivity as a function of applied DC bias field obtained using a 1 µT AC field for DC (red) and RF (blue) deposited FeCoSiB films with compressive film stress. Markings in the DC and RF curves highlight the magnetic bias field in which the best LOD measurement has been achieved. All curves in (**a**,**b**) were recorded from -10 mT to $+10$ mT and vice versa. The measurement step size in the range from -1 mT to $+1$ mT is 25 µT. AC and DC fields were applied perpendicular to the SAW propagation direction.

With 510° for the DC sputtered sensor and 450° for RF the total phase changes are very similar for both deposition techniques. In fact, several sensors have been measured to confirm the obtained results and all show maximum phase changes between 430° and 510°. The hysteresis characteristics are in accordance with the magnetometric analysis of the obtained bias curves, i.e., the large hysteresis between the two field scanning directions for each deposition technique. While in an ideal hysteresis-free case both phase minima are expected to be at 0 mT bias field [5], they are around the coercive fields at −0.125 mT and +0.125 mT for the DC deposition case and −0.2 mT and +0.175 mT for the RF deposited films. Nevertheless, the bias curve hysteresis is very similar for both deposition techniques and therefore independent of the type of deposition bias.

In Figure 11b, the measured sensitivity, i.e., the derivative of the phase change, shows two characteristic sensitivity maxima for each measurement direction and each sensor. The maxima indicate the positions of the highest slope in the static bias curves in Figure 11a. These points of highest sensitivity for the RF deposition case are at −0.175 mT and +0.475 mT coming from negative saturation with maximum sensitivities of 907°/mT and 996°/mT, respectively, and at +0.15 mT and −0.5 mT coming from positive saturation with maximum sensitivities of 900°/mT and 978°/mT, respectively. For the DC deposited films, these respective points are at −0.175 mT and +0.475 mT coming from negative saturation and +0.175 mT and −0.475 mT coming from positive saturation with sensitivity of 1081°/mT, 1138°/mT, 1144°/mT and 1188°/mT, respectively. The slight differences in total phase change and sensitivities originate from slightly varying FeCoSiB thicknesses [9] and tilted orientation of magnetic anisotropy [8]. However, the values presented here are in good agreement with previously published results covering a sensitivity range from 500°/mT to 2000°/mT [5,8,9,32].

While the sensitivity is a useful sensor characteristic, the measure which ultimately quantifies the performance of a magnetic field sensor, however, is the limit of detection (LOD). The LOD results from dividing the sensors noise, in this case phase noise, by its sensitivity. As shown in Figures 11b and 12a, while the maximum sensitivity of the sensor with DC deposited FeCoSiB is higher than the phase noise is, in general, it is lower for the RF deposited films.

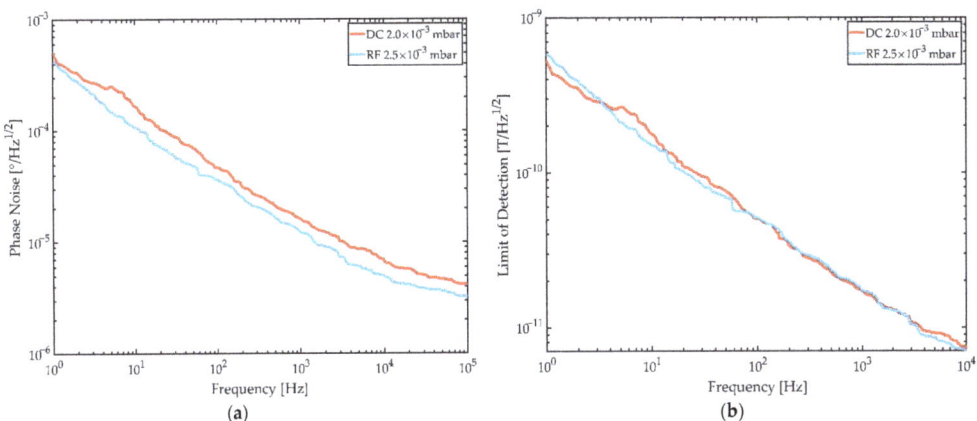

Figure 12. (a) Phase noise as a function of frequency from the SAW excitation carrier for DC (red) and RF (blue) deposited FeCoSiB films with compressive film stress. The shown phase noise spectra are measured at magnetic bias fields and excitation powers, which lead to the lowest limits of detection. In the DC deposition case this corresponds to a bias field of −0.3 mT (from negative saturation) and an excitation power of 1 dBm, and in the RF deposition case this is +0.35 mT (from positive saturation) and 2 dBm. (b) Limit of detection as a function of frequency from the SAW excitation carrier for DC (red) and RF (blue) deposited FeCoSiB films with compressive film stress. The limit of detection is the quotient of a sensors phase noise and its sensitivity.

The phase noise spectra in Figure 12a are both recorded under sensor operating parameters, which lead to the lowest LOD for each sensor. More precisely, these parameters are the excitation power and $\mu_0 H_{bias}$. Phase noise spectra were measured for various DC bias fields and it turned out that the lowest LODs are achieved when applying a bias field slightly higher, i.e., slightly off the point of highest sensitivity. For the DC deposition case, this is at -0.3 mT coming from negative saturation and $+0.35$ mT coming from positive saturation for the RF case. For the excitation power, it was shown before that there exists an optimum excitation power where the phase noise is the lowest in SAW magnetic field sensors [32]. For the sensor design considered in Ref. [32] and here, this optimum excitation power is 1 dBm for the DC deposition case and for the RF case it is 2 dBm. Finally, this optimum operation leads to detection limits as low as 179 $pT/Hz^{1/2}$ for DC deposited films and 152 $pT/Hz^{1/2}$ for RF deposited films at 10 Hz, respectively. At 100 Hz the LODs are 51 $pT/Hz^{1/2}$ (DC) and 52 $pT/Hz^{1/2}$ (RF).

However, it should be noted that the LODs of all sensors investigated for this study vary for both deposition types between 152 $pT/Hz^{1/2}$ and 190 $pT/Hz^{1/2}$ at 10 Hz and can therefore be considered equivalent. This shows that in terms of SAW magnetic field sensor performance with DC deposited films, the same LODs can be achieved while providing lower deposition times and less heat generation.

4. Conclusions

FeCoSiB thin films using DC magnetron sputtering for magnetic field SAW sensors were shown to give equivalent results in terms of magnetic properties as well as sensor performance as compared to commonly used RF deposited layers.

Due to the higher deposition rate in DC operation, it becomes possible to reduce the total deposition time by a factor of 3, which is a significant improvement and advantage over the common RF sputter process. Furthermore, the chemical composition measured by ERDA and the structure measured by XRD of the prepared films are identical for DC and RF deposited FeCoSiB. Major factors influencing the magnetic performance of the SAW sensors such as film stress can be controlled and show that compressively stressed FeCoSiB films provide better control of the soft-magnetic film properties. On the SAW device level, it was shown that the magnetic layer properties and the resulting detection limits of the sensors with a single layer of 200 nm FeCoSiB are virtually the same with 51 $pT/Hz^{1/2}$ (DC) and 52 $pT/Hz^{1/2}$ (RF) at 100 Hz and with 179 $pT/Hz^{1/2}$ on average at 10 Hz. Through further development of DC mode deposited magnetic films it will be possible to further improve the magnetic anisotropy of the thin film as well as to reduce the film variance. Thus, it will be possible to further improve the sensitivity, as well as the resulting LOD, with a significantly accelerated preparation process.

Another approach to improve the sensor performance is to use other structuring mechanisms, such as ion beam etching, to gain better control over the edge profile of the FeCoSiB, while also reducing possible defect structures caused by the resist needed for the lift-off process. Alternative magnetic layer systems, such as exchange bias stacks, show significant improvements in sensor performances [39–41]. However, due to the complexity of the exchange bias systems, the deposition times would increase significantly when using the RF mode, which can now be ideally compensated by DC film deposition.

Author Contributions: Conceptualization, L.T., D.S., V.S., F.M., J.M. and D.M.; methodology, L.T., V.S. and D.S.; validation, L.T.; formal analysis, L.T., D.S. and V.S.; investigation, L.T., D.S., V.S., F.M.; resources, L.T., V.S., F.M. and D.S.; data curation, L.T.; writing—original draft preparation, L.T., V.S., D.S., F.M. and D.M., J.M.; writing—review and editing, D.M., J.M. and F.M.; visualization, L.T., D.S. and V.S.; supervision, D.M. and J.M.; project administration, D.M. and J.M.; funding acquisition, D.M. and J.M. All authors have read and agreed to the published version of the manuscript.

Funding: This research was funded by the German Research Foundation (Deutsche Forschungsgemeinschaft, DFG) through the project A1 of the Collaborative Research Centre CRC 1261 'Magnetoelectric Sensors: From Composite Materials to Biomagnetic Diagnostics'. The chemical analysis was funded by Helmholtz-Zentrum Dresden-Rossendorf (HZDR) through the proposal 2002176.

Data Availability Statement: The data presented in this study are available on request from the corresponding author.

Acknowledgments: We thank Katrin Brandenburg and Sabrina Curt is for proof-reading of the manuscript. Also, the authors would like to thank DFG and HZDR for funding.

Conflicts of Interest: The authors declare no conflict of interest. The funders had no role in the design of the study; in the collection, analyses, or interpretation of data; in the writing of the manuscript, or in the decision to publish the results.

Appendix A

Table A1. Chronological comparison of selected SAW magnetic field sensors from 2018 to 2021 based on substrate and magnetic layer.

Source	Substrate	IDT	SAW Guiding/Top Layer	Magnetic Layer	Wave Form	Operating Frequency	Sensitivity
This work	ST-cut quartz x + 90°	10 nm Ta/200 nm Au/10 nm Cr	4.5 µm SiO$_2$	10 nm Ta/200 nm FeCoSiB/10 nm Ta	Love wave	144 MHz	1188°/m (2373 kHz/mT *)
Fahim 2021 [42]	ST-cut quartz (Murata RO3101)	Al	-	PVA + Ni/Fe nanopowder	-	433.92 MHz	5.83 kHz/mT
Mishra 2020 [43]	ST-cut quartz x + 90°	Al	400 nm ZnO	100 nm CoFeB	Love wave	421 MHz	0.75 kHz/mT
Yang 2020 [44]	ST-cut quartz x + 90°	100 nm Al	333 nm ZnO/400 nm SiO$_2$	5 nm Ta/100 nm Co$_{40}$Fe$_{40}$B$_{20}$/5 nm Pt	Love wave	433 MHz	−170.4 kHz/mT
Schell 2020 [8]	ST-cut quartz x + 90°	10 nm Cr/300 nm Au/C10 nm Cr	4.5 µm SiO$_2$	8 nm Ta/200 nm FeCoSiB/5 Ta	Love wave	147–148 MHz	2000°/mT (3999 kHz/mT *)
Mishra 2020 [27]	ST-cut quartz x + 90°	100 nm Al	510 nm ZnO	100 nm CoFeB	Love wave	433 MHz	15.53 kHz/mT
Jia 2020 [13]	128°YX LiNbO$_3$	300 nm Al	50 nm SiO$_2$	50 nm Cr/500 nm FeCo (dot array)	-	150 MHz	6 kHz/mT
Xiangli 2018 [4]	ST-cut quartz x + 90°	Ta	SiO$_2$	FeCoSiB	Love wave	221.76 MHz	663.98 kHz/mT
Kittman 2018 [5]	ST-cut quartz x + 90°	12 nm Cr/300 nm Au/12 nm Cr	4.5 µm SiO$_2$	10 nm Ta/200 nm FeCoSiB/10 nm Ta	Love wave	147.2 MHz	504°/mT (1008 kHz/mT *)

* The specification in kHz/mT represents an approximation. For data conversion a delay time of 1389 ns has been assumed. This delay time was reported for a similar sensor design as considered here [24].

References

1. Ruppel, C.W.; Dill, R.; Fischerauer, A.; Fischerauer, G.; Gawlik, A.; Machui, J.; Muller, F.; Reindl, L.; Ruile, W.; Scholl, G.; et al. SAW devices for consumer communication applications. *IEEE Trans. Ultrason. Ferroelectr. Freq. Control* **1993**, *40*, 438–452. [CrossRef]
2. Malave, A.; Schlecht, U.; Gronewold, T.M.A.; Perpeet, M.; Tewes, M. Lithium Tantalate Surface Acoustic Wave Sensors for Bio-Analytical Applications. In Proceedings of the 5th IEEE Conference on Sensors (IEEE Sensors 2006), Daegu, Korea, 22–25 October 2006; pp. 604–607. [CrossRef]
3. Schlensog, M.D.; Gronewold, T.M.A.; Tewes, M.; Famulok, M.; Quandt, E. A Love-wave biosensor using nucleic acids as ligands. *Sens. Actuators B Chem.* **2004**, *101*, 308–315. [CrossRef]
4. Yang, Y.; Mengue, P.; Mishra, H.; Floer, C.; Hage-Ali, S.; Petit-Watelot, S.; Lacour, D.; Hehn, M.; Han, T.; Elmazria, O. Wireless Multifunctional Surface Acoustic Wave Sensor for Magnetic Field and Temperature Monitoring. *Adv. Mater. Technol.* **2021**, 2100860. [CrossRef]
5. Kittmann, A.; Durdaut, P.; Zabel, S.; Reermann, J.; Schmalz, J.; Spetzler, B.; Meyners, D.; Sun, N.X.; McCord, J.; Gerken, M.; et al. Wide Band Low Noise Love Wave Magnetic Field Sensor System. *Sci. Rep.* **2018**, *8*, 278. [CrossRef]
6. Liu, X.; Tong, B.; Ou-Yang, J.; Yang, X.; Chen, S.; Zhang, Y.; Zhu, B. Self-biased vector magnetic sensor based on a Love-type surface acoustic wave resonator. *Appl. Phys. Lett.* **2018**, *113*, 82402. [CrossRef]
7. Polewczyk, V.; Dumesnil, K.; Lacour, D.; Moutaouekkil, M.; Mjahed, H.; Tiercelin, N.; Petit Watelot, S.; Mishra, H.; Dusch, Y.; Hage-Ali, S.; et al. Unipolar and Bipolar High-Magnetic-Field Sensors Based on Surface Acoustic Wave Resonators. *Phys. Rev. Appl.* **2017**, *8*, 024001. [CrossRef]

8. Schell, V.; Müller, C.; Durdaut, P.; Kittmann, A.; Thormählen, L.; Lofink, F.; Meyners, D.; Höft, M.; McCord, J.; Quandt, E. Magnetic anisotropy controlled FeCoSiB thin films for surface acoustic wave magnetic field sensors. *Appl. Phys. Lett.* **2020**, *116*, 73503. [CrossRef]
9. Kittmann, A.; Müller, C.; Durdaut, P.; Thormählen, L.; Schell, V.; Niekiel, F.; Lofink, F.; Meyners, D.; Knöchel, R.; Höft, M.; et al. Sensitivity and noise analysis of SAW magnetic field sensors with varied magnetostrictive layer thicknesses. *Sens. Actuators A Phys.* **2020**, *311*, 111998. [CrossRef]
10. Kadota, M.; Ito, S. Sensitivity of Surface Acoustic Wave Magnetic Sensors Composed of Various Ni Electrode Structures. *Jpn. J. Appl. Phys.* **2012**, *51*, 07GC21. [CrossRef]
11. Li, W.; Dhagat, P.; Jander, A. Surface Acoustic Wave Magnetic Sensor using Galfenol Thin Film. *IEEE Trans. Magn.* **2012**, *48*, 4100–4102. [CrossRef]
12. Mishra, H.; Hehn, M.; Hage-Ali, S.; Petit-Watelot, S.; Mengue, P.W.; Zghoon, S.; M'Jahed, H.; Lacour, D.; Elmazria, O. Microstructured Multilayered Surface-Acoustic-Wave Device for Multifunctional Sensing. *Phys. Rev. Appl.* **2020**, *14*, 014053. [CrossRef]
13. Jia, Y.; Wang, W.; Sun, Y.; Liu, M.; Xue, X.; Liang, Y.; Du, Z.; Luo, J. Fatigue Characteristics of Magnetostrictive Thin-Film Coated Surface Acoustic Wave Devices for Sensing Magnetic Field. *IEEE Access* **2020**, *8*, 38347–38354. [CrossRef]
14. Mishra, H.; Hehn, M.; Lacour, D.; Elmazria, O.; Tiercelin, N.; Mjahed, H.; Dumesnil, K.; Petit Watelot, S.; Polewczyk, V.; Talbi, A.; et al. Intrinsic versus shape anisotropy in micro-structured magnetostrictive thin films for magnetic surface acoustic wave sensors. *Smart Mater. Struct.* **2019**, *28*, 12LT01. [CrossRef]
15. McCord, J.; Schäfer, R.; Frommberger, M.; Glasmachers, S.; Quandt, E. Stress-induced remagnetization in magnetostrictive films. *J. Appl. Phys.* **2004**, *95*, 6861–6863. [CrossRef]
16. Peng, B.; Zhang, W.L.; Zhang, W.X.; Jiang, H.C.; Yang, S.Q. Effects of stress on the magnetic properties of the amorphous magnetic films. *Phys. B Condens. Matter* **2006**, *382*, 135–139. [CrossRef]
17. O'Handley, R.C. *Modern Magnetic Materials: Principles and Applications*; Wiley: New York, NY, USA, 2000; ISBN 0-471-15566-7.
18. Fiorillo, F. *Measurement and Characterization of Magnetic Materials*; Elsevier: Amsterdam, The Netherlands, 2004; ISBN 978-0-12-257251-7.
19. Wang, Y.; Nastasi, M.A. (Eds.) *Handbook of Modern Ion Beam Materials Analysis*, 2nd ed.; Materials Research Soc.: Warrendale, PA, USA, 2009; ISBN 978-1-60511-215-2.
20. Kreissig, U.; Grigull, S.; Lange, K.; Nitzsche, P.; Schmidt, B. In situ ERDA studies of ion drift processes during anodic bonding of alkali-borosilicate glass to metal. *Nucl. Instrum. Methods Phys. Res. Sect. B Beam Interact. Mater. At.* **1998**, *136-138*, 674–679. [CrossRef]
21. Barradas, N.P.; Jeynes, C.; Webb, R.P. Simulated annealing analysis of Rutherford backscattering data. *Appl. Phys. Lett.* **1998**, *71*, 291. [CrossRef]
22. Janssen, G.C.A.M.; Abdalla, M.M.; van Keulen, F.; Pujada, B.R.; van Venrooy, B. Celebrating the 100th anniversary of the Stoney equation for film stress: Developments from polycrystalline steel strips to single crystal silicon wafers. *Thin Solid Film.* **2009**, *517*, 1858–1867. [CrossRef]
23. Piorra, A.; Jahns, R.; Teliban, I.; Gugat, J.L.; Gerken, M.; Knöchel, R.; Quandt, E. Magnetoelectric thin film composites with interdigital electrodes. *Appl. Phys. Lett.* **2013**, *103*, 32902. [CrossRef]
24. Durdaut, P.; Kittmann, A.; Rubiola, E.; Friedt, J.-M.; Quandt, E.; Knochel, R.; Hoft, M. Noise Analysis and Comparison of Phase- and Frequency-Detecting Readout Systems: Application to SAW Delay Line Magnetic Field Sensor. *IEEE Sens. J.* **2019**, *19*, 8000–8008. [CrossRef]
25. Devkota, J.; Ohodnicki, P.R.; Greve, D.W. SAW Sensors for Chemical Vapors and Gases. *Sensors* **2017**, *17*, 801. [CrossRef] [PubMed]
26. Wang, W.; Jia, Y.; Xue, X.; Liang, Y.; Du, Z. Grating-patterned FeCo coated surface acoustic wave device for sensing magnetic field. *AIP Adv.* **2018**, *8*, 15134. [CrossRef]
27. Mishra, H.; Streque, J.; Hehn, M.; Mengue, P.; M'Jahed, H.; Lacour, D.; Dumesnil, K.; Petit-Watelot, S.; Zhgoon, S.; Polewczyk, V.; et al. Temperature compensated magnetic field sensor based on love waves. *Smart Mater. Struct.* **2020**, *29*, 45036. [CrossRef]
28. McCord, J. Progress in magnetic domain observation by advanced magneto-optical microscopy. *J. Phys. D Appl. Phys.* **2015**, *48*, 333001. [CrossRef]
29. Lee, S.L.; Windover, D.; Lu, T.-M.; Audino, M. In situ phase evolution study in magnetron sputtered tantalum thin films. *Thin Solid Film.* **2002**, *420–421*, 287–294. [CrossRef]
30. Thornton, J.A.; Hoffman, D.W. Stress-related effects in thin films. *Thin Solid Film.* **1989**, *171*, 5–31. [CrossRef]
31. Hubert, A.; Schäfer, R. *Magnetic Domains: The Analysis of Magnetic Microstructures*; Corr. Print., [Nachdr.]; Springer: Berlin, Germany, 2011; ISBN 978-3-540-64108-7.
32. Durdaut, P.; Müller, C.; Kittmann, A.; Schell, V.; Bahr, A.; Quandt, E.; Knöchel, R.; Höft, M.; McCord, J. Phase Noise of SAW Delay Line Magnetic Field Sensors. *Sensors* **2021**, *21*, 5631. [CrossRef]
33. Urs, N.O.; Golubeva, E.; Röbisch, V.; Toxvaerd, S.; Deldar, S.; Knöchel, R.; Höft, M.; Quandt, E.; Meyners, D.; McCord, J. Direct Link between Specific Magnetic Domain Activities and Magnetic Noise in Modulated Magnetoelectric Sensors. *Phys. Rev. Appl.* **2020**, *13*, 28. [CrossRef]
34. Urs, N.O.; Teliban, I.; Piorra, A.; Knöchel, R.; Quandt, E.; McCord, J. Origin of hysteretic magnetoelastic behavior in magnetoelectric 2-2 composites. *Appl. Phys. Lett.* **2014**, *105*, 202406. [CrossRef]

35. Ludwig, A.; Quandt, E. Optimization of the ΔE effect in thin films and multilayers by magnetic field annealing. *IEEE Trans. Magn.* **2002**, *38*, 2829–2831. [CrossRef]
36. O'Handley, R.; Mendelsohn, L.; Nesbitt, E. New non-magnetostrictive metallic glass. *IEEE Trans. Magn.* **1976**, *12*, 942–944. [CrossRef]
37. McCord, J.; Schäfer, R.; Mattheis, R.; Barholz, K.-U. Kerr observations of asymmetric magnetization reversal processes in CoFe/IrMn bilayer systems. *J. Appl. Phys.* **2003**, *93*, 5491–5497. [CrossRef]
38. Hubert, A.; Schäfer, R. *Magnetic Domains: The Analysis of Magnetic Microstructures*; Softcover Reprint of the Hardcover 1st ed. 1998, Corrected Print. 2000; Springer: Berlin, Germany, 2014; ISBN 978-3-540-85054-0.
39. Lage, E.; Kirchhof, C.; Hrkac, V.; Kienle, L.; Jahns, R.; Knöchel, R.; Quandt, E.; Meyners, D. Exchange biasing of magnetoelectric composites. *Nat. Mater.* **2012**, *11*, 523–529. [CrossRef] [PubMed]
40. Röbisch, V.; Salzer, S.; Urs, N.O.; Reermann, J.; Yarar, E.; Piorra, A.; Kirchhof, C.; Lage, E.; Höft, M.; Schmidt, G.U.; et al. Pushing the detection limit of thin film magnetoelectric heterostructures. *J. Mater. Res.* **2017**, *32*, 1009–1019. [CrossRef]
41. Jovičević Klug, M.; Thormählen, L.; Röbisch, V.; Toxværd, S.D.; Höft, M.; Knöchel, R.; Quandt, E.; Meyners, D.; McCord, J. Antiparallel exchange biased multilayers for low magnetic noise magnetic field sensors. *Appl. Phys. Lett.* **2019**, *114*, 192410. [CrossRef]
42. Fahim; Mainuddin; Rajput, P.; Kumar, J.; Nimal, A.T. A simple and novel SAW magnetic sensor with PVA bound magnetostrictive nanopowder film. *Sens. Actuators A Phys.* **2021**, *331*, 112926. [CrossRef]
43. Mishra, H.; Hehn, M.; Hage-Ali, S.; Petit-Watelot, S.; Mengue, P.W.; M'Jahed, H.; Lacour, D.; Elmazria, O.; Zghoon, S. Multifunctional sensor (Magnetic field and temperature) based on Micro-structured and multilayered SAW device. In Proceedings of the 2020 IEEE International Ultrasonics Symposium (IUS), Las Vegas, NV, USA, 7–11 September 2020; pp. 1–4, ISBN 978-1-7281-5448-0.
44. Yang, Y.; Mishra, H.; Mengue, P.; Hage-Ali, S.; Petit-Watelot, S.; Lacour, D.; Hehn, M.; M'Jahed, H.; Han, T.; Elmazria, O. Enhanced Performance Love Wave Magnetic Field Sensors With Temperature Compensation. *IEEE Sens. J.* **2020**, *20*, 11292–11301. [CrossRef]

Communication

Thin-Film-Based SAW Magnetic Field Sensors

Jana Marie Meyer [1,*], Viktor Schell [2], Jingxiang Su [1], Simon Fichtner [1,2], Erdem Yarar [1], Florian Niekiel [1], Thorsten Giese [1], Anne Kittmann [2], Lars Thormählen [2], Vadim Lebedev [3], Stefan Moench [3], Agnė Žukauskaitė [3], Eckhard Quandt [2] and Fabian Lofink [1]

1. Fraunhofer Institute for Silicon Technology ISIT, Fraunhoferstrasse 1, 25524 Itzehoe, Germany
2. Institute for Materials Science, Kiel University, Kaiserstraße 2, 24143 Kiel, Germany
3. Fraunhofer Institute for Applied Solid State Physics IAF, Tullastrasse 72, 79108 Freiburg, Germany
* Correspondence: jana.meyer@isit.fraunhofer.de

Abstract: In this work, the first surface acoustic-wave-based magnetic field sensor using thin-film AlScN as piezoelectric material deposited on a silicon substrate is presented. The fabrication is based on standard semiconductor technology. The acoustically active area consists of an AlScN layer that can be excited with interdigital transducers, a smoothing SiO_2 layer, and a magnetostrictive FeCoSiB film. The detection limit of this sensor is 2.4 nT/$\sqrt{Hz}$ at 10 Hz and 72 pT/$\sqrt{Hz}$ at 10 kHz at an input power of 20 dBm. The dynamic range was found to span from about ±1.7 mT to the corresponding limit of detection, leading to an interval of about 8 orders of magnitude. Fabrication, achieved sensitivity, and noise floor of the sensors are presented.

Keywords: surface acoustic waves; surface acoustic wave sensor; magnetic field sensor; current sensor; magnetostriction; AlScN; FeCoSiB; MEMS; thin film

1. Introduction

The sensing of magnetic fields has a multitude of use cases ranging from biomedical applications to current sensing in automotive applications [1–5], each having different requirements on the sensor regarding bandwidth, dynamic range, dc capability, size, and price [6,7].

A promising sensing principle of magnetic fields is based on surface acoustic waves (SAW) [8] and the change of the Young's modulus (ΔE effect) of magnetostrictive films [9]. This differs from other sensor approaches, such as using a magnetoelectric composite cantilever suffering from disadvantages such as a small bandwidth, and a good LOD that can only be achieved in resonance [10].

Today, for the fabrication of SAW sensors, the use of piezoelectric single-crystal substrates such as quartz [1,11,12] or $LiNbO_3$ [13,14] is state-of-the-art. For ST-cut quartz sensors, sensitivities of up to 2000°/mT and a limit of detection of 100 pT/$Hz^{1/2}$ at 10 Hz are reached [15] and, for $LiNbO_3$, a variation of the SAW velocity of $\Delta v/v$ = 0.27% at 400 mT [14]. To enable a greater material flexibility, especially in terms of compatibility with CMOS and MEMS technology, a reduction in chip size and the use of cleverly designed multilayers to enhance device performance requires a change to thin-film technology.

For that purpose, thin-film AlN is a promising piezoelectric material due to its high wave velocity, good mechanical and dielectric properties, high thermal conductivity, and high breakdown voltage [16]. Additionally, a SAW sensor operation up to several GHz can be realized, which can significantly increase the sensitivity in many sensor applications [17]. Further, it was shown that alloying AlN with Sc improves the electromechanical coupling significantly without losing the attractive material properties of AlN [18]. The electromechanical coupling in AlScN even increases with increasing frequency, so that its use is particularly interesting for high SAW frequencies [19,20]. The Sc concentration adds an additional parameter for tuning crucial properties of SAW devices, such as the phase velocity and the electromechanical coupling [20]. AlScN as a promising thin-film material

for SAW sensors, as it is described in [21], is studied in this work, and it can be fabricated at reasonable cost with standard semiconductor technology on larger wafer sizes and an easier process integration compared to bulk piezoelectric wafers.

For SAW devices, two common design approaches exist: a delay line, and a resonator configuration [22,23]. In this work, the delay line configuration is chosen, as shown in Figure 1, to increase the interaction volume between the excited wave and the magnetic field sensitive area (magnetostrictive film). For this purpose, two inter-digital transducers (IDTs) are structured on the acoustic layer (thin film AlScN) to excite and readout the SAW signal via the piezoelectric effect [24]. The delay line of length l is located between the two IDTs. To prevent a short circuit between the magnetostrictive film and the IDTs and, more importantly, to reduce the roughness of the underlaying layer of the magnetostrictive film, a SiO$_2$ layer is grown on top of the piezoelectric layer. The topmost layer in this area of the delay line is the magnetostrictive material FeCoSiB. The acoustic wave passing through the delay line couples to an external magnetic field via the induced change in the Young's modulus of the magnetostrictive film [9].

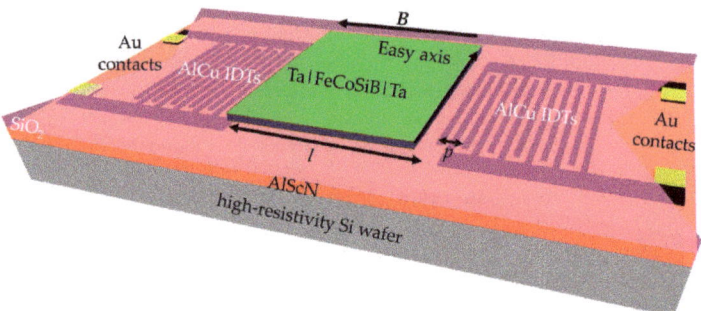

Figure 1. Schematic sketch of the SAW thin-film magnetic field sensor. Magnetostrictive FeCoSiB on top of the silicon dioxide layer of length l is in between the AlCu IDTs. FeCoSiB is capped with Ta to avoid corrosion. The easy axis of the magnetostrictive film defines the sensitive direction of the sensor against an external applied magnetic field **B** and is chosen to be perpendicular to the direction of SAW propagation. As the piezoelectric material, AlScN is chosen.

As the change of Young's modulus ΔE alters the phase velocity v of the acoustic wave [25], a **B**-field-induced phase change $\Delta \varphi = 2\pi\, l\, f/(v(B_1) - v(B_0))$ can be detected at the output IDTs via the direct piezoelectric effect [26]. The sensitivity of the sensor, which is defined as the phase change per change in magnetic field $S = \partial \varphi / \partial H$, can be written as the product of its individual contributions: magnetic layer sensitivity S_{mag} (change in Young's modulus with magnetic field), structural sensitivity S_{str} (change of the wave velocity with change in Young's modulus), and geometric sensitivity S_{geo} (phase change with change in wave velocity) [1]:

$$S = \frac{\partial \varphi}{\partial H} = \frac{\partial G}{\partial H} \cdot \frac{\partial v}{\partial G} \cdot \frac{\partial \varphi}{\partial v} = S_{mag} \cdot S_{str} \cdot S_{geo} \qquad (1)$$

By means of S and the power spectral density S_φ of the random phase fluctuations of the sensor, the limit of detection (LOD) of the sensor can be calculated by [27]:

$$\mathrm{LOD} = \frac{\sqrt{S_\varphi}}{S} \qquad (2)$$

The logarithmic presentation of the power spectral density $10 \log_{10}(S_\varphi)$ is referred to as phase noise.

2. Materials and Methods

2.1. Sensor Fabrication

On a 200 mm, 725 µm thick, single-side polished high-resistivity Si (001) wafer, a 1 µm $Al_{0.77}Sc_{0.23}N$ layer is sputtered as described in [28].

Afterwards, 200 nm thick AlCu IDTs are sputtered and patterned by dry chloride etching to a design with a delay line length of $l = 3.8$ mm, a split-finger structure [29] of 25 pairs, a periodicity of $p = 16$ µm, and a finger width of 2 µm, resulting in a theoretical phase velocity of the Rayleigh-like mode of 283 MHz (Figure 2(1)). Three-hundred-nanometer-thick gold contacts with a 40 nm WTi adhesion layer are sputter-deposited and structured with a wet etching step (Figure 2(2)). A 1.5 µm thick, low-stress SiO_2 interlayer is deposited with plasma-enhanced chemical vapor deposition (PECVD) at 400 °C and smoothed and thinned with a chemical mechanical polishing (CMP) step to a thickness of 1 µm. An atomic force microscopy analysis showed that this reduces the surface roughness from 2 nm of the AlScN layer to a roughness of below 1 nm. Such a reduction significantly enhances the soft magnetic properties of the FeCoSiB thin film [30]. Afterwards, the layer is structured with dry etching (Figure 2(3)).

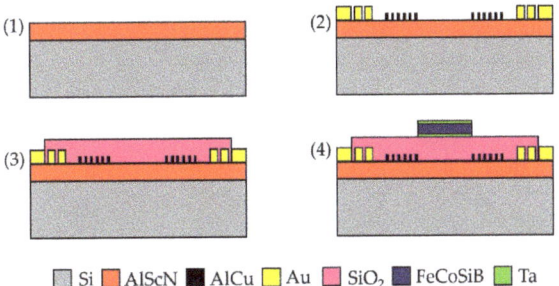

Figure 2. Schematic cross-sections of the processing steps of the thin film SAW sensor. (**1**) A layer of 1 µm AlScN is sputter-deposited on top of a high-resistance silicon (001) wafer, followed by 200 nm AlCu IDTs and 300 nm gold contacts with a 40 nm WTi adhesion layer that are patterned afterwards (**2**). A 1.5 µm SiO_2 layer is deposited via PECVD and thinned with CMP to a thickness of 1 µm (**3**). On top, the magnetostrictive FeCoSiB film with a thickness of 200 nm is deposited with an additional layer of 10 nm Ta on the top and on the bottom (**4**).

The magnetostrictive layer consisting of 200 nm $(Fe_{90}Co_{10})_{78}Si_{12}B_{10}$ is deposited via RF magnetron sputtering on top of the SiO_2 layer and structured with ion beam etching to realize steep and straight edges. To improve adhesion and prevent oxidation, 10 nm Ta is deposited on top and below the FeCoSiB film (Figure 2(4)). To induce a uniaxial magnetic anisotropy in the soft magnetic film, an annealing step at 250 °C for 30 min is performed while applying a magnetic field of 0.2 T. Thereby, the easy axis of the FeCoSiB film is aligned perpendicular to the SAW propagation direction (see Figure 1). The simple process of thermal alignment of the magnetization is an example of the integration-related advantages of the silicon substrate-based thin-film concept. When using single crystal piezoelectric substrates, such a simple thermal imprint is not possible due to anisotropic thermal expansion in the piezoelectric substrate and would result in a significant reduction of the soft magnetic film properties. Instead, a more complex, low-temperature deposition with an applied magnetic field must be applied to achieve a proper alignment of the magnetization [26]. Finally, the sensor is glued and wire-bonded on top of a printed circuit board (PCB), on which there are balun devices to symmetrize the signal. The final sensor is shown in the inset of Figure 3.

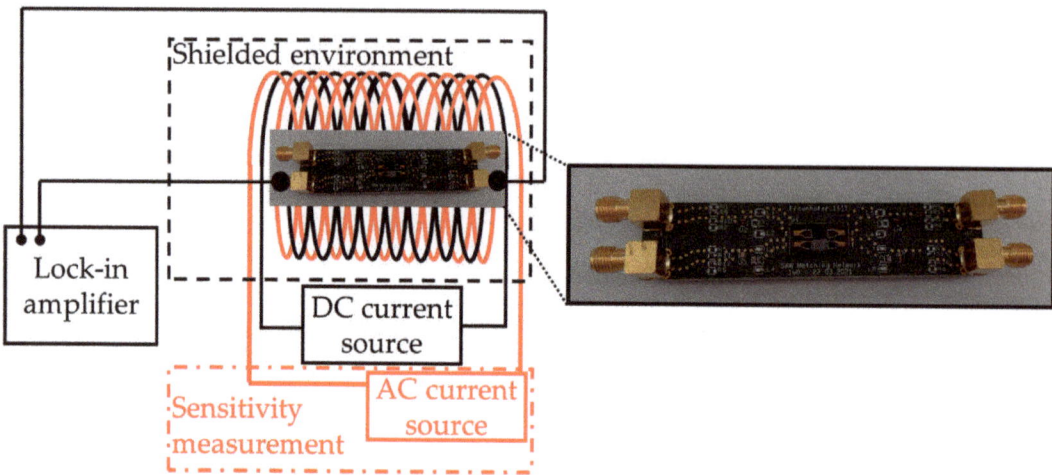

Figure 3. Sketch of the measurement setup. The SAW sensor is placed in a magnetically, electrically, and acoustically shielded measurement chamber inside of two solenoids. For the phase shift measurements, only a dc current source is used to apply a homogeneous magnetic field. A lock-in amplifier is used to apply the synchronous SAW frequency and measure the phase change. The sensitivity S is measured by applying an additional ac magnetic field using the second solenoid that is supplied with another current source using a test amplitude and frequency. The inset shows a zoom-in of the ready-to-use sensor with a balun attached to symmetrize the signal.

2.2. Experimental Setup

The sensor's two-port scattering parameters (S-parameters) are characterized with a vector network analyzer E8361A from Agilent Technologies. A signal power of $p = 0$ dBm is used throughout the experiments in this paper, except for the noise measurements. For all measurements in a magnetic field, the sensor is placed in the center of two axially stacked coils, which are used to generate ac and dc magnetic signals by means of a programmable current source (KEPCO BOP20-10ML) for the dc magnetic field. The solenoids are placed inside a magnetically, electrically, and acoustically shielded measurement chamber. The magnetic field shielding is provided by a mu-metal cylinder ZG1 from Aaronia AG to prevent external influences. The magnetically induced phase shift of the sensor is measured in the homogeneous magnetic field region of the solenoids. To then record the sensor behavior in the magnetic field, the magnetic flux density is swept from negative to positive values and reversed. A lock-in amplifier (UHFLI from Zurich Instruments) is used to apply the synchronous SAW frequency determined by the measurement of the S-parameters and to measure the static phase response $\varphi(B)$ of the sensor at a chosen input power (here 0 dBm).

In order to determine the sensor's optimum working point with the highest sensitivity, the phase φ is analyzed as a function of a dc bias field H. In principle, a numerical calculation of the sensitivity $S = \partial \varphi / \partial H$ should be sufficient to determine the point of steepest slope, which refers to the point of highest sensitivity, but often small phase jumps related to domain wall movement can give the appearance of incorrectly high sensitivities. Therefore, a dynamic phase detection measurement is performed to accurately determine $\partial \varphi / \partial H$ at every single measurement point. For this, one solenoid generates the static magnetic bias field that is superimposed with an ac test-field generated with the second solenoid powered by a current source (Keithley 6221) with a defined amplitude of 10 μT and a frequency of 10 Hz. By choosing the amplitude of the ac field that is large enough, the phase fluctuations can be neglected. The sensor's output signal and the phase reference are fed into the UHFLI lock-in amplifier that is used as a phase demodulator. The phase sensitivity is obtained by the evaluation of the amplitude spectrum of the demodulated phase signal [26].

The phase noise measurements are performed with the Rohde & Schwarz FSWP phase noise analyzer at the sensor's magnetic working point and at magnetic saturation at a different sensor's input power. The SAW sensor is placed in the electrically, magnetically, and acoustically shielded chamber during these measurements. To minimize external noise sources, especially those appearing in common dc current sources, a battery-based current source controlled by a potentiometer is applied in series with the solenoids for the generation of the dc magnetic bias field. The internal generator of the phase noise analyzer excites the sensor at the synchronous SAW frequency determined in previous measurements. The LOD can be determined from the measured noise floor with Equation (2). A more detailed description of the measurement setup can be found in [27].

3. Sensor Characterization

A finite element method (FEM) analysis and spectra analysis were performed using COMSOL Multiphysics® software [31] based on the acoustic and electromagnetic parameters of the constituent layers (AlScN, SiO$_2$, FeCoSiB). The parameters for AlScN were taken from [21] and for FeCoSiB from [1]. The simulated admittance and the displacement of the SAW modes are shown in Figure 4a. A Rayleigh-like thin-film mode is simulated to be at 283 MHz with a high admittance and relatively high displacement that are defined on the surface with some energy losses in the direction of the Si substrate (see Figure 4b).

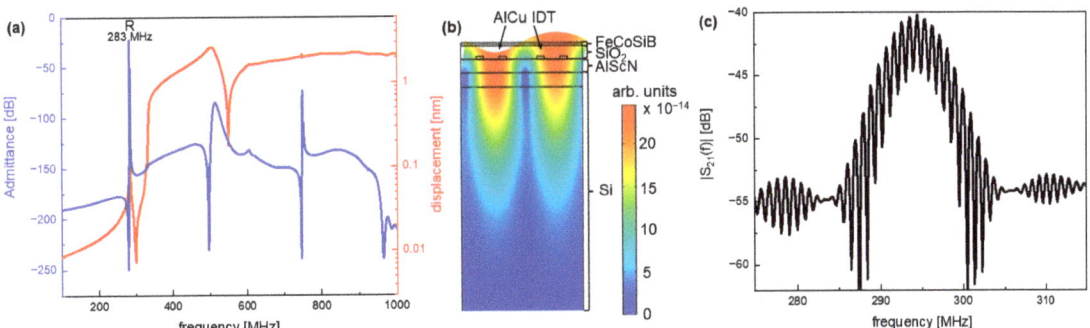

Figure 4. (a) FEM-simulated displacement (red) and admittance (blue) for the presented sensor design. (b) Colored map of absolute deflection for the Rayleigh-like mode at 283 MHz. The deflection into the FeCoSiB layer, the SiO$_2$ intermediate layer, the IDTs, the AlScN layer, and the Si substrate are displayed. (c) Measured transmission behavior (scattering parameter S$_{21}$) of the presented sensor. The synchronous frequency of the sensor is determined to be 294.2 MHz with a return loss of 40 dB.

The measured transmission behavior of the thin-film magnetic field sensor is shown in Figure 4c, exhibiting a synchronous frequency of 294.2 MHz at zero flux density, which is very close to the simulated value. The deviation can be explained by the material parameters in the simulation deviating from the experimental parameters in the real sensor, or imperfections in the fabrication, and is assessed as low.

The performance of the sensor in a magnetic field is measured as described above by applying the synchronous frequency of 294.2 MHz with the lock-in amplifier and measuring the static phase response of the sensor shown in Figure 5a from negative to positive field values (black) and reverse (grey). A slight hysteresis is observable, as was expected, resulting from the magnetic material. The linear region of the static phase response, which determines the dynamic range, is marked with a blue line in Figure 5a.

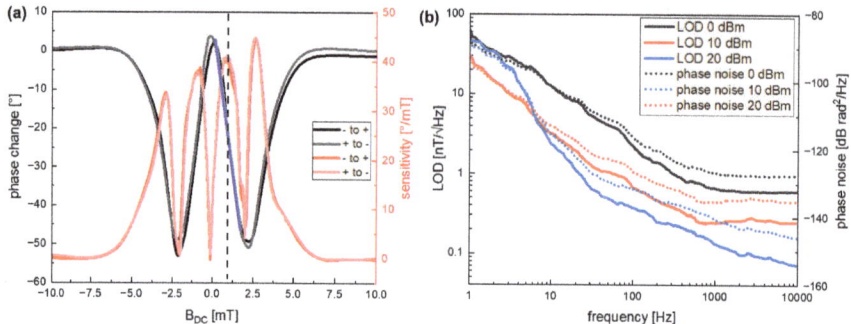

Figure 5. (a) The induced phase shift in the sensor with an external magnetic flux density (black) and the direct measurement of the sensitivity with an ac test signal (red) is shown. The highest slope in the phase change occurs at about 0.85 mT and 2.65 mT, resulting in the highest values of sensitivity of up to 45°/mT. A value of 0.85 mT is chosen as a working point (indicated with the dotted line) due to the high sensitivity and the lower field value compared to 2.65 mT. The dynamic range of the sensor is marked with the blue line, showing the linear region of the sensor. The ac signal has an amplitude of 10 µT and a frequency of 10 Hz. (b) Measured phase noise (dotted line) and calculated limit of detection (solid line) as a function of the frequency at magnetic saturation for 0 dBm (black), 10 dBm (red), and 20 dBm (blue) input power.

The sensitivity that is given by the derivative of the phase change as described above is important for the performance of the sensor. The dynamically measured sensitivity via an ac test signal depending on the bias field is shown in Figure 5a (red). The two regions with the maximum phase change are observed at about 0.85 mT and 2.65 mT (Figure 5a) and are the most interesting for sensor application. The highest sensitivity of about 45°/mT is reached at 2.65 mT.

Besides a high sensitivity, a good LOD is an important sensor parameter of the SAW sensor [15]. The LOD dependency on frequency from the carrier and sensor input power is shown in Figure 5b. The measurements are performed at the sensor's working point at H_{bias} = 0.85 mT, which is chosen due to the technological limitation of the LOD measurement setup and at magnetic saturation. As both measurements are almost identical, only the measurement at saturation is shown in Figure 5b.

Up to a frequency of 1 kHz, a regime of flicker (1/f) noise dominates the spectra. This noise is related to defects in the substrate and the SiO_2 layer, as well as random fluctuations of the magnetization and magnetic hysteresis losses [27]. In the specific case of SiO_2, additional surface roughness is introduced during the ion beam etching step to pattern the FeCoSiB layer. A possible way to reduce this roughness would be to add a lift-off process, though this would have the disadvantage of less defined edges of the magnetostrictive film.

In the 1/f noise regime, a LOD of 3.2 $nT/Hz^{1/2}$ is achieved at 10 Hz and an input power of 10 dBm. Above 1 kHz, the noise is dominated by white noise, which is additive noise and decreases with increasing signal power [27]. Here, a LOD of 246 $pT/Hz^{1/2}$ is reached for 10 kHz. When the input power is increased to 20 dBm, the LOD can be decreased even further at higher frequencies above 10 Hz so that a LOD of 2.4 $nT/Hz^{1/2}$ can be reached at 10 Hz and 72 $pT/Hz^{1/2}$ at 10 kHz.

The dynamic range of the sensor is given by the linear region around the working point of the sensor and is indicated in Figure 5a with the blue line. It spans from about 1.7 mT to the corresponding LOD, leading to an interval of 8 orders of magnitude. The hysteretic behavior could be compensated as is done in AMR sensors (anisotropic magnetoresistance) with controlled current pulses [32].

With these characteristics, our sensor has already high potential for sensing a wide range of technically relevant electrical currents via the generated magnetic field. In contrast, other current sensor concepts, such as Hall sensors, are limited in the bandwidth in the needed dynamic range and cannot achieve the measurement of fast signals [33]. AMR sensors also have a limit of the bandwidth at 1 MHz and a low dynamic range [34]. The presented SAW sensor is dc-compatible with a moderately high bandwidth of about 1.2 MHz limited by the delay line length and a high dynamic range of about 8 orders of magnitude.

4. Conclusions

The first thin-film SAW magnetic field sensor using AlScN as piezoelectric material on a silicon substrate is presented. The limit of detection is 2.4 nT/Hz$^{1/2}$ at 10 Hz, which probably can be lowered further with an impedance matching, higher input power values, and further insights on the sensor design and noise sources. This will also have a high impact on the phase change and the resulting sensitivity of the sensors. At higher frequencies above 10 kHz, the LOD is found to be as low as 72 pT/Hz$^{1/2}$. Additionally, the magnetic layer can be optimized by an exchange bias [35] to eliminate the need for an external bias field. Due to the possibility to measure galvanically isolated values from dc up to MHz with a high dynamic of up to 8 orders of magnitude, the presented sensor is very interesting for a variety of modern measuring tasks, such as the control of modern power switches, where it is well suited for monolithic wafer-level integration circuits. This clearly sets it apart from the competition in the segment of galvanically isolated magnetic field sensors for power transformers in the field of electromobility, which include Hall sensors and AMR sensors. Thus, this sensor concept has the potential to manage the rising requirements on current sensors regarding bandwidth, dynamic range, precision, and compactness.

Author Contributions: Conceptualization, J.M.M., J.S., S.F., V.L., S.M., A.Ž. and F.L.; methodology, F.N., F.L., V.L. and J.M.M.; software, J.M.M., T.G. and F.N.; validation, J.M.M., V.S. and T.G.; formal analysis, J.M.M., S.F., V.L., S.M. and F.L.; investigation, J.M.M., V.S., T.G., V.L., S.M. and E.Y.; resources, A.Ž., A.K., L.T., E.Q. and F.L.; data curation, J.M.M.; writing—original draft preparation, J.M.M.; writing—review and editing, F.L., V.S., S.F., F.N., V.L., A.Ž. and S.M.; visualization, J.M.M.; supervision, F.L. and E.Q.; project administration, J.M.M., F.L., V.S., E.Q. and A.Ž.; funding acquisition, F.L., E.Q. and A.Ž. All authors have read and agreed to the published version of the manuscript.

Funding: This research was funded by the German Research Foundation (Deutsche Forschungsgemeinschaft, DFG) through the project A9 of the Collaborative Research Centre CRC 1261 'Magnetoelectric Sensors: From Composite Materials to Biomagnetic Diagnostics'. This work was partly funded by the BMBF under the project reference numbers 16FMD01K, 16FMD02, 16FMD03. Additionally, this work was supported by the Fraunhofer Internal Programs under Grand No. MAVO 840 173.

Institutional Review Board Statement: Not applicable.

Informed Consent Statement: Not applicable.

Data Availability Statement: The data presented in this study are available on reasonable request from the corresponding author.

Conflicts of Interest: The authors declare no conflict of interest. The funders had no role in the design of the study; in the collection, analyses, or interpretation of data; in the writing of the manuscript; or in the decision to publish the results.

References

1. Kittmann, A.; Durdaut, P.; Zabel, S.; Reermann, J.; Schmalz, J.; Spetzler, B.; Meyners, D.; Sun, N.X.; McCord, J.; Gerken, M.; et al. Wide band low noise love wave magnetic field sensor system. *Sci. Rep.* **2018**, *8*, 278. [CrossRef] [PubMed]
2. Lenz, J.; Edelstein, S. Magnetic sensors and their applications. *IEEE Sens. J.* **2006**, *6*, 631–649. [CrossRef]
3. Sternickel, K.; Braginski, A.I. Biomagnetism using SQUIDs: Status and perspectives. *Supercond. Sci. Technol.* **2006**, *19*, S160. [CrossRef]
4. Williamson, S.J.; Hoke, M. *Advances in Biomagnetism*; Springer Science & Business Media: Berlin, Germany, 2012.
5. Zuo, S.; Schmalz, J.; Ozden, M.O.; Gerken, M.; Su, J.; Niekiel, F.; Lofink, F.; Nazarpour, K.; Heidari, H. Ultrasensitive magnetoelectric sensing system for pico-tesla magnetomyography. *IEEE Trans. Biomed. Circuits Syst.* **2020**, *14*, 971–984. [CrossRef]
6. Ziegler, S.; Woodward, R.C.; Iu, H.H.C.; Borle, L.J. Current sensing techniques: A review. *IEEE Sens. J.* **2009**, *9*, 354–376. [CrossRef]
7. Labrenz, J.; Bahr, A.; Durdaut, P.; Höft, M.; Kittmann, A.; Schell, V.; Quandt, E. Frequency Response of SAW Delay Line Magnetic Field/Current Sensor. *IEEE Sens. Lett.* **2019**, *3*, 1500404. [CrossRef]
8. Webb, D.; Forester, D.; Ganguly, A.; Vittoria, C. Applications of amorphous magnetic-layers in surface-acoustic-wave devices. *IEEE Trans. Magn.* **1979**, *15*, 1410–1415. [CrossRef]

9. Ludwig, A.; Quandt, E. Optimization of the Delta E effect in thin films and multilayers by magnetic field annealing. *IEEE Trans. Magn.* **2002**, *38*, 2829–2831. [CrossRef]
10. Su, J.; Niekiel, F.; Fichtner, S.; Kirchhof, C.; Meyners, D.; Quandt, E.; Wagner, B.; Lofink, F. Frequency tunable resonant magnetoelectric sensors for the detection of weak magnetic field. *J. Micromech. Microeng.* **2020**, *30*, 075009. [CrossRef]
11. Liu, X.; Tong, B.; Ou-Yang, J.; Yang, X.; Chen, S.; Zhang, Y.; Zhu, B. Self-biased vector magnetic sensor based on a Love-type surface acoustic wave resonator. *Appl. Phys. Lett.* **2018**, *113*, 082402. [CrossRef]
12. Yokokawa, N.; Tanaka, S.; Fujii, T.; Inoue, M. Love-type surface-acoustic waves propagating in amorphous iron-boron films with multilayer structure. *J. Appl. Phys.* **1992**, *72*, 360–366. [CrossRef]
13. Ganguly, A.; Davis, K.; Webb, D.; Vittoria, C.; Forester, D. Magnetically tuned surface-acoustic-wave phase shifter. *Electron. Lett.* **1975**, *11*, 610–611. [CrossRef]
14. Yamaguchi, M.; Hashimoto, K.; Kogo, H.; Naoe, M. Variable SAW delay line using amorphous TbFe2 film. *IEEE Trans. Magn.* **1980**, *16*, 916–918. [CrossRef]
15. Kittmann, A.; Müller, C.; Durdaut, P.; Thormählen, L.; Schell, V.; Niekiel, F.; Lofink, F.; Meyners, D.; Knöchel, R.; Höft, M.; et al. Sensitivity and noise analysis of SAW magnetic field sensors with varied magnetostrictive layer thicknesses. *Sens. Actuators A Phys.* **2020**, *311*, 111998. [CrossRef]
16. Dubois, M.A.; Muralt, P. Properties of aluminum nitride thin films for piezoelectric transducers and microwave filter applications. *Appl. Phys. Lett.* **1999**, *74*, 3032–3034. [CrossRef]
17. Caliendo, C.; Imperatori, P. High-frequency, high-sensitivity acoustic sensor implemented on ALN/Si substrate. *Appl. Phys. Lett.* **2003**, *83*, 1641–1643. [CrossRef]
18. Fichtner, S.; Wolff, N.; Krishnamurthy, G.; Petraru, A.; Bohse, S.; Lofink, F.; Chemnitz, S.; Kohlstedt, H.; Kienle, L.; Wagner, B. Identifying and overcoming the interface originating c-axis instability in highly Sc enhanced AlN for piezoelectric microelectromechanical systems. *J. Appl. Phys.* **2017**, *122*, 035301. [CrossRef]
19. Ding, A. Surface Acoustic Wave Devices Based on C-Plane and A-Plane AlScN. Ph.D. Thesis, Albert-Ludwigs-Universität Freiburg, Freiburg im Breisgau, Germany, 2020.
20. Kurz, N.; Ding, A.; Urban, D.F.; Lu, Y.; Kirste, L.; Feil, N.M.; Žukauskaitė, A.; Ambacher, O. Experimental determination of the electro-acoustic properties of thin film AlScN using surface acoustic wave resonators. *J. Appl. Phys.* **2019**, *126*, 075106. [CrossRef]
21. Wang, W.; Mayrhofer, P.M.; He, X.; Gillinger, M.; Ye, Z.; Wang, X.; Bittner, A.; Schmid, U.; Luo, J. High performance AlScN thin film based surface acoustic wave devices with large electromechanical coupling coefficient. *Appl. Phys. Lett.* **2014**, *105*, 133502. [CrossRef]
22. Slobodnik, A.J. Surface acoustic waves and SAW materials. *Proc. IEEE* **1976**, *64*, 581–595. [CrossRef]
23. Bell, D.T.; Li, R.C. Surface-acoustic-wave resonators. *Proc. IEEE* **1976**, *64*, 711–721. [CrossRef]
24. White, R.M.; Voltmer, F.W. Direct piezoelectric coupling to surface elastic waves. *Appl. Phys. Lett.* **1965**, *7*, 314–316. [CrossRef]
25. Smole, P.; Ruile, W.; Korden, C.; Ludwig, A.; Quandt, E.; Krassnitzer, S.; Pongratz, P. Magnetically tunable SAW-resonator. In Proceedings of the IEEE International Frequency Control Symposium and PDA Exhibition Jointly with the 17th European Frequency and Time Forum, Tampa, FL, USA, 4–8 May 2003; pp. 903–906.
26. Schell, V.; Müller, C.; Durdaut, P.; Kittmann, A.; Thormählen, L.; Lofink, F.; Meyners, D.; Höft, M.; McCord, J.; Quandt, E. Magnetic anisotropy controlled FeCoSiB thin films for surface acoustic wave magnetic field sensors. *Appl. Phys. Lett.* **2020**, *116*, 073503. [CrossRef]
27. Durdaut, P.; Müller, C.; Kittmann, A.; Schell, V.; Bahr, A.; Quandt, E.; Knöchel, R.; Höft, M.; McCord, J. Phase Noise of SAW Delay Line Magnetic Field Sensors. *Sensors* **2021**, *21*, 5631. [CrossRef]
28. Lu, Y.; Reusch, M.; Kurz, N.; Ding, A.; Christoph, T.; Kirste, L.; Lebedev, V.; Žukauskaitė, A. Surface morphology and microstructure of pulsed DC magnetron sputtered piezoelectric AlN and AlScN thin films. *Phys. Status Solidi A* **2018**, *215*, 1700559. [CrossRef]
29. Holland, M.G.; Claiborne, L.T. Practical surface acoustic wave devices. *Proc. IEEE* **1974**, *62*, 582–611. [CrossRef]
30. Piorra, A.; Jahns, R.; Teliban, I.; Gugat, J.; Gerken, M.; Knochel, R.; Quandt, E. Magnetoelectric thin film composites with interdigital electrodes. *Appl. Phys. Lett.* **2013**, *103*, 032902. [CrossRef]
31. COMSOL Multiphysics®®v. 5.6. COMSOL AB, Stockholm, Sweden. Available online: https://www.comsol.com (accessed on 11 November 2020).
32. Xie, F.; Weiss, R.; Weigel, R. Hysteresis compensation based on controlled current pulses for magnetoresistive sensors. *IEEE Trans. Ind. Electron.* **2015**, *62*, 7804–7809. [CrossRef]
33. Popovic, R.S.; Randjelovic, Z.; Manic, D. Integrated Hall-effect magnetic sensors. *Sens. Actuators A Phys.* **2001**, *91*, 46–50. [CrossRef]
34. Ripka, P. Electric current sensors: A review. *Meas. Sci. Technol.* **2010**, *21*, 112001. [CrossRef]
35. Spetzler, B.; Bald, C.; Durdaut, P.; Reermann, J.; Kirchhof, C.; Teplyuk, A.; Meyners, D.; Quandt, E.; Höft, M.; Schmidt, G.; et al. Exchange biased delta-E effect enables the detection of low frequency pT magnetic fields with simultaneous localization. *Sci. Rep.* **2021**, *11*, 5269. [CrossRef] [PubMed]

Article

Modeling and Parallel Operation of Exchange-Biased Delta-E Effect Magnetometers for Sensor Arrays

Benjamin Spetzler *, Patrick Wiegand, Phillip Durdaut, Michael Höft, Andreas Bahr, Robert Rieger and Franz Faupel

Institute of Materials Science, Faculty of Engineering, Kiel University, 24143 Kiel, Germany; pw@tf.uni-kiel.de (P.W.); pd@tf.uni-kiel.de (P.D.); michael.hoeft@tf.uni-kiel.de (M.H.); andreas.bahr@tf.uni-kiel.de (A.B.); rri@tf.uni-kiel.de (R.R.); ff@tf.uni-kiel.de (F.F.)
* Correspondence: benjamin.spetzler@tu-ilmenau.de

Abstract: Recently, Delta-E effect magnetic field sensors based on exchange-biased magnetic multilayers have shown the potential of detecting low-frequency and small-amplitude magnetic fields. Their design is compatible with microelectromechanical system technology, potentially small, and therefore, suitable for arrays with a large number N of sensor elements. In this study, we explore the prospects and limitations for improving the detection limit by averaging the output of N sensor elements operated in parallel with a single oscillator and a single amplifier to avoid additional electronics and keep the setup compact. Measurements are performed on a two-element array of exchange-biased sensor elements to validate a signal and noise model. With the model, we estimate requirements and tolerances for sensor elements using larger N. It is found that the intrinsic noise of the sensor elements can be considered uncorrelated, and the signal amplitude is improved if the resonance frequencies differ by less than approximately half the bandwidth of the resonators. Under these conditions, the averaging results in a maximum improvement in the detection limit by a factor of $\sqrt{N}$. A maximum $N \approx 200$ exists, which depends on the read-out electronics and the sensor intrinsic noise. Overall, the results indicate that significant improvement in the limit of detection is possible, and a model is presented for optimizing the design of delta-E effect sensor arrays in the future.

Keywords: magnetometer; delta-E effect; sensor array; magnetoelectric; cantilever; exchange bias

Citation: Spetzler, B.; Wiegand, P.; Durdaut, P.; Höft, M.; Bahr, A.; Rieger, R.; Faupel, F. Modeling and Parallel Operation of Exchange-Biased Delta-E Effect Magnetometers for Sensor Arrays. *Sensors* **2021**, *21*, 7594. https://doi.org/10.3390/s21227594

Academic Editor: Iren E. Kuznetsova

Received: 9 October 2021
Accepted: 14 November 2021
Published: 16 November 2021

Publisher's Note: MDPI stays neutral with regard to jurisdictional claims in published maps and institutional affiliations.

Copyright: © 2021 by the authors. Licensee MDPI, Basel, Switzerland. This article is an open access article distributed under the terms and conditions of the Creative Commons Attribution (CC BY) license (https://creativecommons.org/licenses/by/4.0/).

1. Introduction

The detection of small amplitude magnetic fields is of interest for various fields of application, e.g., in magnetic recording, geomagnetism, and aerospace engineering [1]. Specific engineering and development challenges arise for biomedical applications, such as cell and particle mapping [2,3], magnetomyography [4,5], or magnetocardiography [6–9]. Such applications are often connected to inverse solution problems that benefit from large arrays with many sensor elements and the possibility of quick spatial field mapping [10,11]. Magnetic flux densities in this field of application are of the order of tens of picotesla and less [12] with frequency components often well below 1 kHz [5,13]. Therefore, research on sensor systems for biomedical applications is devoted to improving the minimum detectable field at low frequencies while minimizing critical parameters such as size, power consumption, and cost [13].

The gold standard for detecting such small magnetic fields is superconducting quantum interference device (SQUID) magnetometry [14,15]. These sensors must be cooled and magnetically well-shielded during operation, which makes them expensive and extensive to operate. Such setups are limited in the number of sensor elements and their minimum distance to the magnetic source. Atomic magnetometers [16–18] have been investigated as an affordable alternative to SQUIDs and have achieved limits of detection (LOD) in the fT/$\sqrt{\text{Hz}}$ regime at low signal frequencies between 1–200 Hz [16]. Despite this progress, atomic magnetometers often require magnetic shielding, and their limited CMOS integrability and downsizing reduce the number and density of sensor elements that can be used in array applications. Miniaturization and MEMS fabrication of atomic magnetometers is currently an active field of research [19,20]. Many magnetometers are being investigated to overcome such limitations [4,15], and an overview and comparison of magnetic field sensors for biomedical applications can be found [13].

In this work, we study magnetic field sensors based on magnetoelectric composite resonators. Previously, sensor systems utilizing the direct magnetoelectric effect were discussed for magnetocardiography [21] and magnetomyography [5], and limits of detection in the low and sub-pT/$\sqrt{Hz}$ regime have been reached with a linear response over several orders of magnitude [22]. Magnetoelectric sensors can be produced on a large scale with standard micro-electro-mechanical system (MEMS) technology and dimensions down to the micrometer range [23]. They are potentially cost-efficient, feature low power consumption, and are integrable with CMOS electronics. These aspects make magnetoelectric sensors promising candidates for sensor arrays. On the other hand, detecting small magnetic flux densities is limited to a narrow bandwidth of a few hertz around the resonance frequency, which is usually in the kilohertz regime for millimeter-sized resonators or the megahertz regime for micrometer-sized devices. Such high and narrow frequency regimes are not suitable for many applications [21]. Shifting them down increases the contributions of 1/f noise and requires large resonators with low resonance frequencies, which are susceptible to mechanical vibrations and reduce the spatial resolution.

Delta-E effect magnetometers extend the measurement range of magnetoelectric sensors and shift it to low frequencies while avoiding 1/f noise and keeping the advantages of magnetoelectric composites and the MEMS fabrication technology. In contrast to sensors based on the direct magnetoelectric effect, delta-E effect sensors benefit from high resonance frequencies because they operate on a modulation scheme. The higher resonance frequencies permit miniaturization and render the devices robust against mechanical disturbances. The modulation occurs via the magnetoelastic delta-E effect [24–26], i.e., the magnetization induced change of the stiffness of the material, which leads to a detuning of the resonance frequency upon the application of a magnetic field. This detuning can be measured as a change of the electrical admittance of the sensor and causes a modulation of the current through the sensor [27]. Although precursor steps towards the delta-E effect sensor concept were already made in the 1990s [28], it took another two decades until fully integrable delta-E effect sensors [29] were developed based on microelectromechanical magnetoelectric composite cantilevers [26,30–34], plate resonators [35,36], or other designs [37], including macroscopic laminate structures [38,39]. MEMS cantilever sensors achieved LODs < 100 pT/$\sqrt{Hz}$ in the frequency range from approximately 10–100 Hz [32]. This is currently of a similar order of magnitude as the LODs of some magnetoresistive sensors [40,41]. As an application example of delta-E effect sensors, magnetic particle mapping was recently demonstrated for cell localization [42]. In this setup, the sensor was operated under a magnetic bias field provided by a permanent magnet. Most studies rely on an external magnetic bias field to operate the sensor at an optimum signal-to-noise ratio. Instead of a permanent magnet, the magnetic bias field is often created with external coils. For delta-E effect sensor arrays with many sensor elements, coils and permanent magnets can be inconvenient because their stray fields shift the operation points of adjacent sensor elements, and the additional electrical components increase the volume of the sensor system.

Recently, we demonstrated a first delta-E effect magnetometer based on exchange biased magnetic multilayers that circumvents such complications and still achieves a minimum detection limit of 350pT/$\sqrt{Hz}$ at 25 Hz [34]. The exchange bias provides an internal bias field and thereby paves the way to flexible and compact delta-E effect sensor arrays with many sensor elements. Only recently were sensor arrays based on magnetoelectric sensor elements reported [43–47], and were limited to direct magnetoelectric detection and were mostly based on macroscopic resonators. A CMOS integrated array of magnetoelastic sensor elements was presented for vector magnetometry [48], but the sensor elements were only characterized individually and without a signal and noise analysis. No attempts of parallel operating delta-E effect sensor elements in array configurations or thorough signal and noise analyses of such have yet been presented.

In this study, we explore the operation of delta-E effect sensor elements in arrays to improve the signal, noise, and limit of detection. Instead of measuring the magnetic field at different locations, spatial resolution can be sacrificed by averaging the outputs of several sensors operating simultaneously. However, the large number of hardware channels required to achieve the desired improvement in the LOD increases the size of the setup and limits the number and density of sensors. As a solution, sensor elements are connected in parallel and operated and read out simultaneously with one set of electronics. This method of parallel operation is accompanied by other complications, and they are analyzed here to identify the potential of such a setup. After presenting the sensor system, which is based on exchange-biased delta-E effect sensors, a signal-and-noise model is developed and validated with measurements. The model is used to analyze the sensor characteristics as functions of the number of sensor elements and variations in the resonance frequency that can occur during fabrication. Implications for the design of delta-E effect sensor arrays are derived and requirements on the reproducibility are identified and discussed.

2. Sensor System

In this study, two MEMS fabricated sensor elements are used, based on 50 μm thick poly-Si cantilevers with a length of 3 mm and a width of 1 mm. They are covered with a 4 μm thick exchange-biased magnetic multilayer [49] and a 2 μm thick piezoelectric AlN layer [50] on the top. The AlN layer is contacted via two Ta-Pt electrodes on its top and rear-side for excitation and read-out. The magnetic multilayer is based on alternating anti-ferromagnetic $Mn_{70}Ir_{30}$ (8 nm) and soft ferromagnetic $Fe_{70.2}Co_{7.8}Si_{12}B_{10}$ (200 nm) layers. In this configuration, the antiferromagnetic layer provides an exchange bias that serves as an internal bias field for the ferromagnetic layer to ensure a nonzero sensitivity without an externally applied magnetic field. Hence, all measurements shown in this study are performed without a magnetic bias field. Details about the layer structure and fabrication process and a comprehensive analysis of sensors with a similar geometry can be found elsewhere [34]. Two sensor elements are mounted on a printed circuit board, respectively, as shown in Figure 1. They are connected in parallel to each other and connected to the input of a low-noise JFET-based charge amplifier [51]. A high-resolution A/D and D/A converter (*Fireface UFX+*, *RME*, Chemnitz, Germany) is used for excitation and read-out (24 bit, 32 kHz). For the measurements, the sensors are placed in a magnetically and electrically shielded setup [52], based on a mu-metal shielding cylinder (ZG1, *Aaronia AG*, Strickscheid, Germany), and are located in a copper fleece coated box that is mechanically decoupled to reduce the impact of mechanical vibrations. All magnetic flux densities are applied along the long axis of the cantilever.

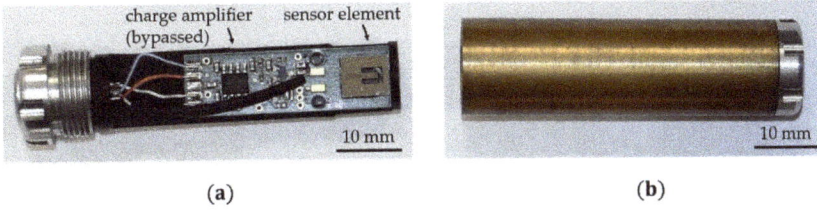

Figure 1. (a) Example sensor (without encapsulation) used in this study; it comprises a MEMS-fabricated cantilever resonator as a sensing element mounted on a printed circuit board (PCB). The JFET charge amplifier on the PCB was used in a previous study and is bypassed here and replaced by an external one. (b) Brass encapsulation for mechanical protection and electrical shielding during the measurements. Further details are reported in Ref. [34].

3. Array Modeling

In an alternating magnetic field, the delta-E effect causes an oscillation of the mechanical stiffness of the cantilever. The response of the cantilever to this stiffness change is damped with increasing magnetic field frequencies because of its mass inertia. Previously, this behavior was modeled with a first-order Bessel filter [27,53], applied to the demodulated simulated output signal of the charge amplifier. Later, a dynamic sensitivity was introduced [54] to consider the low-pass filter characteristics of the sensor as a function of the magnetic field frequency. The dynamic sensitivity was derived from the frequency response of a simple damped harmonic oscillator; however, it is only fully valid if the sensor is excited at its mechanical resonance frequency. For many previously analyzed sensors [33,53,55], this approximation was well justified, as their resonance frequency was close to their optimum excitation frequency, i.e., the excitation frequency with the largest signal-to-noise ratio. This is not a general property of such sensors but depends on their geometry, material, and electrical capacitance. Significant quantitative and qualitative deviations between measurements and simulations can occur if the excitation is not in mechanical resonance [56] (p. 139). In an array, not all sensor elements can be excited in mechanical resonance because of variations in their resonance frequencies that occur during fabrication. In this section, a signal and noise model is developed based on an altered approach, and it permits describing the output signal of an array of N sensor elements excited at an arbitrary excitation frequency.

3.1. Signal Model

During operation of the sensor array, a sinusoidal voltage $u_{ex}(t)$ with amplitude $\hat{u}_{ex}$ and frequency f_{ex} is applied. It excites the magnetoelectric resonators of each sensor element at, or close to, its respective resonance frequency $f_{r,n}$. In linear approximation, the voltage at the charge amplifier's output can be described by:

$$u_{co}(t) \approx -Z_f(f_{ex}) \cdot i_s(t). \tag{1}$$

In this equation, the time is denoted by t and the feedback impedance of the charge amplifier by Z_f. The current i_s through the array of parallel-connected sensor elements can be expressed as the sum of all individual currents $i_{s,n}$ that flow through the respective sensor element n. To describe $i_{s,n}$, we use a modified Butterworth-van Dyke (mBvD) equivalent circuit representation, illustrated in detail in Appendix A. It consists of a series resonant circuit with a resistance $R_{r,n}$, inductance $L_{r,n}$, and capacitance $C_{r,n}$ that consider the resonant behavior of the cantilever. The electrodes of each sensor element form a capacitor with the piezoelectric layer. It is described by a capacitance $C_{p,n}$ and resistance $R_{p,n}$, both in parallel to the series LCR-circuit. Further, the current $i_{s,n}$ can be separated into a current $i_{r,n}$, which passes through the resonant LCR circuit, and a current $i_{p,n}$, which passes through the parallel pathway. A sketch of the circuit model is provided in Figure A1 (Appendix A). For N parallel-connected sensor elements, i_s can be described by:

$$i_s(t) = \sum_{n=1}^{N} i_{s,n} = \sum_{n=1}^{N} \left(i_{p,n} + i_{r,n} \right). \tag{2}$$

The current $i_{p,n}$ is described as a function of the magnitude $|Y_{p,n}|$ and the phase angle $\phi_{p,n}\text{angle}\{Y_{p,n}\}$ of the electrical admittance $Y_{p,n}$ of the parallel pathway, and results in:

$$i_{p,n} = \hat{u}_{ex} \cdot |Y_{p,n}(f_{ex})| \cdot \cos\left(2\pi f_{ex} t + \phi_{p,n}(f_{ex})\right). \tag{3}$$

This current is independent of the magnetic field, and the corresponding electrical admittance $Y_{p,n}(f) = R_{p,n}^{-1} + 2\pi f C_{p,n}$ as a function of frequency f is entirely determined by the capacitance $C_{p,n}$ of the respective piezoelectric layer-electrode configuration and its resistance $R_{p,n}$. Similarly, the current $i_{r,n}$ can be described as a function of the magnitude $|Y_{r,n}|$ and the phase angle $\phi_{r,n}\text{angle}\{Y_{r,n}\}$ of the magnetic-field and frequency-dependent admittance $Y_{r,n}$ of the resonant circuit of a sensor element n. The current $i_{r,n}$ is filtered in the time domain to consider the frequency response of the resonator. We use a second-order digital peaking (resonator) filter with a rational transfer function $\mathcal{H}_r$ that is determined by the resonance frequency f_r and the quality factor Q (Appendix B). It is given by:

$$i_{r,n} = \mathcal{H}_r\{\hat{u}_{ex} \cdot |Y_{r,n}(f_{ex}, B, t)| \cdot \cos(2\pi f_{ex} t + \phi_{r,n}(f_{ex}, B, t))\}. \tag{4}$$

In contrast to $i_{p,n}$, the resonant current $i_{r,n}$ depends on the magnetic flux density $B = B_0 + B_{ac}(t)$, which can be expressed as a static magnetic flux density B_0, superposed by a small, time t dependent contribution $B_{ac}(t)$. For small amplitudes $\hat{B}_{ac}$ of $B_{ac}(t)$, the admittance $Y_{r,n}(f, B)$ around B_0 and at $f = f_{ex}$ can be approximated by a first-order Taylor series:

$$|Y_{r,n}(f_{ex}, B, t)| \approx |Y_{r,n}(f_r, B_0)| + \left.\frac{d|Y_{r,n}(f, B_0)|}{df}\right|_{f=f_{ex}} \left.\frac{df_{r,n}(B)}{dB}\right|_{B=B_0} \cdot B_{ac}(t), \tag{5}$$

and

$$\phi_{r,n}(f_{ex}, B, t) \approx \phi_{r,n}(f_r, B_0) + \left.\frac{d\phi_{r,n}(f, B_0)}{df}\right|_{f=f_{ex}} \left.\frac{df_{r,n}}{dB}\right|_{B=B_0} \cdot B_{ac}(t). \tag{6}$$

Because the damping of the carrier relative to its maximum value at $f_{ex} = f_r$ is already considered by $\mathcal{H}_r$, the zero-order element in the series expansion is taken at $f = f_r$ instead of $f = f_{ex}$. If not stated differently, we always use $B_0 = 0$ because the exchange bias sensors used here do not require an externally applied magnetic bias field.

3.2. Definition of Sensitivities

The derivatives in the previous two equations describe the influence of the magnetic field on the electrical admittance and can be referred to as sensitivities. A magnetic sensitivity can be defined as:

$$S_{mag,n} = \left.\frac{df_{r,n}}{dB}\right|_{B=B_0}, \tag{7}$$

and two electrical sensitivities $S_{el,am,n}$ and $S_{el,pm,n}$ as:

$$S_{el,am,n} = \left.\frac{d|Y_{r,n}(f, B_0)|}{df}\right|_{f=f_{ex}}, \quad S_{el,pm,n} = \left.\frac{d\phi_{r,n}(f, B_0)}{df}\right|_{f=f_{ex}}. \tag{8}$$

These definitions of electrical sensitivities differ from previous work [26,53,57], which is further discussed at the end of this section. A normalization, as in Refs. [26,57], is still required to compare the electric and magnetic sensitivity of sensors with different resonance frequencies. After amplification by the charge amplifier, the output signal $u_{co}(t)$ is fed into a quadrature amplitude demodulator to obtain the demodulated signal $u(t)$. The amplitude spectrum $\hat{U}(f)$ of $u(t)$ can then be used to define the voltage sensitivity $S_V(f)$ as a normalized measure for the sensor's signal response:

$$S_V(f_{ac}) = \frac{\hat{U}(f_{ac})}{\hat{B}_{ac}} \text{ with } [S_V] = \frac{V}{T}. \tag{9}$$

The voltage sensitivity $S_V(f_{ac})$ can be estimated by applying a sinusoidal magnetic test signal $B_{ac} = \hat{B}_{ac} \sin(2\pi f_{ac} t)$, with well-defined amplitude $\hat{B}_{ac}$ and frequency f_{ac}, to obtain $U(f_{ac})$ from the measurement. With $S_V(f_{ac})$, a measure for the smallest detectable magnetic field can be defined. This measure is frequently referred to as limit of detection (LOD) [22,27], equivalent magnetic noise [58,59], or detectivity [40]:

$$\text{LOD}(f_{ac}) = \frac{E(f_{ac})}{S_V(f_{ac})} \text{ with } [\text{LOD}] = \frac{T}{\sqrt{Hz}}, \tag{10}$$

where $E(f_{ac})$ is the voltage noise density of $u(t)$ at f_{ac}, after demodulation and measured without any magnetic field applied. The response of Delta-E effect magnetometers to magnetic fields depends on the mutual orientation of the magnetic field and sensor element. Consequently, the sensitivities and LOD (Equations (7)–(10)) are generally functions of the orientation of the magnetic field. Details about the signal-and-noise characterization of ΔE-effect magnetometers can be found elsewhere [27,53,54].

The definitions of the electrical sensitivities (Equation (8)) differ from previous formulations [26,53,57] limited to the special case of one sensor element excited in resonance ($f_{ex} = f_r$). The electrical sensitivities defined within those models use the total sensor admittance $Y_s Y_r + Y_p$ instead of Y_r to form the derivatives with respect to the frequency. Here, the parallel admittance Y_p is considered in the total sensor current i_s. This definition arises naturally from separating the sensor current into the resonator current and the current though the capacitor, and it is used to consider the response of the resonator to the alternating magnetic field.

3.3. Noise Model

In the following, we modify and extend the model presented in Ref. [54] to analyze the noise of the array sensor system and how it is influenced by adding parallel sensor elements. Additional sensor elements are considered and minor noise sources, e.g., of the cables, are omitted. The equivalent circuit noise model is shown in Figure 2 and a summary of the parameters is given in Table A1 in Appendix C. The noise of the excitation source is described by a thermal-electrical noise source E_{ex} of the D/A converter's output resistance R_{ex} and the D/A converter's quantization noise E_{Vex}. Similarly, E_{AD} describes the noise that occurs during the analog-digital conversion. The noise source of the JFET charge amplifier is calculated from the model in [51] and is summarized in E_{JCA}. Each sensor element of the array is described by the mBvD equivalent circuit (Figure 2b). For the sensor intrinsic noise of the nth sensor element, we consider the thermal-electrical noise source $E_{p,n}$ of the piezoelectric layer and the thermal-mechanical noise source $E_{r,n}$ of the resonator. The value of the thermal-electrical noise sources can be calculated from:

$$E_x = \sqrt{4k_B T R_x} \text{ with } x \in \{ex, p, r\}, \tag{11}$$

with Boltzmann's constant $k_B = 1.38 \times 10^{-23}$ J/K and the temperature $T = 290$ K. The noise source E_{Vex} and E_{AD} were obtained from measurements. Here, we consider small excitation amplitudes $\hat{u}_{ex} < 100$ mV only and obtain $E_{Vex} = 26.8$ nV/$\sqrt{\text{Hz}}$ and $E_{AD} = 6.9$ nV/$\sqrt{\text{Hz}}$ in this case. Each noise source is transformed to the output of the charge amplifier to analyze its contribution to the total noise density at the charge amplifier's output. To simplify the final expressions, the following impedances are defined. The impedance $Z_{s,n}$ of the nth sensor is obtained from:

$$Z_{s,n} = Z_{r,n} \| Z_{p,n}, \tag{12}$$

$$Z_{r,n} = R_{r,n} + j\omega L_{r,n} + \frac{1}{j\omega C_{r,n}}, \tag{13}$$

$$Z_{p,n} = \frac{1}{j\omega C_{p,n}} \| R_{p,n}, \tag{14}$$

where || denotes the parallel operator (a || b = $(a^{-1} + b^{-1})^{-1}$).

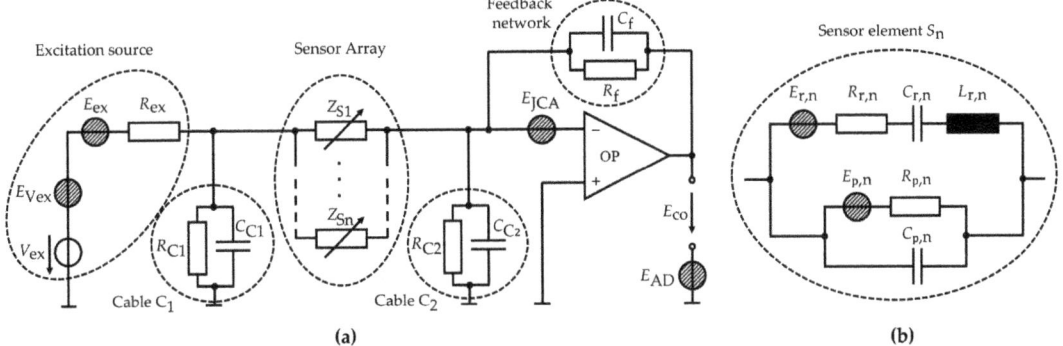

Figure 2. (a) Noise model of the sensor system comprising an excitation source, charge amplifier, and N sensor elements S_n with impedances Z_{Sn}, connected in parallel. (b) Equivalent circuit noise model of each sensor element S_n, with resonator intrinsic noise source $E_{r,n}$ and piezoelectric noise source $E_{p,n}$.

The total impedance of all N sensor elements connected in parallel is:

$$Z_s = \left[\sum_{n=1}^{N} Z_{s,n}^{-1}\right]^{-1}. \quad (15)$$

The impedance Z_{C2} of the cable with capacitance C_{C2} and resistance R_{C2} between the sensor elements and the charge amplifier is given by:

$$Z_{C2} = \frac{1}{j\omega C_{C2}} \| R_{C2}, \quad (16)$$

and the feedback impedance of the charge amplifier by:

$$Z_f = \frac{1}{j\omega C_f} \| R_f, \quad (17)$$

with the capacitance C_f and the resistance R_f. The total voltage noise density at the output of the charge amplifier is obtained from the superposition of the individual output referred noise sources,

$$E_{co}^2 = E_{co,JCA}^2 + E_{co,Vex}^2 + E_{co,AD}^2 + E_{co,s}^2, \quad (18)$$

$E_{co,JCA}$ of the charge amplifier, $E_{co,Vex}$ of the D/A converter, $E_{co,AD}$ of the A/D converter, and the contribution $E_{co,s}$ of the parallel sensor elements. These noise contributions are given by:

$$E_{co,JCA}^2 = E_{JCA}^2 \left|1 + \frac{Z_f}{Z_s + Z_{C2}}\right|^2, \quad (19)$$

$$E_{co,Vex}^2 = E_{Vex}^2 \left|\frac{Z_f}{Z_s}\right|^2, \quad (20)$$

$$E_{co,AD}^2 = E_{AD}^2, \quad (21)$$

$$E_{co,s}^2 = \sum_{n=1}^{N} \left(E_{r,n}^2 \left|\frac{Z_f}{Z_{r,n}}\right| + E_{p,n}^2 \left|\frac{Z_f}{R_{p,n}}\right|\right) E_{co,r}^2 + E_{co,p}^2. \quad (22)$$

4. Characterization and Validation of the Signal-and-Noise Model

In this section, the sensor elements and the array are characterized regarding their impedance, signal, and noise as well as their frequency response. The measurements are compared with simulations to demonstrate the validity of the model. Details on the implementation of the model are given in Appendix D.

4.1. Electrical Sensitivity and Admittance Characterization

To eventually compare simulations with measurements, the admittance of the sensor elements is characterized. Measurements of the admittance magnitude $|Y_s|$ as functions of frequency f are shown in Figure 3a (top) for the two sensor elements S_1, S_2, and both connected in parallel ($S_1 || S_2$). All measurements were made at $B_0 = 0$ and an excitation voltage amplitude of $\hat{u}_{ex} = 10$ mV. An mBvD equivalent circuit as described earlier and illustrated in Appendix A is fitted to the magnitude data. The parameters obtained from the fit are given in Table A1 in Appendix C. From the mBvD parameters, we obtain resonance frequencies of $f_{r,1} = 7674.9$ Hz and $f_{r,2} = 7676.5$ Hz and quality factors of $Q_1 = 642$ and $Q_2 = 558$ (equations in Appendix A). This results in resonator bandwidths $f_{BW,n} \approx f_{r,n}/Q_n$ of $f_{BW,1} \approx 12$ Hz and $f_{BW,2} \approx 14$ Hz. Hence, the difference in resonance frequencies $\Delta f_r = |f_{r,2} - f_{r,1}| = 1.6$ Hz is significantly smaller than the bandwidth of the sensor elements.

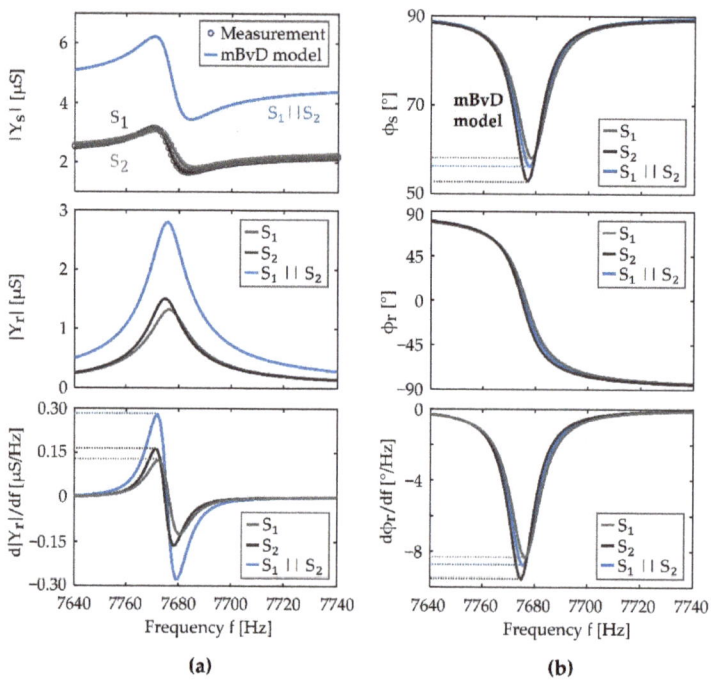

Figure 3. (a) Top: magnitudes $|Y_s|$ of the admittance of the sensor elements S_1, S_2 and both connected in parallel ($S_1 || S_2$) measured at an applied magnetic flux density of $B = 0$ and an excitation voltage amplitude of $\hat{u}_{ex} = 10$ mV, compared with a modified Butterworth-van Dyke (mBvD) equivalent circuit fit; middle: magnitude $|Y_r|$ of the electrical admittance of the LCR series circuit of the mBvD model; bottom: derivative of $|Y_r|$ with respect to the frequency f, which we refer to as electrical amplitude sensitivity. (b) Top: corresponding phase angles ϕ_s of the sensor elements obtained from the mBvD model; middle: phase angles ϕ_r of the admittance of the LCR series circuits; bottom: their derivates, which we refer to as electrical phase sensitivities.

With the mBvD parameters, the phase angle $\phi_{s,n}$ is calculated and plotted in Figure 3b (top). It shows the typical minimum of an electromechanical resonator that is caused by the superposition of the current through the resonator and the current through the parallel capcitance $C_{p,n}$ and resistance $R_{p,n}$. The values of $Y_{s,n}$ and $\phi_{s,n}$ are similar to other electromechanical resonators that have been operated as delta-E effect sensors (e.g., [32,34,54]). Hence, the chosen sensor elements are representative examples. The admittance magnitude $|Y_{r,n}|$ and phase angle $\phi_{r,n}$ of the series resonance circuit are obtained from the mBvD model by omitting the parallel current $i_{p,n}$ and are plotted in Figure 3a,b (middle). They exhibit the behavior expected from a linear resonator and the main difference between the two sensor elements is the small difference in their resonance frequencies. The electrical sensitivities $S_{el,am,n}$ and $S_{el,pm,n}$ are calculated following the definitions in Equation (8) from the derivatives of $|Y_{r,n}|$ and $\phi_{r,n}$ with respect to the frequency. They are plotted in Figure 3a,b (bottom). Both sensor elements have similar electrical sensitivities with extrema of $S_{el,am,1}^{max} \approx S_{el,am,2}^{max} \approx \pm 0.15\ \mu\text{S/Hz}$ and $S_{el,pm,1}^{max} \approx S_{el,pm,2}^{max} \approx -8.5/\text{Hz}$. Note that $S_{el,am,n} = 0$ at $f_{ex} = f_{r,n}$, but $S_{el,pm,n} = S_{el,pm,n}^{max}$. Because the two sensor elements have very similar resonance frequencies, their total electrical admittance $Y_s = Y_{s,1} + Y_{s,2}$ in parallel connection ($S_1||S_2$) shows qualitatively the same behavior but with a much increased admittance magnitude and electrical amplitude sensitivity by approximately a factor of two compared to the single sensor elements. The corresponding plots are shown in Figure 3. Comparing the magnitude and phase of $Y_s(S_1||S_2)$ emphasizes that an improvement in the sensitivity is only expected for the electrical amplitude sensitivity $S_{el,am}$, because the magnitudes $|Y_{r,n}|$ add up. In contrast, the electrical phase sensitivity $S_{el,pm}$ does not improve, as it results from averaging $S_{el,pm,1}$ and $S_{el,pm,2}$. For a more comprehensive and general discussion, signal and noise must be considered and, in particular, their dependencies on the magnetic field frequency and the differences in resonance frequency. For that, the signal model is validated in the following section.

4.2. Frequency Response of the Sensor

The electrical sensitivities and sensor parameters found in the previous section are used here in the signal model and the simulations are compared with measurements. In Figure 4a, the spectrum $\hat{U}_{co}$ of the modulated signal is shown from a measurement of the sensor element S_1 (top) and $S_1||S_2$ (bottom) using an excitation signal with a voltage amplitude of $\hat{u}_{ex} = 25$ mV, a frequency $f_{ex} = 7680$ Hz and a sinusoidal magnetic test signal with an arbitrarily chosen frequency of $f_{ac} = 5.8$ Hz, and an amplitude of $\hat{B}_{ac} = 1\ \mu$T. Besides the carrier peak at f_{ex}, both spectra show one pair of peaks at $f_{ex} \pm f_{ac}$, which corresponds to the modulating signal caused by the magnetic field. Following the magnitude-frequency response of the transfer function of the resonator, the side peak closest to the resonance frequency at $f_{r,1} = 7674.9$ Hz (S_1) is the largest. The signal model fits the measurements very well for magnetic sensitivities of $S_{mag,1} \approx S_{mag,2} = 24$ Hz/mT. Considering the normalization required for a comparison [26,57], $S_{mag,n}/f_{r,n}$ is in the typical range expected from similar sensor elements [34,57].

Several $f_{ex} \neq f_r$ are chosen to analyze the sensor's magnitude-frequency response for operating out of resonance. They are indicated in Figure 4b for S_1 (top) and $S_1||S_2$ (bottom) as the difference $\Delta f_{ex,1}\ f_{ex} - f_{r,1}$ of f_{ex} to the resonance frequency $f_{r,1}$ of S_1, and the difference $\Delta f_{ex,2}\ f_{ex} - f_{r,2}$ of f_{ex} to $f_{r,2}$, respectively. For each excitation frequency, the voltage sensitivity S_V as a function of the magnetic field frequency f_{ac} was measured four times, averaged, and plotted in Figure 4c. As expected, the measurements of both configurations (S_1 and $S_1||S_2$) show qualitatively the same behavior. For excitation frequencies close to f_r, the sensor's voltage sensitivity S_V exhibits a low-pass behavior with a maximum voltage sensitivity at the lowest magnetic field frequency f_{ac}.

With an increasing deviation of f_{ex} from f_r, the maximum shifts to larger values of f_{ac} and further reduces its value. The reduction of the voltage sensitivity is caused by a change of the electrical sensitivities as well as the transfer function of the resonator. The model matches the measurements well with deviations mostly smaller than a factor of two and well within the estimated errors of the measurements. In line with the estimation based on the electrical sensitivity in the previous section, the simulations, and measurements of $S_1||S_2$, show an overall improved voltage sensitivity compared to S_1. A more detailed analysis of this is given in Section 5, where the model is used to estimate the influence of a resonance mismatch for otherwise identical sensor elements.

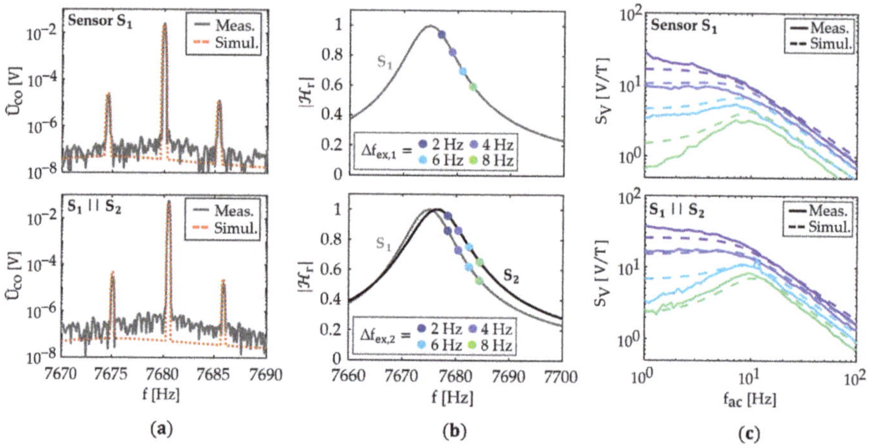

Figure 4. Comparison of simulations with measurements. (**a**) Example amplitude spectrum of the measured and simulated output signal using only the sensor element S_1 (top) and both sensor elements in parallel $S_1||S_2$ (bottom) ($\hat{u}_{ex} = 25$ mV). (**b**) Magnitude $|\mathcal{H}_r|$ of the transfer function $\mathcal{H}_r$ of the resonator used to indicate several excitation frequencies f_{ex} by $\Delta f_{ex,1}$ $f_{ex} - f_{r,1}$, relative to the resonance frequency $f_{r,1} = 7674.9$ Hz of the sensor element S_1 (top), and $\Delta f_{ex,2}$ $f_{ex} - f_{r,2}$, relative to the resonance frequency $f_{r,2} = 7676.5$ Hz of the sensor element S_2 (bottom). (**c**) Measured and simulated voltage sensitivity S_V (Equation (9)) as a function of the magnetic field frequency f_{ac} for the excitation frequencies indicated in (**b**) for the sensor element S_1 (top) and both sensor elements in parallel $S_1||S_2$ (bottom) ($\hat{u}_{ex} = 10$ mV).

4.3. Validation of the Noise Model

We omit the effect of $\hat{u}_{ex}$ on the quality factor and noise floor for the small $\hat{u}_{ex}$ used in this work, in line with previous investigations [53,54]. Noise measurements are performed for $\hat{u}_{ex} = 0$, i.e., the sensor's input is shortened to ground potential, and data are recorded for 5 min with a sample rate of 32 kHz. The measured noise density spectra are compared with the simulations in Figure 5. The contributions of the sensor intrinsic thermal-mechanical noise $E_{co,r}$, and piezoelectric thermal-electric noise $E_{co,p}$, as well as the operational amplifier's noise $E_{co,JCA}$, are shown. The measurements and simulations match well and show what is expected for no excitation, or small excitation amplitudes. Thermal-mechanical noise dominates the noise floor around the resonance frequency and, further away, thermal-electrical noise of the piezoelectric resistance. The maximum noise density peak in Figure 5 is increased by a factor of approximately 1.3 compared to the single sensor elements. This is slightly less than the maximum increase by a factor of $\sqrt{2}$ expected from Equation (22), and it is likely caused by the resonance mismatch.

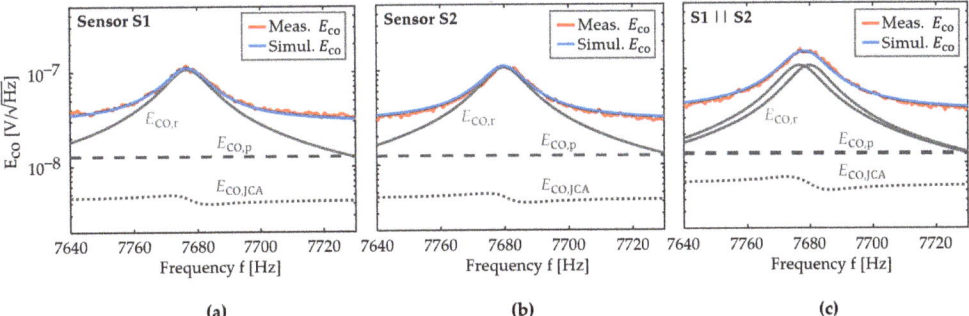

Figure 5. Comparison of the simulated and measured total voltage noise density E_{co} (Equation (18)) around the sensor's resonance frequency. The simulated contributions of the thermal-mechanical noise density $E_{co,r}$, the thermal-electrical noise density $E_{co,p}$ of the piezoelectric layer, and the operational amplifier's noise density $E_{co,JCA}$ are shown as well. Measurements and simulations are compared for a sensor system with (**a**) a single sensor element S_1, (**b**) a single sensor element S_2, and (**c**) two sensor elements connected in parallel ($S_1 \mid\mid S_2$).

5. Implications for Sensor Arrays

5.1. Influence of the Number of Sensor Elements

The noise model is used to estimate the influence of the number N of sensor elements on the minimum detectable magnetic flux density. First, we consider the ideal case of identical sensor elements described with the mBvD parameters of the sensor S_1. In this case, the signal magnitude increases linearly with N. The change of the total voltage noise density is less trivial because the various noise contributions depend differently on N. A simulation of the voltage noise density at the resonance frequency is shown in Figure 6 as a function of N. While the sensor intrinsic thermal-mechanical noise and thermal-electrical noise increase $\propto \sqrt{N}$, the noise of the JFET charge amplifier is $\propto N$. This linear relationship can be explained with the expression for the noise gain $|1 + Z_f/(Z_s + Z_{C2})|$ of the amplifier in Equation (19). According to this expression, the noise gain is in good approximation ($Z_s \gg Z_{C2}, 1$) inversely proportional to the impedance Z_s of the array. Each additional sensor element increases the capacitance and reduces Z_s (Equation (15)), and therefore, the noise gain increases linearly with N if all sensor elements are identical. The thermal-mechanical noise and the thermal-electrical noise of the amplifier dominate the noise floor at different N owing to their different dependencies on N. At small N, the thermal-mechanical noise dominates the noise floor and the LOD $\propto 1/\sqrt{N}$ can be improved by adding sensor elements. At large N, the noise contribution of the amplifier dominates and no improvement in the LOD can be achieved because signal and noise are both $\propto N$. A transition region exists at intermediate values of N where the improvement in LOD decreases continuously with N. This transition region is approximately around $N = 200 N_{max}$ for the set of sensor parameters considered.

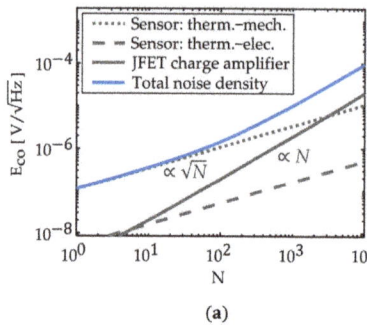 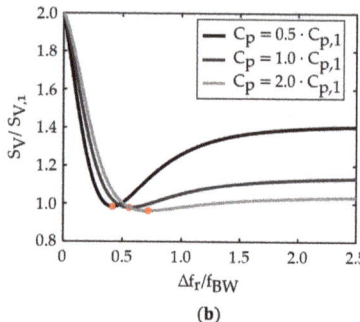

Figure 6. (a) Simulation of the voltage noise density at the resonance frequency as a function of the number N of the parallel connected sensor elements. Below approximately $N = 200$, the noise level is dominated by the sensor intrinsic thermal-mechanical noise and increases $\propto \sqrt{N}$. At approximately $N > N_{\max} \approx 200$, the noise of the charge amplifier dominates the noise floor and is $\propto N$. No significant improvement in the signal-to-noise ratio is expected for $N > N_{\max} \approx 200$ because the signal amplitude increases $\propto N$ as well. (b) Simulated gain $S_V/S_{V,1}$ in the voltage sensitivity S_V of a system with two parallel connected sensor elements relative to the voltage sensitivity $S_{V,1}$ of a single sensor element as a function of the difference Δf_r in their resonance frequencies, normalized to the bandwidth $f_{BW} f_r/Q$ of the resonator. Examples are shown for three different parallel capacities, expressed as multiples of the parallel capacitance $C_{p,1}$ of the sensor element S_1. The minima are indicated with red dots.

5.2. Influence of Resonance Frequency Mismatch

The reproducibility of sensor elements can be considerably impaired by the relaxation of small stress during fabrication [60] and by small variations in the resonator geometry. The condition $f_{ex} = f_r$ cannot be fulfilled simultaneously for all N sensor elements because both mechanisms cause a distribution in resonance frequency f_r. At this point, it remains unclear to what extent such a distribution impairs signal, noise, and LOD. However, knowing the tolerable variation in resonance frequency is important for the design of sensor arrays because it imposes limitations on the resonator geometry and on the tolerances of the fabrication process.

First, the voltage sensitivity S_V is calculated as a function of the resonance frequency mismatch because it must be known to estimate the LOD (Equation (10)). The model sensor system considered comprises two sensor elements connected in parallel. Both sensor elements have identical resonance frequencies and sensitivities, and they are described by the same set of mBvD equivalent circuit parameters of the sensor element S_1. The resonance frequency $f_r \propto 1/\sqrt{L_r C_r}$ of one sensor element is altered by increasing the mBvD parameters L_r and C_r in equal ratios. This keeps the quality factor $Q \propto L_r/C_r$ constant (Appendix A), and it causes only a negligible change in the bandwidth for the range of resonance frequencies tested. For each difference Δf_r in the two resonance frequencies, we simulate the output signal using a magnetic test signal with a frequency of $f_{ac} = 1$ Hz, and calculate the voltage sensitivity S_V (Equation (9)). This procedure is repeated for three different example capacities C_p with values that are multiples of the parallel capacitance $C_{p,1}$ of the sensor element S_1. We define the sensitivity gain $S_V/S_{V,1}$ by the voltage sensitivity S_V of the two parallel sensor elements, normalized to the voltage sensitivity $S_{V,1}$ of the single sensor element. The results are plotted in Figure 6b as functions of Δf_r normalized to the bandwidth $f_{BW} = f_r/Q$ of S_1.

For all three values of C_p, the sensitivity gain reaches a maximum value of $S_V/S_{V,1} = 2$, when the resonance frequencies are identical $\Delta f_r/f_{BW} = 0$. It decreases to a minimum of around $S_V/S_{V,1} = 1$ at roughly $\Delta f_r/f_{BW} = 0.5$, indicated in Figure 6b with red dots. For larger $\Delta f_r/f_{BW}$, $S_V/S_{V,1}$ increases slightly but it does not reach its maximum value again. The influence of the parallel capacitance C_p on $S_V/S_{V,1}$ is distinct but it does not change the curves qualitatively in the considered range. A larger C_p reduces $S_V/S_{V,1}$ at high $\Delta f_r/f_{BW}$ and shifts the location of the minimum to a larger $\Delta f_r/f_{BW}$; hence, it slightly broadens the peak around the maximum of $S_V/S_{V,1}$. Consequently, the condition of approximately $\Delta f_r/f_{BW} < 0.5$ (depending on C_p) should be fulfilled to increase the signal magnitude in an array with two sensor elements. This condition can be expressed as:

$$\Delta f_{r,BW} \frac{\Delta f_r}{f_{BW}} \approx \frac{\Delta f_r}{f_r} \cdot Q < 0.5. \tag{23}$$

The simulations in Figure 6b demonstrate that Equation (23) is not a strict criterium because the locations of the minima on the $\Delta f_{r,BW}$-axis vary by up to 50% for different tested values of C_p (e.g., for $C_p = 2C_{p,1}$ the minimum is at $\Delta f_{r,BW} \approx 0.75$). The exact location of the minimum depends on the contribution of the current through the LCR pathway to the total sensor current, relative to the capacitive contribution of C_p. For all practical purposes, these two contributions can hardly be varied independently because changing the parallel capacitance is accompanied by a change of the excitation efficiency, e.g., by altering the electrode geometry [32] or the piezoelectric material [33].

Not only for the voltage sensitivity, but also the sensor intrinsic thermal-mechanical noise referred to as the output follows the transfer function of the resonator; this is demonstrated with the measurements and simulations in Figure 5. Therefore, the LOD is constant if the sensor intrinsic thermal-mechanical noise dominates the noise floor, which is typically fulfilled for excitation frequencies f_{ex} within the bandwidth of the resonator and sufficiently small magnetic field frequencies f_{ac}. This conclusion is in line with other experimental results [55] (Figure 6) and does still hold for two parallel operating sensor elements with different resonance frequencies. Consequently, it is also $\text{LOD}(\Delta f_{r,BW}) = const.$ if the thermal-mechanical noise dominates the voltage noise density at $f_{ex} + f_{ac}$. The frequency band around the resonance frequency where the LOD is constant depends on the difference between the thermal-electric noise level and the resonance-amplified thermal-mechanical noise level and changes with N. For the sensors analyzed, this range is approximately $< f_{BW}$ around f_r, as shown in Figure 5.

6. Discussion and Conclusions

The signal-and-noise model developed matches well with measurements on exchange-biased sensor elements operated separately and in parallel in a setup with a single oscillator and amplifier. The model does still hold for excitation frequencies out of resonance and more than one sensor element. Hence, two major limitations of previous models have been solved and a tool is presented that can further support the design of delta-E effect sensors and sensor arrays. From the good match of the model and consistency with noise measurements, we find that the sensor intrinsic noise in our setup can be considered as uncorrelated, despite the parallel connection of sensor elements and their operation and read-out by a single oscillator and single amplifier. This is an essential precondition for improving the sensor performance by operating in parallel while using fewer electronic elements to keep the setup compact. Additional requirements were identified, which must be fulfilled to improve the signal and the limit of detection by operating many sensor elements in parallel. For the given sensor system, no significant improvement in the limit of detection can be achieved if a maximum number $N_{max} \approx 200$ of sensor elements is exceeded. Above this number, the noise contribution of the amplifier dominates the noise floor and increases, like the signal amplitude $\propto N$. Below, the sensor intrinsic noise dominates around the resonance and increases merely $\propto \sqrt{N}$, which results in $LOD \propto 1/\sqrt{N}$. With the given N_{max}, this would correspond to an LOD improvement by a factor of approximately 14. The value of N_{max} depends on the contribution of the thermal-mechanical noise relative to the thermal-mechanical noise. Therefore, N_{max} can potentially be improved by reducing the noise contribution of the charge amplifier. The proportionalities found do only hold strictly if all sensor elements are identical. Simulations of the voltage sensitivity confirmed that the improvement in signal amplitude depends significantly on the difference in the resonance frequencies of the sensor elements. It vanishes at a bandwidth normalized resonance frequency difference of approximately $\Delta f_{r,BW} \approx 0.5$, depending on the value of the parallel capacitance of the sensor element. Consequently, a large signal improvement by parallel operation requires tight tolerances on the resonance frequency and, therefore, on the reproducibility provided by the fabrication process. Because the sensor intrinsic noise follows the same resonator transfer function as the signal, we expect the LOD to be constant with $\Delta f_{r,BW}$ for sufficiently small $\Delta f_{r,BW}$, and here at approximately $\Delta f_{r,BW} < 2$. This value depends on the level of the thermal-mechanical noise, relative to the thermal-electrical noise of the piezoelectric layer and the noise contribution of the amplifier.

In conclusion, a model was presented that overcomes previous limitations and can be used to explore the signal and noise characteristics of delta-E effect sensor arrays. The results from measurements and simulations indicate that large arrays of parallel operating sensor elements can be an option to improve the signal and limit of detection in the future.

Author Contributions: Conceptualization, B.S.; methodology, B.S. and P.D.; formal analysis, B.S. and P.W.; investigation, B.S., P.W. and P.D.; data curation, B.S. and P.W.; writing—original draft preparation, B.S.; writing—review and editing, B.S., P.W., P.D., M.H., A.B., R.R. and F.F.; visualization, B.S.; project administration, M.H., A.B., R.R. and F.F.; funding acquisition, M.H., A.B., R.R. and F.F. All authors have read and agreed to the published version of the manuscript.

Funding: This research was funded by the German Research Foundation (DFG) via the collaborative research center CRC 1261.

Institutional Review Board Statement: Not applicable.

Informed Consent Statement: Not applicable.

Data Availability Statement: The data presented in this study are available on reasonable request from the corresponding author.

Acknowledgments: The authors gratefully thank Christine Kirchhof for the fabrication of the sensor elements.

Conflicts of Interest: The authors declare no conflict of interest.

Appendix A. Equivalent Circuit Model

The equivalent circuit model used to describe the electrical admittance of each sensor element is illustrated in Figure A1a and the structure of the modeled sensor array in Figure A1b.

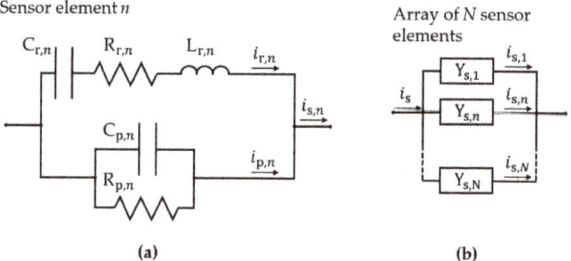

Figure A1. Illustration of (a) the modified Butterworth-van Dyke equivalent-circuit model used to describe the sensor elements, with the current $i_{r,n}$ through the resonator LCR-circuit and $i_{p,n}$ as the parallel capacitive pathway, and (b) all N parallel-connected sensor elements of the array, with the current $i_{s,n}$ through the nth sensor element and the total current i_s through the array.

The following equations are used to estimate the resonance frequency $f_{r,n}$ and the quality factor Q_n of the nth sensor element:

$$Q_n = \frac{1}{R_{r,n}} \sqrt{\frac{L_{r,n}}{C_{r,n}}}, \tag{A1}$$

and
$$f_{r,n} = \frac{1}{2\pi\sqrt{L_{r,n}C_{r,n}}}. \tag{A2}$$

Appendix B. Transfer Function of the Resonator

The frequency response of the resonator is modeled as second-order infinite impulse response (IIR) peaking filter with the transfer function $\mathcal{H}_r(z)$ [61]:

$$\mathcal{H}_r(z) = \frac{n_1 + n_2 z^{-1} + n_3 z^{-2}}{d_1 + d_2 z^{-1} + d_3 z^{-2}}, \tag{A3}$$

and the components n_i of the numerator coefficients as well as d_i of the denominator coefficients, which are functions of the quality factor Q and the resonance frequency f_r [62]:

$$n = \begin{pmatrix} 1-g \\ 0 \\ g-1 \end{pmatrix}, \quad d = \begin{pmatrix} 1-g \\ -2g\cdot\cos(\pi f_r) \\ 2g-1 \end{pmatrix}. \tag{A4}$$

To ensure a gain of -3 dB at the bandwidth, the factor g is set to:

$$g = \left[1 + \sqrt{2}\cdot\tan\left(\frac{\pi}{2}\frac{f_r}{Q}\right)\right]^{-1}. \tag{A5}$$

Appendix C. System Parameters

In the following Table A1, the model parameters of the sensor system and the equivalent circuit parameters of the two sensor elements S_1 and S_2 are summarized.

Table A1. Parameters of the equivalent circuit noise model and the modified Butterworth-van Dyke (mBvD) model.

Component	Parameter	Value	Parameter	Value
Excitation	R_{ex}	75 Ω		
Cable C_1	R_{C1}	147 MΩ	C_{C1}	208 pF
Cable C_2	R_{C2}	184 MΩ	C_{C2}	36 pF
Sensor element S_1	$R_{p,1}$ $R_{r,1}$ $L_{r,1}$	295 MΩ 663.47 kΩ 8.826 kH	$C_{ME,1}$ $C_{r,1}$ $f_{r,1}$	47.157 pF 48.725 fF 7674.9 Hz
Sensor element S_2	$R_{p,2}$ $R_{r,2}$ $L_{r,2}$	310 MΩ 755.96 kΩ 8.753 kH	$C_{ME,2}$ $C_{r,2}$ $f_{r,2}$	48.568 pF 49.112 fF 7676.4 Hz
Amplifier	R_f	5 GΩ	C_f	30 pF

Appendix D. Implementation of the Model

The equations, which describe the signal-and-noise model (Equations (1)–(22)), are implemented in MATLAB (The MathWorks, Inc., Natick, MA, USA). The voltage at the charge amplifier's output $u_{co}(t)$ is calculated in the time domain using Equations (1)–(6), and the electric sensitivities (Equations (7) and (8)) obtained from the impedance measurements in Section 4.1. Estimated magnetic sensitivities of $S_{mag,1} \approx S_{mag,2} = 24$ Hz/mT are used, which is in the typical range expected from similar sensor elements [34,57]. The simulated time domain signal is demodulated with a quadrature amplitude demodulator and subsequently converted to the frequency domain using Welch's method [63]. From the power spectral density estimate, we calculate the amplitude spectrum $\hat{U}(f)$ of the demodulated signal $u(t)$ and the voltage sensitivity following the definition provided by

Equation (9). For the noise simulations, Equations (11)–(22) are implemented to obtain the voltage noise density at the output of the charge amplifier in the frequency domain.

References

1. Lenz, J.; Edelstein, S. Magnetic Sensors and Their Applications. *IEEE Sens. J.* **2006**, *6*, 631–649. [CrossRef]
2. Gleich, B.; Weizenecker, J. Tomographic Imaging Using the Nonlinear Response of Magnetic Particles. *Nature* **2005**, *435*, 1214–1217. [CrossRef]
3. Friedrich, R.-M.; Zabel, S.; Galka, A.; Lukat, N.; Wagner, J.-M.; Kirchhof, C.; Quandt, E.; McCord, J.; Selhuber-Unkel, C.; Siniatchkin, M.; et al. Magnetic Particle Mapping Using Magnetoelectric Sensors as an Imaging Modality. *Sci. Rep.* **2019**, *9*, 2086. [CrossRef]
4. Cohen, D.; Givler, E. Magnetomyography: Magnetic Fields around the Human Body Produced by Skeletal Muscles. *Appl. Phys. Lett.* **1972**, *21*, 114–116. [CrossRef]
5. Zuo, S.; Schmalz, J.; Özden, M.-Ö.; Gerken, M.; Su, J.; Niekiel, F.; Lofink, F.; Nazarpour, K.; Heidari, H. Ultrasensitive Magnetoelectric Sensing System for Pico-Tesla MagnetoMyoGraphy. *IEEE Trans. Biomed. Circuits Syst.* **2020**, *14*, 971–984. [CrossRef]
6. Koch, H. Recent Advances in Magnetocardiography. *J. Electrocardiol.* **2004**, *37*, 117–122. [CrossRef]
7. Fenici, R.; Brisinda, D.; Meloni, A.M. Clinical Application of Magnetocardiography. *Expert Rev. Mol. Diagn.* **2005**, *5*, 291–313. [CrossRef]
8. Kwong, J.S.W.; Leithäuser, B.; Park, J.-W.; Yu, C.-M. Diagnostic Value of Magnetocardiography in Coronary Artery Disease and Cardiac Arrhythmias: A Review of Clinical Data. *Int. J. Cardiol.* **2013**, *167*, 1835–1842. [CrossRef] [PubMed]
9. Duez, L.; Beniczky, S.; Tankisi, H.; Hansen, P.O.; Sidenius, P.; Sabers, A.; Fuglsang-Frederiksen, A. Added Diagnostic Value of Magnetoencephalography (MEG) in Patients Suspected for Epilepsy, Where Previous, Extensive EEG Workup Was Unrevealing. *Clin. Neurophysiol.* **2016**, *127*, 3301–3305. [CrossRef] [PubMed]
10. Fenici, R.; Brisinda, D.; Nenonen, J.; Fenici, P. Phantom Validation of Multichannel Magnetocardiography Source Localization. *Pacing and Clin. Electrophysiol.* **2003**, *26*, 426–430. [CrossRef] [PubMed]
11. Bertero, M.; Piana, M. Inverse problems in biomedical imaging: Modeling and methods of solution. In *Complex Systems in Biomedicine*; Quarteroni, A., Formaggia, L., Veneziani, A., Eds.; Springer: Milano, Italy, 2006; pp. 1–33. ISBN 978-88-470-0396-5.
12. Yang, J.; Poh, N. *Recent Application in Biometrics*; IntechOpen: London, UK, 2011; ISBN 978-953-307-488-7.
13. Murzin, D.; Mapps, D.J.; Levada, K.; Belyaev, V.; Omelyanchik, A.; Panina, L.; Rodionova, V. Ultrasensitive Magnetic Field Sensors for Biomedical Applications. *Sensors* **2020**, *20*, 1569. [CrossRef] [PubMed]
14. Kleiner, R.; Koelle, D.; Ludwig, F.; Clarke, J. Superconducting Quantum Interference Devices: State of the Art and Applications. *Proc. IEEE* **2004**, *92*, 1534–1548. [CrossRef]
15. Robbes, D. Highly Sensitive Magnetometers—A Review. *Sens. Actuators A Phys.* **2006**, *129*, 86–93. [CrossRef]
16. Griffith, W.C.; Knappe, S.; Kitching, J. Femtotesla Atomic Magnetometry in a Microfabricated Vapor Cell. *Opt. Express* **2010**, *18*, 27167–27172. [CrossRef] [PubMed]
17. Johnson, C.N.; Schwindt, P.D.D.; Weisend, M. Multi-Sensor Magnetoencephalography with Atomic Magnetometers. *Phys. Med. Biol.* **2013**, *58*, 6065–6077. [CrossRef]
18. Osborne, J.; Orton, J.; Alem, O.; Shah, V. Fully Integrated Standalone Zero Field Optically Pumped Magnetometer for Biomagnetism. In Proceedings of the Steep Dispersion Engineering and Opto-Atomic Precision Metrology XI, San Francisco, CA, USA, 29 January–1 February 2018; Volume 10548, pp. 89–95.
19. Mhaskar, R.R.; Knappe, S.; Kitching, J. Low-Frequency Characterization of MEMS-Based Portable Atomic Magnetometer. In Proceedings of the 2010 IEEE International Frequency Control Symposium, Newport Beach, CA, USA, 1–4 June 2010; pp. 376–379.
20. Oelsner, G.; IJsselsteijn, R.; Scholtes, T.; Krüger, A.; Schultze, V.; Seyffert, G.; Werner, G.; Jäger, M.; Chwala, A.; Stolz, R. Integrated Optically Pumped Magnetometer for Measurements within Earth's Magnetic Field. *arXiv* **2021**, arXiv:2008.01570.
21. Reermann, J.; Durdaut, P.; Salzer, S.; Demming, T.; Piorra, A.; Quandt, E.; Frey, N.; Höft, M.; Schmidt, G. Evaluation of Magnetoelectric Sensor Systems for Cardiological Applications. *Measurement* **2018**, *116*, 230–238. [CrossRef]
22. Yarar, E.; Salzer, S.; Hrkac, V.; Piorra, A.; Höft, M.; Knöchel, R.; Kienle, L.; Quandt, E. Inverse Bilayer Magnetoelectric Thin Film Sensor. *Appl. Phys. Lett.* **2016**, *109*, 022901. [CrossRef]
23. Tu, C.; Chu, Z.-Q.; Spetzler, B.; Hayes, P.; Dong, C.-Z.; Liang, X.-F.; Chen, H.-H.; He, Y.-F.; Wei, Y.-Y.; Lisenkov, I.; et al. Mechanical-Resonance-Enhanced Thin-Film Magnetoelectric Heterostructures for Magnetometers, Mechanical Antennas, Tunable RF Inductors, and Filters. *Materials* **2019**, *12*, 2259. [CrossRef]
24. Lee, E.W. Magnetostriction and Magnetomechanical Effects. *Rep. Prog. Phys.* **1955**, *18*, 184–229. [CrossRef]
25. Spetzler, B.; Golubeva, E.V.; Müller, C.; McCord, J.; Faupel, F. Frequency Dependency of the Delta-E Effect and the Sensitivity of Delta-E Effect Magnetic Field Sensors. *Sensors* **2019**, *19*, 4769. [CrossRef] [PubMed]
26. Spetzler, B.; Golubeva, E.V.; Friedrich, R.-M.; Zabel, S.; Kirchhof, C.; Meyners, D.; McCord, J.; Faupel, F. Magnetoelastic Coupling and Delta-E Effect in Magnetoelectric Torsion Mode Resonators. *Sensors* **2021**, *21*, 2022. [CrossRef] [PubMed]
27. Reermann, J.; Zabel, S.; Kirchhof, C.; Quandt, E.; Faupel, F.; Schmidt, G. Adaptive Readout Schemes for Thin-Film Magnetoelectric Sensors Based on the Delta-E Effect. *IEEE Sens. J.* **2016**, *16*, 4891–4900. [CrossRef]
28. Osiander, R.; Ecelberger, S.A.; Givens, R.B.; Wickenden, D.K.; Murphy, J.C.; Kistenmacher, T.J. A Microelectromechanical-based Magnetostrictive Magnetometer. *Appl. Phys. Lett.* **1996**, *69*, 2930–2931. [CrossRef]

29. Gojdka, B.; Jahns, R.; Meurisch, K.; Greve, H.; Adelung, R.; Quandt, E.; Knöchel, R.; Faupel, F. Fully Integrable Magnetic Field Sensor Based on Delta-E Effect. *Appl. Phys. Lett.* **2011**, *99*, 223502. [CrossRef]
30. Jahns, R.; Zabel, S.; Marauska, S.; Gojdka, B.; Wagner, B.; Knöchel, R.; Adelung, R.; Faupel, F. Microelectromechanical Magnetic Field Sensor Based on ΔE Effect. *Appl. Phys. Lett.* **2014**, *105*, 052414. [CrossRef]
31. Zabel, S.; Kirchhof, C.; Yarar, E.; Meyners, D.; Quandt, E.; Faupel, F. Phase Modulated Magnetoelectric Delta-E Effect Sensor for Sub-Nano Tesla Magnetic Fields. *Appl. Phys. Lett.* **2015**, *107*, 152402. [CrossRef]
32. Zabel, S.; Reermann, J.; Fichtner, S.; Kirchhof, C.; Quandt, E.; Wagner, B.; Schmidt, G.; Faupel, F. Multimode Delta-E Effect Magnetic Field Sensors with Adapted Electrodes. *Appl. Phys. Lett.* **2016**, *108*, 222401. [CrossRef]
33. Spetzler, B.; Su, J.; Friedrich, R.-M.; Niekiel, F.; Fichtner, S.; Lofink, F.; Faupel, F. Influence of the Piezoelectric Material on the Signal and Noise of Magnetoelectric Magnetic Field Sensors Based on the Delta-E Effect. *APL Mater.* **2021**, *9*, 031108. [CrossRef]
34. Spetzler, B.; Bald, C.; Durdaut, P.; Reermann, J.; Kirchhof, C.; Teplyuk, A.; Meyners, D.; Quandt, E.; Höft, M.; Schmidt, G.; et al. Exchange Biased Delta-E Effect Enables the Detection of Low Frequency PT Magnetic Fields with Simultaneous Localization. *Sci. Rep.* **2021**, *11*, 5269. [CrossRef]
35. Nan, T.; Hui, Y.; Rinaldi, M.; Sun, N.X. Self-Biased 215 MHz Magnetoelectric NEMS Resonator for Ultra-Sensitive DC Magnetic Field Detection. *Sci. Rep.* **2013**, *3*, 1985. [CrossRef]
36. Li, M.; Matyushov, A.; Dong, C.; Chen, H.; Lin, H.; Nan, T.; Qian, Z.; Rinaldi, M.; Lin, Y.; Sun, N.X. Ultra-Sensitive NEMS Magnetoelectric Sensor for Picotesla DC Magnetic Field Detection. *Appl. Phys. Lett.* **2017**, *110*, 143510. [CrossRef]
37. Staruch, M.; Matis, B.R.; Baldwin, J.W.; Bennett, S.P.; van't Erve, O.; Lofland, S.; Bussmann, K.; Finkel, P. Large Non-Saturating Shift of the Torsional Resonance in a Doubly Clamped Magnetoelastic Resonator. *Appl. Phys. Lett.* **2020**, *116*, 232407. [CrossRef]
38. Zhuang, X.; Sing, M.L.C.; Dolabdjian, C.; Wang, Y.; Finkel, P.; Li, J.; Viehland, D. Sensitivity and Noise Evaluation of a Bonded Magneto(Elasto) Electric Laminated Sensor Based on In-Plane Magnetocapacitance Effect for Quasi-Static Magnetic Field Sensing. *IEEE Trans. Magn.* **2015**, *51*, 1–4. [CrossRef]
39. Staruch, M.; Yang, M.-T.; Li, J.F.; Dolabdjian, C.; Viehland, D.; Finkel, P. Frequency Reconfigurable Phase Modulated Magnetoelectric Sensors Using ΔE Effect. *Appl. Phys. Lett.* **2017**, *111*, 032905. [CrossRef]
40. Stutzke, N.A.; Russek, S.E.; Pappas, D.P.; Tondra, M. Low-Frequency Noise Measurements on Commercial Magnetoresistive Magnetic Field Sensors. *J. Appl. Phys.* **2005**, *97*, 10Q107. [CrossRef]
41. Deak, J.G.; Zhou, Z.; Shen, W. Tunneling Magnetoresistance Sensor with PT Level 1/f Magnetic Noise. *AIP Adv.* **2017**, *7*, 056676. [CrossRef]
42. Lukat, N.; Friedrich, R.-M.; Spetzler, B.; Kirchhof, C.; Arndt, C.; Thormählen, L.; Faupel, F.; Selhuber-Unkel, C. Mapping of Magnetic Nanoparticles and Cells Using Thin Film Magnetoelectric Sensors Based on the Delta-E Effect. *Sens. Actuators A Phys.* **2020**, *309*, 112023. [CrossRef]
43. Chu, Z.; Shi, W.; Shi, H.; Chen, Q.; Wang, L.; PourhosseiniAsl, M.J.; Xiao, C.; Xie, T.; Dong, S. A 1D Magnetoelectric Sensor Array for Magnetic Sketching. *Adv. Mater. Technol.* **2019**, *4*, 1800484. [CrossRef]
44. Cuong, T.D.; Viet Hung, N.; Le Ha, V.; Tuan, P.A.; Duong, D.D.; Tam, H.A.; Duc, N.H.; Giang, D.T.H. Giant Magnetoelectric Effects in Serial-Parallel Connected Metglas/PZT Arrays with Magnetostrictively Homogeneous Laminates. *J. Sci. Adv. Mater. Devices* **2020**, *5*, 354–360. [CrossRef]
45. Xi, H.; Lu, M.-C.; Yang, Q.X.; Zhang, Q.M. Room Temperature Magnetoelectric Sensor Arrays For Application of Detecting Iron Profiles in Organs. *Sens. Actuators A Phys.* **2020**, *311*, 112064. [CrossRef] [PubMed]
46. Lu, Y.; Cheng, Z.; Chen, J.; Li, W.; Zhang, S. High Sensitivity Face Shear Magneto-Electric Composite Array for Weak Magnetic Field Sensing. *J. Appl. Phys.* **2020**, *128*, 064102. [CrossRef]
47. Li, H.; Zou, Z.; Yang, Y.; Shi, P.; Wu, X.; Ou-Yang, J.; Yang, X.; Zhang, Y.; Zhu, B.; Chen, S. Microbridge-Structured Magnetoelectric Sensor Array Based on PZT/FeCoSiB Thin Films. *IEEE Trans. Magn.* **2020**, *56*, 1–4. [CrossRef]
48. Kim, H.J.; Wang, S.; Xu, C.; Laughlin, D.; Zhu, J.; Piazza, G. Piezoelectric/Magnetostrictive MEMS Resonant Sensor Array for in-Plane Multi-Axis Magnetic Field Detection. In Proceedings of the 2017 IEEE 30th International Conference on Micro Electro Mechanical Systems (MEMS), Las Vegas, NV, USA, 22–26 January 2017; pp. 109–112.
49. Lage, E.; Kirchhof, C.; Hrkac, V.; Kienle, L.; Jahns, R.; Knöchel, R.; Quandt, E.; Meyners, D. Exchange Biasing of Magnetoelectric Composites. *Nat. Mater.* **2012**, *11*, 523–529. [CrossRef]
50. Yarar, E.; Hrkac, V.; Zamponi, C.; Piorra, A.; Kienle, L.; Quandt, E. Low Temperature Aluminum Nitride Thin Films for Sensory Applications. *AIP Adv.* **2016**, *6*, 075115. [CrossRef]
51. Durdaut, P.; Penner, V.; Kirchhof, C.; Quandt, E.; Knöchel, R.; Höft, M. Noise of a JFET Charge Amplifier for Piezoelectric Sensors. *IEEE Sens. J.* **2017**, *17*, 7364–7371. [CrossRef]
52. Jahns, R.; Knöchel, R.; Greve, H.; Woltermann, E.; Lage, E.; Quandt, E. Magnetoelectric Sensors for Biomagnetic Measurements. In Proceedings of the 2011 IEEE International Symposium on Medical Measurements and Applications, Bari, Italy, 30–31 May 2011; pp. 107–110.
53. Durdaut, P.; Reermann, J.; Zabel, S.; Kirchhof, C.; Quandt, E.; Faupel, F.; Schmidt, G.; Knöchel, R.; Höft, M. Modeling and Analysis of Noise Sources for Thin-Film Magnetoelectric Sensors Based on the Delta-E Effect. *IEEE Trans. Instrum. Meas.* **2017**, *66*, 2771–2779. [CrossRef]

54. Durdaut, P.; Rubiola, E.; Friedt, J.-M.; Müller, C.; Spetzler, B.; Kirchhof, C.; Meyners, D.; Quandt, E.; Faupel, F.; McCord, J.; et al. Fundamental Noise Limits and Sensitivity of Piezoelectrically Driven Magnetoelastic Cantilevers. *J. Microelectromech. Syst.* **2020**, *29*, 1347–1361. [CrossRef]
55. Spetzler, B.; Kirchhof, C.; Reermann, J.; Durdaut, P.; Höft, M.; Schmidt, G.; Quandt, E.; Faupel, F. Influence of the Quality Factor on the Signal to Noise Ratio of Magnetoelectric Sensors Based on the Delta-E Effect. *Appl. Phys. Lett.* **2019**, *114*, 183504. [CrossRef]
56. Durdaut, P. Ausleseverfahren Und Rauschmodellierung Für Magnetoelektrische Und Magnetoelastische Sensorsysteme. Ph.D. Thesis, Kiel University, Kiel, Germany, 2019.
57. Spetzler, B.; Kirchhof, C.; Quandt, E.; McCord, J.; Faupel, F. Magnetic Sensitivity of Bending-Mode Delta-E-Effect Sensors. *Phys. Rev. Appl.* **2019**, *12*, 064036. [CrossRef]
58. Wang, Y.J.; Gao, J.Q.; Li, M.H.; Shen, Y.; Hasanyan, D.; Li, J.F.; Viehland, D. A Review on Equivalent Magnetic Noise of Magnetoelectric Laminate Sensors. *Philos. Trans. R. Soc. A Math. Phys. Eng. Sci.* **2014**, *372*, 20120455. [CrossRef] [PubMed]
59. Ding, L.; Saez, S.; Dolabdjian, C.; Melo, L.G.C.; Yelon, A.; Menard, D. Equivalent Magnetic Noise Limit of Low-Cost GMI Magnetometer. *IEEE Sens. J.* **2009**, *9*, 159–168. [CrossRef]
60. Matyushov, A.D.; Spetzler, B.; Zaeimbashi, M.; Zhou, J.; Qian, Z.; Golubeva, E.V.; Tu, C.; Guo, Y.; Chen, B.F.; Wang, D.; et al. Curvature and Stress Effects on the Performance of Contour-Mode Resonant ΔE Effect Magnetometers. *Adv. Mater. Technol.* **2021**, *6*, 2100294. [CrossRef]
61. Oppenheim, A.V.; Schafer, R.W.; Buck, J.R. *Discrete-Time Signal Processing*, 2nd ed.; Prentice Hall: Upper Saddle River, NJ, USA, 1998.
62. Orfanidis, S.J. *Introduction to Signal Processing*; Prentice Hall: Hoboken, NJ, USA, 1996; ISBN 978-0-13-240334-4.
63. Welch, P.D. The Use of Fast Fourier Transform for the Estimation of Power Spectra: A Method Based on Time Averaging Over a Short, Modified Periodograms. *IEEE Trans. Audio Electroacoust.* **1967**, *15*, 70–73. [CrossRef]

Article

Magnetoelastic Coupling and Delta-E Effect in Magnetoelectric Torsion Mode Resonators

Benjamin Spetzler, Elizaveta V. Golubeva, Ron-Marco Friedrich, Sebastian Zabel, Christine Kirchhof, Dirk Meyners, Jeffrey McCord and Franz Faupel *

Institute of Materials Science, Faculty of Engineering, Kiel University, 24143 Kiel, Germany; besp@tf.uni-kiel.de (B.S.); elgo@tf.uni-kiel.de (E.V.G.); rmfr@tf.uni-kiel.de (R.-M.F.); seza@tf.uni-kiel.de (S.Z.); cki@tf.uni-kiel.de (C.K.); dm@tf.uni-kiel.de (D.M.); jmc@tf.uni-kiel.de (J.M.)
* Correspondence: ff@tf.uni-kiel.de; Tel.: +49-431-880-6229

Abstract: Magnetoelectric resonators have been studied for the detection of small amplitude and low frequency magnetic fields via the delta-E effect, mainly in fundamental bending or bulk resonance modes. Here, we present an experimental and theoretical investigation of magnetoelectric thin-film cantilevers that can be operated in bending modes (BMs) and torsion modes (TMs) as a magnetic field sensor. A magnetoelastic macrospin model is combined with an electromechanical finite element model and a general description of the delta-E effect of all stiffness tensor components C_{ij} is derived. Simulations confirm quantitatively that the delta-E effect of the C_{66} component has the promising potential of significantly increasing the magnetic sensitivity and the maximum normalized frequency change Δf_r. However, the electrical excitation of TMs remains challenging and is found to significantly diminish the gain in sensitivity. Experiments reveal the dependency of the sensitivity and Δf_r of TMs on the mode number, which differs fundamentally from BMs and is well explained by our model. Because the contribution of C_{11} to the TMs increases with the mode number, the first-order TM yields the highest magnetic sensitivity. Overall, general insights are gained for the design of high-sensitivity delta-E effect sensors, as well as for frequency tunable devices based on the delta-E effect.

Keywords: delta-E effect; magnetoelectric; magnetoelastic; resonator; torsion mode; bending mode; magnetic modeling; MEMS; FEM

1. Introduction

In recent years, thin-film magnetoelectric sensors have been studied, frequently envisioning biomedical applications in the future [1,2]. Such applications often require the measurement of small amplitude and low frequency magnetic fields [1–3]. With the direct magnetoelectric effect, such small detection limits are only obtained at high frequencies and in small-signal bandwidths of a few Hz [2,4]. One way to overcome these limitations is by using a modulation scheme based on the delta-E effect. The delta-E effect is the change of the effective elastic properties with magnetization due to magnetoelastic coupling [5–8]. It results from inverse magnetostriction that adds additional stress-induced magnetostrictive strain to the purely elastic Hookean strain. The delta-E effect can occur generally in various elastic moduli and several components of the elastic stiffness tensor C [9,10]. Hence, it is sometimes referred to as the delta-C effect [11]. Typically, delta-E effect sensors are based on magnetoelectric resonators that are electrically excited via the piezoelectric layer at or close to the resonance frequency f_r. Upon the application of a magnetic field, the magnetization changes and the delta-E effect alters the mechanical stiffness tensor of the magnetostrictive layer. If the altered stiffness tensor components contribute to the resonance frequency of the excited mode, the resonance frequency changes, which can be read-out electrically. The delta-E effect of the Young's modulus has especially been studied thoroughly in soft magnetic amorphous materials [12–18]. It was used for magnetic field sensing with magnetoelectric plate resonators [19–22] and beam structures [23–32]. Such resonators are operated in bending or bulk modes and some have achieved limits of detection down to the sub-nT regime at low frequencies. Microelectromechanical systems (MEMS) cantilever sensors based on the delta-E effect were recently used for the mapping of magnetically labeled cells [33], and have shown promising properties for sensor array applications [34].

In contrast to the delta-E effect of the Young's modulus, the delta-E effect of the shear modulus has been studied less extensively [35] and mainly in amorphous wires [36,37]. It has been used for a different kind of delta-E effect sensors where shear waves, traveling through the magnetoelastic material, are influenced by the delta-E effect. This concept was realized with bulk acoustic shear waves in amorphous ribbons [38] and recently with surface acoustic shear waves in magnetic thin film devices [10,39–42]. Only very few studies investigate torsion modes in beam structures [43,44], either with electrostatically actuated cantilevers [43] or double-clamped beams [44]. Both studies are limited to specific configurations of the magnetic system and consider neither the full tensor relations of the mechanics and the delta-E effect nor higher resonance modes. Until now, a comprehensive experimental and theoretical analysis has been missing as well as a discussion of implications for the design of delta-E effect-based devices.

2. MEMS Torsion Mode Sensors

In this study, all measurements and models are made for a microelectromechanical system (MEMS) technology-fabricated cantilever with an electrode design that permits the excitation of torsion modes. A sketch including dimensions and layer structure and a top-view photograph of the design are shown in Figure 1. The approximately 3.1 mm-long and 2.15 mm-wide cantilever consists of a ≈ 2 µm-thick piezoelectric layer of AlN [45] on a 50 µm-thick poly-Si substrate. A 2 µm-thick amorphous magnetostrictive multilayer is deposited on the rear side. A magnetic field is applied during the deposition to induce a magnetic easy axis along the short cantilever axis. For actuation and read-out, three top electrodes (E_1, E_2 and E_3) of 100 nm-thick Au with lengths $L_1 = L_2 \approx 1$ mm and $L_3 \approx 0.6$ mm and widths $W_1 = W_2 \approx 0.5$ mm and $W_3 \approx 1$ mm contact the AlN layer on the top. The counter electrode (150 nm Pt) covers the whole beam area and is located between the AlN layer and the substrate. All measurements are performed with electrode E_1. As a magnetostrictive material, we use a 2 µm multilayer of 20 × (100 nm $(Fe_{90}Co_{10})_{78}Si_{12}B_{10}$ and 6 nm Cr). It is covered by a top Cr-layer that serves as a protection against corrosion. More information about the layer structures and the fabrication process can be found elsewhere [27]. In contrast to the sensors in Ref. [27], the sensor presented here is significantly wider and the adapted electrode design additionally permits the excitation of torsion modes. Details on the geometry are given in the appendix.

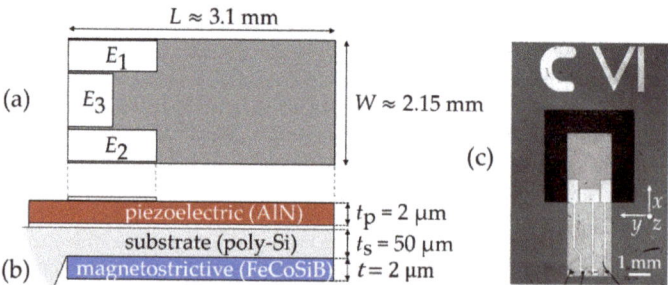

Figure 1. Delta-E effect sensor analyzed in this study: (**a**) schematic top view of the cantilever, with three different electrodes E_1, E_2 and E_3 of lengths $L_1 = L_2 \approx 1$ mm, $L_3 \approx 0.6$ mm and widths $W_1 = W_2 \approx 0.5$ mm and $W_3 \approx 1$ mm; (**b**) schematic side-view of the cantilever with the thickness of the functional layers and the poly-Si substrate; (**c**) top-view photograph of the fabricated structure.

3. Sensitivity

3.1. Definition of the Sensitivity

An important parameter that characterizes a magnetic field sensor is its sensitivity. During sensor operation, an alternating voltage is applied to excite the cantilever at its mechanical resonance frequency f_r. Applying a magnetic field, shifts f_r via the delta-E effect and correspondingly the sensor's admittance characteristic on its frequency axis f. Hence, the magnitude $|Y| = \text{abs}\{Y\}$ and phase angle $\phi = \arg\{Y\}$ of the sensor admittance Y depend on the magnetic field. Consequently, the ac magnetic field to be measured causes an amplitude modulation (am) and phase modulation (pm) of the current through the sensor. Detailed information on the operation and read-out can be found elsewhere [46–48]. The linearized change of $|Y|$ and ϕ with the magnetic field can be described by the amplitude sensitivity $S_{am} = S_{Y,r} \cdot S_{H,r}$ and the phase sensitivity $S_{pm} = S_{\phi,r} \cdot S_{H,r}$ [49], respectively. Both sensitivities have a magnetic part $S_{H,r}$ that includes the delta-E effect and an electric part $S_{Y,r}$ or $S_{\phi,r}$, which can be determined from the admittance. We refer to the three sensitivities as relative sensitivities, because they are normalized to the excitation frequency $f_{ex} = f_r$. The normalization is required to eventually compare the electrical and magnetic sensitivities of sensors with different geometries operated at different f_r or in different resonance modes. Usually a magnetic bias field H_0 is applied to operate the sensor at optimum conditions. The relative sensitivities are then defined in linear approximation as derivatives [49]:

$$S_{Y,r} \frac{\partial |Y|}{\partial f}\bigg|_{f=f_r,\, H=H_0} \cdot f_r; \quad S_{\phi,r} \frac{\partial \phi}{\partial f}\bigg|_{f=f_r, H=H_0} \cdot f_r; \quad S_{H,r} \frac{1}{f_r}\frac{\partial f_r}{\partial \mu_0 H}\bigg|_{H=H_0}, \quad (1)$$

with the magnetic vacuum permeability $\mu_0 \approx 4\pi \cdot 10^{-7}$ N/A^2. From Equation (1), the relative magnetic sensitivity $S_{H,r}$ is the linearized and normalized change of the resonance frequency f_r with the applied magnetic flux density $\mu_0 H$.

3.2. Magnetic Sensitivity of Arbitrary Resonance Modes

The delta-E effect is included in the relative magnetic sensitivity $S_{H,r}$ because the resonance frequency $f_r = f_r(C_{ij})$ is a function of the stiffness tensor components C_{ij}. Depending on the respective resonance mode, different C_{ij} dominate f_r and depending on the magnetoelastic properties they might result in non-zero $S_{H,r}$. To describe $S_{H,r}$ for arbitrary resonance modes, it can be separated into a purely mechanical part $f_r^{-1} \partial f_r / \partial C_{ij}$ that contains the resonance properties of the structure and a purely magnetoelastic part $\partial C_{ij}/\partial \mu_0 H$:

$$S_{H,r} \frac{1}{f_r}\frac{\partial f_r}{\partial \mu_0 H}\bigg|_{H=H_0} = \sum_{i=1}^{3}\sum_{j=1}^{3} \frac{1}{f_r}\frac{\partial f_r}{\partial C_{ij}}\frac{\partial C_{ij}}{\partial \mu_0 H}\bigg|_{H=H_0} \sum_{i=1}^{3}\sum_{j=1}^{3} \partial_C f_{r,ij}\, \partial_H C_{ij}. \quad (2)$$

If treated separately, the factors $\partial f_r/\partial C_{ij}$ and $\partial C_{ij}/\partial \mu_0 H$ must be normalized to remove the dependency on the absolute value of C_{ij} that cancels out in $S_{H,r}$. We define:

$$\partial_C f_{r,ij} \frac{C_{ij}}{f_r}\frac{\partial f_r}{\partial C_{ij}}\bigg|_{H=H_0}; \quad \partial_H C_{ij}\frac{1}{C_{ij}}\frac{\partial C_{ij}}{\partial \mu_0 H}\bigg|_{H=H_0}. \quad (3)$$

From Equation (3), the factor $\partial_C f_{r,ij}$ represents a normalized measure for the influence of the stiffness tensor component C_{ij} on the resonance frequency f_r of the considered resonance mode. It is a purely mechanical quantity and hence determined by the geometry, the resonance mode, and the effective mechanical properties of the resonator. The second factor $\partial_H C_{ij}$, includes the delta-E effect and describes the normalized influence of the applied flux density $\mu_0 H$ on C_{ij}. Hence, the two factors quantify the mechanical and the magnetoelastic parts of the relative magnetic sensitivity $S_{H,r}$. They will be used later to analyze the sensitivity and the frequency detuning of higher bending and torsion modes of the cantilever.

4. Sensor Modelling

The model used to describe and analyze the sensor consists of two parts. With a semi-analytical magnetoelastic macrospin model, the delta-E effect is obtained, i.e., the effective mechanical stiffness tensor $C(H)$ as a function of the applied field H. It is used as an input for an electromechanical finite element mechanics (FEM) model that describes the resonance frequency and the sensor's impedance response. In addition to a macrospin approximation, we assume a quasi-static magnetization behavior. Consequently, it is only valid for operation frequencies and magnetic field frequencies far below the ferromagnetic resonance frequency (FMR). The FMR generally depends on the geometry and the magnetic properties of the thin-film [50] and can cause a frequency dependency of the delta-E effect [51]. For the soft-magnetic material and thin-film geometry used here, it is in the GHz regime [51,52]. Because the operation frequencies are of the order of several kHz, magnetodynamic effects and the frequency dependency of the delta-E effect are neglected [51]. Due to the low frequencies, we assume that also electrodynamic effects can be omitted in the electromechanical model. In the following, both parts of the model are discussed in detail.

4.1. Electromechanical Model

In the electromechanical part of the model, we consider a simplified cantilever geometry, reduced to the poly-Si substrate, the piezoelectric AlN layer, and the magnetic FeCoSiB layer. Details on the geometry used are given in Appendix C. We assume all materials to be mechanically linear, which is a good approximation at sufficiently small excitation voltages. The material parameters used are given in the appendix. The cantilever is oriented in a cartesian coordinate system as illustrated in Figure 2, used throughout this paper. The mechanical equation of motion is given by (e.g., [53])

$$\rho \frac{\partial^2 \overline{u}}{\partial t^2} = \overline{\nabla} \cdot \sigma , \tag{4}$$

if no external forces are present. It includes the displacement vector $\overline{u}$, the time t, the mass density ρ, and the divergence $\overline{\nabla} \cdot \sigma$ of the mechanical stress tensor σ. For sufficiently small excitation frequencies, eddy current effects can be neglected and the electrostatic equations [54]:

$$\begin{aligned} \overline{E} &= -\overline{\nabla} U, \\ \overline{\nabla} \cdot \overline{D} &= \rho_c, \end{aligned} \tag{5}$$

are valid in good approximation. They include the electrical vector field $\overline{E}$, the gradient $\overline{\nabla} U$ of the electrical potential U and the divergence $\overline{\nabla} \cdot \overline{D}$ of the electric flux density $\overline{D}$ with the free charge density ρ_c. The electrostatic equations are coupled to the mechanical equation of motion via the constitutive piezoelectric equations, here in the stress-charge form [54,55]:

$$\begin{aligned} \sigma &= C^* \varepsilon - e_c^T \overline{E} \\ \overline{D} &= e_c \varepsilon - \varepsilon_{el} \overline{E} , \end{aligned} \tag{6}$$

with the linear strain tensor ε and the complex mechanical stiffness tensor $C^* = C(1 + i\eta)$. Its real part is the material's stiffness tensor C and its imaginary part $C\eta$ includes the isotropic loss factor η, which is used to consider damping in the materials [56]. The electromechanical coupling tensor is denoted as e_c and the electrical permittivity tensor as ε_{el}. For the calculation, we set fixed boundary conditions ($\overline{u} = 0$) at the left face of the beam. For the piezoelectric material, we assumed at the boundaries $\overline{n}\overline{D} = 0$ (with surface normal vector $\overline{n}$), and an initial value for the electric potential of $U = 0$, except for the area covered by the electrodes. The electrodes are modeled with a fixed potential boundary condition, where an alternating voltage $U_{app} = U_0 \cdot \exp(i[\omega t + \varphi_v])$ is applied, with amplitude U_0, the angular frequency ω and phase angle φ_v. To calculate the electrical admittance $Y = I/U$ the current I is obtained from integrating the surface charge density over the electrode areas. For the solution, a linear response of the system is assumed, with a displacement of the form $\bar{u} = \hat{u} \cdot \exp(i[\omega t + \varphi_u])$ and a solution for the electrical potential of $U = \hat{U} \cdot \exp(i[\omega t + \varphi_v])$. The equations are solved within a frequency domain study in COMSOL$^{(r)}$ Multiphysics v. 5.3a (COMSOL AB, Stockholm, Sweden) [56]. All material parameters used are given in the appendix.

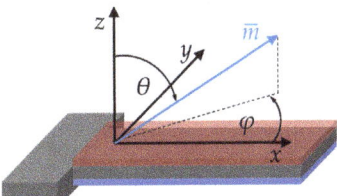

Figure 2. Coordinate system used for the electromechanical and the magnetic model. All three components m_i of the reduced magnetization vector $\bar{m}$ are described by the polar angle θ and the azimuthal angle φ.

4.2. Magnetoelastic Model

For the magnetic model, we consider the enthalpy density function of a macrospin with a uniaxial anisotropy energy density, an external magnetic field, a demagnetizing term, and magnetoelastic energy density. Using Einstein's summation convention, the enthalpy density term we use is:

$$u = K\left(1 - (m_i EA_i)^2\right) - \mu_0 M_s m_i H_i - \frac{1}{2} \mu_0 M_s m_i H_{d,i} - \sigma_j \lambda_j \quad \text{with } i = 1, 2, 3 \quad j = 1, \ldots, 6. \quad (7)$$

In this equation, the components of the reduced magnetization vector are denoted by m_i, the magnitude of the magnetization vector by M_s and the magnetic vacuum permeability by μ_0. The effective easy axis of magnetization is characterized by its orientation vector EA_i and the effective first-order uniaxial anisotropy energy density constant K. The components of the external magnetic field vector are given by H_i and the components of the mean demagnetizing field by $H_{d,i} = -D_{ii} m_i M_s$, with the main diagonal components D_{ii} of the demagnetizing tensor. For the magnetoelastic energy density, we use the coupling term $-\sigma_i \lambda_i$ with the stress tensor components σ_i and the components λ_i of the isotropic magnetostrictive strain tensor. Both are given in Voigt's notation. The coupling term results from omitting magnetostrictive self-energy and incorporating the term constant with stress into K [57]. In the following, the polar angle θ and the azimuthal angle φ of $\bar{m}$ in the spherical coordinate system (Figure 2) are used to define its components m_i. The exact definition of all vector and tensor components is given in the appendix. The linearized change of the elastic compliance components S_{ij} with the magnetic field and stress is derived from the expression

$$S_{ij}(H, \sigma) = \frac{\partial \varepsilon_i}{\partial \sigma_j} = \frac{\partial (e_i + \lambda_i)}{\partial \sigma_j} S_{m,ij} + \Delta S_{ij}. \quad (8)$$

where the first summand $S_{m,ij}$ is the constant, fixed magnetization elastic compliance tensor component. The magnetization dependent part ΔS_{ij} can be obtained from the equilibrium conditions that are given by the first-order derivatives of u:

$$u_\varphi \frac{\partial u}{\partial \varphi} = 0 \quad \text{and} \quad u_\theta \frac{\partial u}{\partial \theta} = 0. \quad (9)$$

From these equilibrium conditions a general expression for the linearized change ΔS_{ij} of the compliance tensor can be derived (Appendix A). Denoting the second-order derivatives as $u_{\varphi\varphi}$ and $u_{\theta\theta}$ it is:

$$\Delta S_{ij} \frac{\partial \lambda_i}{\partial \sigma_j} = -\frac{\partial \lambda_i}{\partial \varphi} \frac{\partial u_\varphi}{\partial \sigma_j} \frac{1}{u_{\varphi\varphi}} - \frac{\partial \lambda_i}{\partial \theta} \frac{\partial u_\theta}{\partial \sigma_j} \frac{1}{u_{\theta\theta}}. \tag{10}$$

This expression permits a quick calculation of the compliance tensor for different magnetic systems described by an enthalpy density u. From Equation (10), the non-zero components of ΔS for in-plane magnetization ($\theta = \pi/2$) are:

$$\Delta S_{11} = \Delta S_{22} = -\Delta S_{12} = \frac{9\lambda_s^2 \cos[\varphi]^2 \sin[\varphi]^2}{u_{\varphi\varphi}}, \tag{11}$$

$$\Delta S_{16} = -\Delta S_{26} = -\frac{9\lambda_s^2 \cos[\varphi] \cos[2\varphi] \sin[\varphi]}{u_{\varphi\varphi}}, \tag{12}$$

$$\Delta S_{44} = \frac{9\lambda_s^2 \sin[\varphi]^2}{u_{\theta\theta}}, \tag{13}$$

$$\Delta S_{45} = \frac{9\lambda_s^2 \cos[\varphi] \sin[\varphi]}{u_{\theta\theta}}, \tag{14}$$

$$\Delta S_{55} = \frac{9\lambda_s^2 \cos[\varphi]^2}{u_{\theta\theta}}, \tag{15}$$

$$\Delta S_{66} = \frac{9\lambda_s^2 \cos[2\varphi]^2}{u_{\varphi\varphi}}. \tag{16}$$

The final compliance tensor for in-plane magnetization as a function of magnetic field and stress is:

$$S(H,\sigma) = \begin{bmatrix} S_{11} & S_{12} & S_{m,13} & 0 & 0 & \Delta S_{16} \\ S_{12} & S_{22} & S_{m,23} & 0 & 0 & \Delta S_{26} \\ S_{m,13} & S_{m,23} & S_{m,33} & 0 & 0 & 0 \\ 0 & 0 & 0 & S_{44} & \Delta S_{45} & 0 \\ 0 & 0 & 0 & \Delta S_{45} & S_{55} & 0 \\ \Delta S_{16} & \Delta S_{26} & 0 & 0 & 0 & S_{66} \end{bmatrix} \text{ with } S_{ij}(H,\sigma) = S_{m,ij} + \Delta S_{ij}. \tag{17}$$

Because in our case both, S_m and ΔS. are symmetric, and S is also symmetric. Note that $S_{m,16} = S_{m,26} = S_{m,45} = 0$ in our isotropic magnetic material and consequently $S_{16} = \Delta S_{16}$, $S_{26} = \Delta S_{26}$ and $S_{45} = \Delta S_{45}$. Finally, the stiffness tensor C is obtained by numerically calculating the inverse $C(H,\sigma) = S(H,\sigma)^{-1}$. It has the same non-zero components and symmetry. All equations (Equations (11)–(17) are obtained from Equation (10) assuming in-plane magnetization ($\theta = \pi/2$) and are valid for the isotropic magnetoelastic coupling used in the enthalpy density function (Equation (7)). For all the following simulations, we additionally assume in-plane magnetic fields ($\theta_H = \pi/2$) and an in-plane easy axis ($\theta_{EA} = \pi/2$).

These two assumptions influence and simplify u and its derivatives, which are given in the appendix.

5. Implications of the Magnetic Model

In the following, results for the C_{ij} of the magnetoelastic model are discussed at the example of a thin-film geometry. For the calculations, we assumed zero static stress ($\sigma_i = 0$) and $D_{33} = 1$. The large shape anisotropy results in $C_{44} \approx C_{m,44}$ and $C_{55} \approx C_{m,55}$. We limit the discussion to the C_{11}, C_{12} and C_{66} components as they are most relevant for torsion and bending modes.

In Figure 3a, the normalized C_{11}, C_{12} and C_{66} components are plotted for a macrospin and $\varphi_{EA} = 90°$. Because $u_{\varphi\varphi}(H = H_K) = 0$ and so $\Delta S_{66}(H \to H_K) \to \infty$ (Equation (4)) it is $C_{66}(H = H_K) = 0$. At $|H| > |H_K|$, it is $C_{66} < C_{m,66}$ with $C_{66} = C_{m,66}$ only for $H \to \infty$. Hence, for finite H even a small shear stress σ_6 can always tilt the magnetization vector out of the applied magnetic field direction. It occurs, because the magnetoelastic energy density contribution $-\sigma_6 \lambda_6$ of the shear stress σ_6 is asymmetric around $\varphi = 0°$. Its minimum is shifted by 45° compared to the minimum of the one-component at $\varphi = 0°$. Consequently, at the two local maxima it is $C_{66}(\varphi = 45°, 135°) = C_{m,66}$. The C_{11} component shows two distinct minima but unlike the delta-E effect in the Young's modulus (e.g., [6,14,49]) no discontinuities at $|H| = |H_K|$. Although the discontinuities are present in S_{11} (not shown), they vanish during the inversion due to contributions of other S_{ij} components to C_{11}. In contrast to C_{11}, C_{12} stiffens with applied magnetic bias field because $\Delta S_{12} = -\Delta S_{11}$. The signs are a direct consequence of the positive isotropic magnetoelastic coupling. As the macrospin rotates towards the x axis, magnetostrictive expansion occurs along the x axis, but contraction occurs along the y axis. Compared with C_{11}, the maximum relative change of C_{12} is larger because $S_{m,12} < S_{m,11}$, which results in a different weighting in Equation (8). In Figure 3b, C_{66} is shown for three different angles of the easy axis $\varphi_{EA} = 90°, 85°$, and 75°. It is apparent that a change of φ_{EA} strongly influences C_{66}. Relative to $\varphi_{EA} = 90°$, the two minima at $H = \pm H_K$ shift to a larger $|H|$ and the minimum value increases strongly by more than 85% at $\varphi_{EA} = 85°$ and about 95% at $\varphi_{EA} = 75°$. The center minimum shifts due to the single domain hysteresis and decreases slightly with decreasing φ_{EA}. A singularity occurs at $\varphi_{EA} = 85°$ due to the magnetic discontinuity at the switching field of the single-spin model. Due to the strong impact of small deviations from $\varphi_{EA} = 90°$ on $C_{66}(H)$, the magnetic sensitivity is expected to change notably with φ_{EA}.

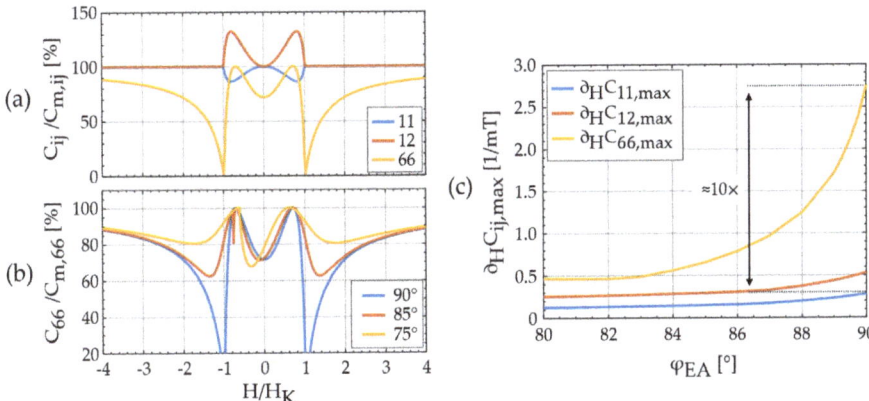

Figure 3. (a) Magnetic field dependent components C_{ij} of the effective stiffness tensor for an ideal hard axis magnetization process of a macrospin. The external magnetic field with magnitude H is applied along the x axis and normalized the anisotropy field H_K; (b) component C_{66} for different angles φ_{EA} of the magnetic easy axis to the x axis; (c) maximum value $\partial_H C_{ij,max}$ of $\partial_H C_{ij}(H)$ (Equation (3)) for the C_{11} and C_{66} components as functions of the easy axis angle φ_{EA}. For the calculations in (c), a distribution of effective anisotropy energy density is used as in Ref. [49] with a standard deviation of $\delta_K = 15\%$ for a more quantitative estimation and to prevent singularities.

In the following, we quantify the influence of the C_{ij} components on the relative magnetic sensitivity $S_{H,r}$ using $\partial_H C_{ij}$ as defined in Equation (3). Calculating $\partial_H C_{ij}$ requires forming the derivate $\partial C_{ij}/\partial H$, which results in singularities for the 66-component at $\varphi_{EA} = 90°$ and $|H| = |H_K|$. A finite derivative can be estimated by including the distribution of the effective anisotropy energy density K in a mean-field approach [15,49,58]. With such a distribution, inhomogeneities in the magnetization response are considered that can occur, e.g., from spatially varying stress or internal stray fields. We use a normal distribution of K with a standard deviation $\delta_K = 15\,\%$ as a representative example value that has been used previously for a similar device [49]. We calculate $\partial_H C_{ij}(H)$ numerically from $C_{ij}(H)$ and extract the maximum $\partial_H C_{ij,\text{max}}(H)$ for $H > 0$, at various angles φ_{EA} of the easy axis. They are plotted in Figure 3c. As a result of the distribution, both $\partial_H C_{ij,\text{max}}$ are finite at $\varphi_{EA} = 90°$ with $\partial_H C_{66,\text{max}} \approx 10 \times \partial_H C_{11,\text{max}} \approx 4.5 \times \partial_H C_{12,\text{max}}$. This is reduced to $\partial_H C_{66,\text{max}} \approx 4 \times \partial_H C_{11,\text{max}} \approx 2 \times \partial_H C_{12,\text{max}}$ at $\varphi_{EA} = 80°$. In conclusion, the C_{66} component potentially offers a significantly larger magnetic sensitivity than the C_{11} and C_{12} components.

6. Results

6.1. Magnetization Measurements

Magneto–optical Kerr effect (MOKE) microscopy [59] was used to analyze the magnetic multilayer. The picture in Figure 4a shows the rear side of the cantilever and is composed of a series of images. For each image, the magnetic multilayer was demagnetized along the x axis and the MOKE sensitivity axis was set along the y axis. The region of the magnetic multilayer is marked with a white frame and the estimated easy axis orientation is indicated with white arrows. In a large region around the left, top, and bottom edge, no magnetic response is visible. A comparison with light microscopy images reveals possibly corroded regions. They might have formed due to incomplete Cr-coverage at the edges. At the time, a particularly thin Cr-layer was deposited to ensure good magnetooptical contrast. Close to the clamping region (blue rectangle in Figure 4a), the layer is partially delaminated. Despite these nonidealities, the overall magnetic response in the magnetically active region is quite homogeneous. The average easy axis orientation is approximately $\varphi_{EA} = -75° \pm 5°$ relative to the x axis. An effective uniaxial anisotropy energy density of $K = (1.2 \pm 0.1)\,\text{kJ}/\text{m}^3$ is estimated with the magnetoelastic model. We used the ballistic demagnetizing tensor [60] in the center of the film and assumed $\sigma_j = 0$. A representative magnetization curve of the center region, recorded along the x axis, is shown in Figure 4b, and compared with one recorded at the clamping region. The difference between these curves indicates a different alignment of the effective anisotropy. However, due to the magnetic multilayer structure and the partial delamination additional effects cannot be excluded. From previous investigations [49], we expect that the deteriorated magnetic properties at the clamping especially reduce the resonance frequency detuning and the magnetic sensitivity of the first bending mode.

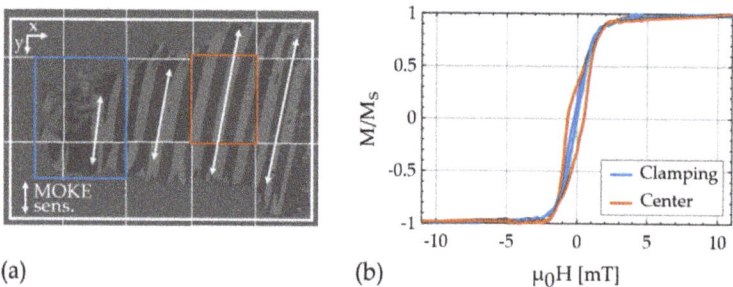

Figure 4. (a) Magneto-optical Kerr effect microscopy image of the analyzed structure, demagnetized along the x axis and composed of a series of different images. The region of the magnetic layer is marked by a white square and the approximated orientation of the magnetic easy axis is indicated with arrows at approximately $-75° \pm 5°$ relative to the x axis; (b) magnetization curve close to the clamping and in the center of the magnetic film. The two evaluated regions are indicated with squares in (a).

6.2. Electromechanical Properties

To analyze the electromechanical properties of the sensor, the sensor admittance $Y(f_{ex})$ is measured over a large range of excitation frequencies f_{ex}. Six resonance modes are characterized in detail by fitting a modified Butterworth van Dyke (mBvD) model (e.g., [61]) with the equivalent circuit configuration from [47] to the measurements. The resonance frequencies f_r and quality factors are calculated from the mBvD parameters of each admittance curve and compared with the eigenfrequencies obtained from the finite element method (FEM) model. With this comparison, the eigenmodes are identified to be the first three bending modes (BM1–3) and the first three torsion modes (TM1–3). The FEM model was fitted to admittance measurements of the first torsion mode (TM1) close to magnetic saturation at $\mu_0 H = -10$ mT. It matches the measurements very well as shown in Figure 5a. The material parameters match excellently with literature values. Details on the material parameters and on the geometry are given in Appendix C. A comparison of the measured resonance frequencies in magnetic saturation with the FEM simulations results in extremely small deviations <2% for all six modes (Appendix B).

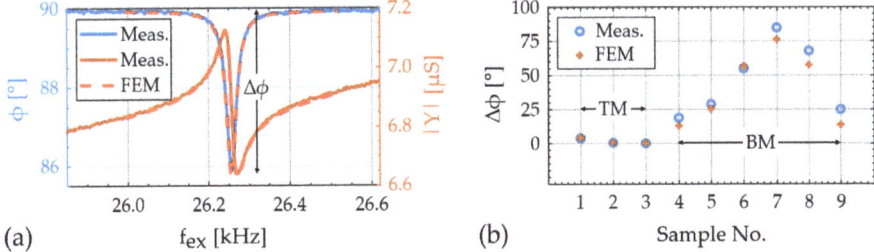

Figure 5. (a) Example comparison of measurement and FEM simulation of the sensor admittance around the first torsion mode (TM1), close to magnetic saturation at -10 mT; (b) comparison of measured and modeled maximum phase shift of the first three torsion modes (TM, samples 1–3) and the first and second bending modes (BMs, samples 4–9) with various electrode configurations. The BM measurements are published in Ref. [28].

The set of material parameters found is used to predict, and compare the impedance characteristic of other cantilever delta-E effect sensors published previously [28]. The sensors differ in their geometry from our torsion mode sensor. They were designed to excite the first and second bending mode with various electrode geometries. For the simulations, we used the same material parameters found for the torsion mode sensors but adjusted the geometry.

As a figure of merit for the electromechanical model, we compared the absolute difference $\Delta\phi = \phi_{max} - \phi_{min}$ of the phase angle ϕ of the electrical admittance Y. The simulation results are plotted in Figure 5b and compared with values of the torsion modes (Appendix B) measured here, and the bending mode from Ref. [28]. The TMs were measured close to magnetic saturation at $\mu_0 H = -10$ mT to reduce the influence of the delta-E effect. Slight deviations between the measurement and simulation might result from effectively different magnetoelectric coupling factors, e.g., due to the slightly different material parameters, geometric inaccuracies, or stress [62]. In conclusion, the model can estimate the electromechanical properties of the device and the effect of different electrode configurations well. For the application of magnetoelastic resonators as delta-E effect sensors, a high $\Delta\phi$ and hence a high electrical sensitivity is desirable. In comparison to the bending modes, the $\Delta\phi$ of the torsion modes is systematically smaller, which is also reflected in the electrical sensitivities. With $S_{Y,r} \approx 0.85$ mS of TM1, the maximum relative electrical amplitude sensitivity $S_{Y,r} \approx 5.8$ mS of sample No. 7 (BM2) [28] is almost a factor of seven larger, despite a similar quality factor. Hence, the large factor potentially gained in the magnetic sensitivity from utilizing the C_{66} component can be diminished by a reduced electrical sensitivity.

Additional simulations show that further optimization of the electrode design and reduction in the parasitic capacity from bond pads and wires could improve $\Delta\phi$ of TM1 to $\Delta\phi = 10°$. Alternatively, the parasitic effect of the sensor capacitance could be neutralized with additional electronics to utilize the phase-modulated signal for magnetic field detection [48]. A further improvement by a factor of two could be obtained by exciting both electrodes E_1 and E_2, phase shifted by 180°. Additionally, alternative piezoelectric materials with larger piezoelectric coefficients, such as AlScN [63–65] could increase the electrical sensitivity significantly and result in $\Delta\phi$ comparable with bending modes.

6.3. Delta-E Effect and Sensitivities

The $f_r(H)$ plots extracted from the modified Butterworth van Dyke (mBvD) fits of the first three bending modes (BM1–3) are shown in Figure 6a (right). They are normalized to $f_{r,max} \Delta f_r(-10\,\mathrm{mT})$ and have a respective minimum resonance frequency $f_{r,min}$. As a measure for the maximum resonance frequency detuning, we defined the normalized resonance frequency change $f_r(f_{r,max} - f_{r,min})/f_{r,max}$. All three curves are w-shaped and Δf_r increases with increasing mode number. This effect was reported previously and explained with a strong weighting of the magnetic properties at the clamping in BM1 [49]. Here, the difference between the BM1 and BM2 is significantly larger, which is consistent with the deteriorated material around the clamping region, visible in the magneto–optical Kerr effect microscopy (MOKE) images (Figure 4a). Correspondingly, the relative magnetic sensitivity $S_{H,r} \approx 3.5\,\mathrm{T}^{-1}$ is smallest in BM1 and increases up to $S_{H,r} \approx 9\,\mathrm{T}^{-1}$ in BM2.

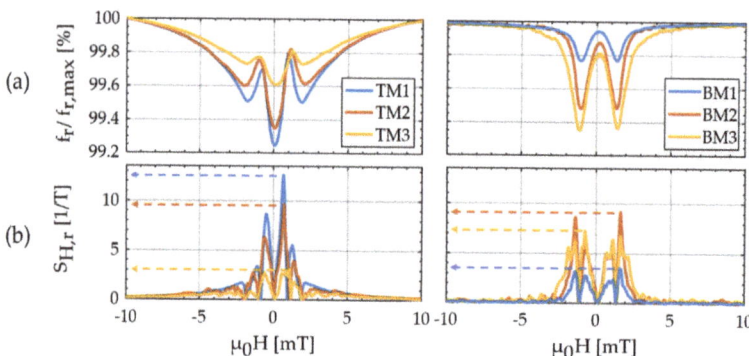

Figure 6. Measurements of the resonance frequency f_r as a function of the applied magnetic flux density $\mu_0 H$ along the long cantilever axis (x axis) starting at $\mu_0 H = -10\,\mathrm{mT}$: (**a**) normalized resonance frequency $f_r/f_{r,max}$ of the first three torsion modes (TMs) (left) and the first three transversal bending modes (BMs) (right). The maximum resonance frequencies $f_{r,max}$ of the TMs are 26.256, 87.478, 175.150 kHz, and of the BMs: 7.649, 47.182, 121.400 kHz; (**b**) relative magnetic sensitivities $S_{H,r} = S_H/f_{r,max}$ calculated from the data in (**a**) with Equation (1).

The normalized $f_r(H)$ plots of the torsion modes (TMs) and their corresponding $S_{H,r}$ are shown in Figure 6 (left). Although the sample is close to magnetic saturation at $\mu_0 H = -10\,\mathrm{mT}$, all three $f_r(H)$ curves still exhibit a non-zero slope as expected from the presented theory. The three $f_r(H)$ curves have a global minimum around $H = 0$, two local minima at around ± 2 mT, and two local maxima at about ± 1 mT. With an increasing mode number, the local maxima are almost unaffected, whereas Δf_r significantly decreases. Consequently, the maximum $S_{H,r}$ also decrease with the increasing mode number, here from $S_{H,r} = 12.6\,\mathrm{T}^{-1}$ in TM1 to $S_{H,r} = 3.0\,\mathrm{T}^{-1}$ in TM3. The trend is notably opposed to the corresponding behavior of the bending modes and will be analyzed and explained in detail in the next section using the magnetoelastic and electromechanical models. Overall, the magnetic sensitivities are in the range of $\approx 10\,\mathrm{T}^{-1}$ also measured with other magnetoelastic resonators in bending and bulk resonance modes [22,49]. At first glance, the similarity of BM and TM in $S_{H,r} \propto \partial_H C_{ij}$, might contradict the magnetoelastic model results in Figure 3c.

To resolve this and explain the dependency of the torsion modes on the mode number, the second factor $\partial_C f_{r,ij}$ that contributes to $S_{H,r}$ must be considered.

6.4. Resonance Frequency Simulations

In the following, we use the stiffness tensor components from the magnetic model as input in the finite element method (FEM) model to describe and analyze the frequency detuning and the magnetic sensitivity of the bending and torsion modes measured before. The demagnetizing tensor is approximated with the ballistic demagnetizing tensor in the center of the magnetic layer [60]. Consistently with the measurements the easy axis angle is set to $\varphi_{EA} = -75°$ and the effective anisotropy energy density constant to $K = 1.2$ kJ/m³, assuming $\sigma_j = 0$. Results for the normalized resonance frequencies $f_r(H)$ of the torsion modes are shown in Figure 7a and of the bending modes in Figure 7b. Despite the simplifying assumptions, a striking similarity with the measurements is apparent. All simulated torsion mode (TM) curves in Figure 7a exhibit two local maxima around one global minimum. Due to the single-domain hysteresis, the local minimum is shifted slightly leftwards away from $\mu_0 H = 0$. The frequency difference between the local maxima and the global minimum decreases significantly with increasing mode number, as also observed in the measurements.

Within the model, this phenomenon can be explained with the mode shapes of the higher torsion modes (Figure 7c). Due to the multiple twisting of the cantilever in higher modes, the resonance nodes are closer together. This results in an increasing contribution of the stiffness tensor components C_{11} and C_{22} to f_r relative to the C_{66} component. Quantitatively, we can explain the contribution of C_{ij} to f_r with the 11- and the 66-components of the normalized frequency factors $\partial_C f_{r,ij}$ (Equation (3)). They are estimated with the FEM model and summarized in Table 1. Whereas $\partial_C f_{r,11}$ increases by almost a factor of three, $\partial_C f_{r,66}$ shows the opposite trend and decreases by a factor of approximately two, from TM1 to TM3. Because the minima and maxima of C_{11} and C_{66} occur at similar magnetic bias fields they increasingly compensate each other in higher torsion modes. This causes similar magnetic sensitivities of TMs and BMs in our sensor, although $\partial_H C_{66,max} > \partial_H C_{11,max}$ in Figure 3c. If the delta-E effect of C_{66} is to be utilized, consequently, the first torsion mode is preferable to higher modes.

Table 1. Normalized frequency factors $\partial_C f_{r,11}$ and $\partial_C f_{r,66}$ (Equation (3)) of the first three torsion modes (TMs) and bending modes (BMs), calculated with the electromechanical finite element model.

Resonance Mode	TM1	TM2	TM3	BM1	BM2	BM3
$\partial_C f_{r,11}$	0.010	0.017	0.029	0.060	0.056	0.052
$\partial_C f_{r,66}$	0.034	0.026	0.016	0	0	0
$\partial_C f_{r,12}$	0	0	0	−0.003	−0.006	−0.001

In contrast to the measured bending mode curves (Figure 6), the corresponding modeled curves (Figure 7b) are almost independent of the mode number. Consistently, the $\partial_C f_{r,11}$ of the BMs are approximately constant with the mode number. The other frequency factor $\partial_C f_{r,12}$ is very small and $\partial_C f_{r,66} \approx 0$. A different effect dominates the mode dependency observed in the measured bending modes. This corroborates the hypothesis stated earlier in Section 6.3 that the reduced maximum normalized resonance frequency change Δf_r (as defined in Section 6.3) of BM1 is likely caused by the deteriorated magnetic layer present around the clamping (Figure 4a).

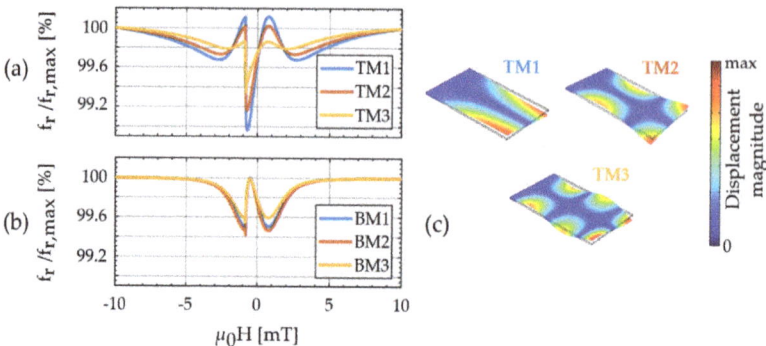

Figure 7. (**a**) Simulated resonance frequency f_r normalized to its maximum, as a function of the magnetic bias flux density $\mu_0 H$ for the first three torsional modes (TMs); and (**b**) the first three bending modes (BMs); and (**c**) the magnitude of the displacement vector of the first three torsion modes plotted and calculated with COMSOL$^{(r)}$ Multiphysics v. 5.3a (COMSOL AB, Stockholm, Sweden).

As shown earlier in Figure 3a, the minima of $C_{11}(H)$ occur at the same magnetic bias fields as the maxima of $C_{12}(H)$. Whereas $C_{11}(H)$ softens upon the application of a magnetic bias field, $C_{12}(H)$ increases. However, upon application of a magnetic field, they both reduce the resonance frequency of bending modes. Consequently, their corresponding frequency factors have opposite signs and $\partial_C f_{r,12} < 0$.

7. Summary and Conclusions

In summary, we provide an experimental and theoretical study on the delta-E effect, the normalized resonance frequency change Δf_r (defined in Section 6.3) and the sensitivity of first and higher-order bending modes (BMs) and torsion modes (TMs). The study was conducted on a magnetoelectric thin-film cantilever with a soft magnetic FeCoSiB–Cr multilayer and an electrode design that enables the excitation of various resonance modes. A general expression was developed that permits the detailed analysis of the magnetic sensitivity of arbitrary resonance modes. An electromechanical finite element method model was set up to describe the resonator and the electrical sensitivity. It was combined with a magnetoelastic macrospin model to include the tensor of the linearized delta-E effect for isotropic magnetostriction in the approximation of negligible magnetostrictive self-energy. The models are valid for moderately high-operation frequencies, where electrodynamic and magnetodynamic effects can be omitted.

The delta-E effect model is discussed in detail for here the most relevant components C_{66}, C_{11}, and C_{12} of the magnetic field-dependent stiffness tensor. Simulation results imply that the C_{66} component potentially offers a ten-fold higher contribution to the magnetic sensitivity than the C_{11} component. With an increasing tilt of the magnetic easy axis, this factor reduces to approximately four at an easy axis angle aligned at 80° relative to the long axis of the cantilever. However, the measurements and simulations of the current design confirm that the TMs exhibit a systematically smaller electromechanical response compared to BMs, which can significantly diminish the potential gain in sensitivity. Possible ways of improvement are sketched out. From simulated and measured resonance frequency curves $f_r(H)$ we found that the maximum normalized resonance frequency change Δf_r and the magnetic sensitivity of TMs reduce with the increasing mode number due to the increasing contribution of C_{11} to the resonance frequency. Hence, the dependency of TMs on the mode number is opposite to the one observed for BMs and caused by a different mechanism.

In conclusion, the delta-E effect of the C_{66} component shows the promising potential of significantly increasing the magnetic sensitivity and the maximum normalized resonance frequency change Δf_r. However, the efficient electrical excitation of TMs remains challenging for achieving high electrical sensitivity. Generally, the results imply that the delta-E effect of different C_{ij} can have opposite effects on Δf_r, depending on the resonance mode. This was demonstrated in the example of torsion modes. Because the contribution of C_{11} increases with the torsion mode number, the first-order torsion mode shows the highest magnetic sensitivity. In addition to fundamental insights on the delta-E effect in higher resonance modes, a model for the electrical and the magnetic sensitivity was presented. The results are not only relevant for the development of magnetoelastic magnetic field sensors, but also for frequency tunable devices based on the delta-E effect.

Author Contributions: Conceptualization, B.S.; methodology, B.S., E.V.G., R.-M.F.; validation, B.S.; formal analysis, B.S., E.V.G.; investigation, B.S., E.V.G., S.Z., C.K.; writing—original draft preparation, B.S.; writing—review and editing, B.S., E.V.G., R.-M.F., S.Z., C.K., D.M., J.M., F.F.; visualization, B.S., E.V.G.; supervision, D.M., J.M., F.F.; project administration, D.M., J.M., F.F.; funding acquisition, D.M., J.M., F.F. All authors have read and agreed to the published version of the manuscript.

Funding: This research was funded by the German Research Foundation (DFG) via the collaborative research center CRC 1261.

Institutional Review Board Statement: Not applicable.

Informed Consent Statement: Not applicable.

Data Availability Statement: Not applicable.

Acknowledgments: The authors thank Christoph Elis for performing the density measurements on FeCoSiB.

Conflicts of Interest: The authors declare no conflict of interest. The funders had no role in the design of the study; in the collection, analyses, or interpretation of data; in the writing of the manuscript, and in the decision to publish the results.

Appendix A. Magnetoelastic Model

Appendix A.1. Definition of Vectors

In the following, a detailed definition of all vectors is given in the spherical coordinate system. All polar angles are denoted by θ and all azimuthal angles by φ, with an index if it is not the angles of the reduced magnetization. The reduced magnetization vector is denoted as

$$\overline{m} = \begin{bmatrix} \cos\varphi\sin\theta & \sin\varphi\sin\theta & \cos\theta \end{bmatrix}^T, \quad (A1)$$

The easy axis unit vector $\overline{EA}$ is given by

$$\overline{EA} = \begin{bmatrix} \cos\varphi_{EA}\sin\theta_{EA} & \sin\varphi_{EA}\sin\theta_{EA} & \cos\theta_{EA} \end{bmatrix}^T. \quad (A2)$$

The vector $\overline{H}$ of the external applied field and $\overline{H}_d$ of the demagnetizing field are given by

$$\overline{H} = H\begin{bmatrix} \cos\varphi_H\sin\theta_H & \sin\varphi_H\sin\theta_H & \cos\theta_H \end{bmatrix}^T, \quad (A3)$$

$$\overline{H}_d = -M_S\begin{bmatrix} D_{11}m_1 & D_{22}m_2 & D_{33}m_3 \end{bmatrix}^T. \quad (A4)$$

For all higher-order tensors, we use the Voigt notation. The magnetostriction tensor is then given by

$$\lambda = \frac{3}{2}\lambda_s \begin{bmatrix} m_1^2 - \frac{1}{3} & m_2^2 - \frac{1}{3} & m_3^2 - \frac{1}{3} & 2m_2m_3 & 2m_1m_3 & 2m_1m_2 \end{bmatrix}^T. \quad (A5)$$

Appendix A.2. General Expression for ΔS_{ij}

From the equilibrium conditions, one can write

$$\Delta S_{ij} = \frac{\partial \lambda_i}{\partial \sigma_j} = \frac{\partial \lambda_i}{\partial \varphi}\frac{\partial \varphi}{\partial \sigma_j} + \frac{\partial \lambda_i}{\partial \theta}\frac{\partial \theta}{\partial \sigma_j} = \frac{\partial \lambda_i}{\partial \varphi}\left(-\frac{\partial u_\varphi}{\partial \sigma_j}\bigg/\frac{\partial u_\varphi}{\partial \varphi}\right) + \frac{\partial \lambda_i}{\partial \theta}\left(-\frac{\partial u_\theta}{\partial \sigma_j}\bigg/\frac{\partial u_\theta}{\partial \theta}\right)$$
$$= -\frac{\partial \lambda_i}{\partial \varphi}\frac{\partial u_\varphi}{\partial \sigma_j}\frac{1}{u_{\varphi\varphi}} - \frac{\partial \lambda_i}{\partial \theta}\frac{\partial u_\theta}{\partial \sigma_j}\frac{1}{u_{\theta\theta}}. \quad (A6)$$

Appendix A.3. Derivatives of the Energy Density Functional

For in-plane magnetization ($\theta = \pi/2$), easy axis ($\theta_{EA} = \pi/2$), and magnetic field ($\theta_H = \pi/2$), the second-order derivatives of u (Equation (7)) are given by

$$u_{\theta\theta} \frac{\partial^2 u}{\partial \varphi^2} = 3\lambda_s[\sigma_{11}\cos(2\varphi) - \sigma_{22}\cos(2\varphi) + 2\sigma_{12}\sin(2\varphi)] + \mu_0 M_s H \cos(\varphi - \varphi_H) + 2K\cos(2[\varphi - \varphi_{EA}]) \\ + \mu_0 M_s^2 (D_{22} - D_{11})\cos(2\varphi) \quad (A7)$$

$$u_{\theta\theta} \frac{\partial^2 u}{\partial \theta^2} = 3\lambda_s[\sigma_{11}\cos^2\varphi + \sigma_{22}\sin^2\varphi - \sigma_{33} + \sigma_{12}\sin(2\varphi)] + \mu_0 M_s H \cos(\varphi - \varphi_H) + 2K\cos^2(\varphi - \varphi_{EA}) \\ - \mu_0 M_s^2 (D_{11}\cos^2\varphi + D_{22}\sin^2\varphi - D_{33}) \quad (A8)$$

Appendix B. Resonance Frequencies and Sensitivities

A summary of the maximum measured resonance frequencies $f_{r,max}$ at $\mu_0 H = -10$ mT is given in Table A1, together with the corresponding quality factor Q_{max} and the maximum magnetic sensitivities from Figure 6. The bending modes (BMs) and the torsion modes (TMs) were all measured using only electrode E1. Hence, the electrodes are not optimized for the bending modes. Additionally, the maximum quality factor of the BMs is a factor of three smaller than in TM1. Due to both factors, the electrical sensitivity of BMs is significantly smaller than that of TMs for our cantilever.

Table A1. Measured resonance frequencies $f_{r,max}$ of the six modes analyzed and the quality factor Q_{max}, both measured in magnetic saturation at $\mu_0 H = -10$ mT. The maximum magnetic sensitivity S_H and the maximum relative magnetic sensitivity $S_{H,r}$ are obtained from Figure 6b. The maximum relative electrical sensitivities found are given by $S_{\phi,r}$ and $S_{Y,r}$.

Mode	$f_{r,max}$ (kHz)	$f_{r,model}$ (kHz)	Q_{max}	S_H (Hz/mT)	$S_{H,r}$ (1/T)	$S_{\phi,r}$ (°)	$S_{Y,r}$ (µS)
TM1	26.26	26.26	900	330	12.6	4780	850
TM2	87.48	88.20	700	837	9.5	59	27.5
TM3	175.15	179.20	280	542	3.0	115	95
BM1	7.65	7.80	300	26.7	3.5	2550	150
BM2	47.18	48.55	300	432	9.2	220	70
BM3	121.40	116.20	300	811	6.7	18	42

Appendix C. Geometry and Material Parameters

Appendix C.1. Geometry

The poly-Si cantilever was measured with an optical microscope to be $L = 3.12$ mm long and $W = 2.15$ mm wide. For the simulations, the length in the model was slightly adjusted within the measurement accuracy to 3.116 mm. The magnetostrictive layer was deposited directly at the clamping on the bottom side of the poly-Si cantilever and has a width of $W_{mag} = 2$ mm and a length of $L_{mag} = 3.05$ mm. The AlN layer is of the same geometry but deposited on top of the poly-Si. The electrodes E_1 and E_2 are positioned at the clamping on top of the AlN layer and the left and right edge, respectively. An additional parallel capacitance of $C_0 = 17.7$ pF is used. It is consistent with the area of the bond pads, the conduction lines, and the relative electrical permittivity used for the AlN.

Appendix C.2. Substrate (Poly-Si)

For the poly-silicon substrate, we use isotropic material parameters, with Young's modulus $E_{Si} = 160$ GPa [66,67] a Poisson's ratio of $v_{Si} = 0.22$ [67] and a mass density of $\rho_{Si} = 2300$ kg/m^3 [68].

Appendix C.3. Magnetic Material (FeCoSiB)

The mass density of FeCoSiB was experimentally determined by estimating the volume with profilometer measurements and the mass with a microbalance. The measurements were performed on a 6-inch wafer with a mean FeCoSiB layer thickness of approximately 1.5 µm. A density of $\rho_{FeCoSiB} = (7870 \pm 1350)$ kg/m^3 was obtained. For the simulation, a corresponding mass density of $\rho_{FeCoSiB} = 7700$ kg/m^3 was used. For the stiffness tensor of the mechanically isotropic magnetic film, we use:

$$C_{m,ij} = \begin{bmatrix} C_{m,11} & C_{m,12} & C_{m,12} & 0 & 0 & 0 \\ C_{m,12} & C_{m,11} & C_{m,12} & 0 & 0 & 0 \\ C_{m,12} & C_{m,12} & C_{m,11} & 0 & 0 & 0 \\ 0 & 0 & 0 & C_{m,44} & 0 & 0 \\ 0 & 0 & 0 & 0 & C_{m,55} & 0 \\ 0 & 0 & 0 & 0 & 0 & C_{m,66} \end{bmatrix} \quad (A9)$$

Using a Young's modulus of $E = 150$ GPa and a Poisson's ratio of $v = 0.3$, both at fixed magnetization, the non-zero components of the fixed magnetization stiffness tensor are:

$$C_{m,11} = C_{m,22} = C_{m,33} = \frac{E(1-v)}{(1+v)(1-2v)} = 201.92 \text{ GPa} \quad (A10)$$

$$C_{m,12} = C_{m,13} = C_{m,23} = \frac{E(v)}{(1+v)(1-2v)} = 86.54 \text{ GPa} \quad (A11)$$

$$C_{m,66} = C_{m,55} = C_{m,44} = \frac{E}{2(1+v)} = 57.69 \text{ GPa} \quad (A12)$$

For the magnetoelastic simulations, we use a saturation magnetic flux density of $\mu_0 M_s = 1.5$ T [17] and saturation magnetostriction of $\lambda_s = 35$ ppm [17].

Appendix C.4. Piezoelectric Material (AlN)

For the stiffness matrix C_{AlN} and the piezoelectric stress-charge coefficient tensor d we use values based on ab initio calculations [69]. Those tend to overestimate d and are here slightly adjusted. The stiffness tensor is:

$$C_{AlN} = \begin{bmatrix} 410.2 & 142.2 & 110.1 & 0 & 0 & 0 \\ 142.2 & 410.2 & 110.1 & 0 & 0 & 0 \\ 110.1 & 110.1 & 385.0 & 0 & 0 & 0 \\ 0 & 0 & 0 & 122.9 & 0 & 0 \\ 0 & 0 & 0 & 0 & 122.9 & 0 \\ 0 & 0 & 0 & 0 & 0 & 134.0 \end{bmatrix} \text{ GPa} \quad (A13)$$

The piezoelectric coupling tensor $c_e = dC_{AlN}$ results as

$$c_e = \begin{bmatrix} 0 & 0 & 0 & 0 & -0.27828 & 0 \\ 0 & 0 & 0 & -0.27828 & 0 & 0 \\ -0.4496 & -0.4496 & 1.41 & 0 & 0 & 0 \end{bmatrix} \frac{\text{pC}}{\text{m}^2}. \quad (A14)$$

For the density, we use $\rho_{AlN} = 3300$ kg/m^3 and for the electrical permittivity $\varepsilon_{el} = \varepsilon_r \varepsilon_0$ with the electrical vacuum permittivity ε_0 and the relative electrical permittivity ε_r, given by

$$\varepsilon_r = \begin{bmatrix} 9.2081 & 0 & 0 \\ 0 & 9.2081 & 0 \\ 0 & 0 & 10.1192 \end{bmatrix} \quad (A15)$$

References

1. Reermann, J.; Durdaut, P.; Salzer, S.; Demming, T.; Piorra, A.; Quandt, E.; Frey, N.; Höft, M.; Schmidt, G. Evaluation of magnetoelectric sensor systems for cardiological applications. *Meas. J. Int. Meas. Confed.* **2018**, *116*, 230–238. [CrossRef]
2. Zuo, S.; Schmalz, J.; Ozden, M.-O.; Gerken, M.; Su, J.; Niekiel, F.; Lofink, F.; Nazarpour, K.; Heidari, H. Ultrasensitive Magnetoelectric Sensing System for pico-Tesla MagnetoMyoGraphy. *IEEE Trans. Biomed. Circuits Syst.* **2020**, *1*. [CrossRef]
3. Kwong, J.S.W.; Leithäuser, B.; Park, J.W.; Yu, C.M. Diagnostic value of magnetocardiography in coronary artery disease and cardiac arrhythmias: A review of clinical data. *Int. J. Cardiol.* **2013**, *167*, 1835–1842. [CrossRef] [PubMed]

4. Röbisch, V.; Salzer, S.; Urs, N.O.; Reermann, J.; Yarar, E.; Piorra, A.; Kirchhof, C.; Lage, E.; Höft, M.; Schmidt, G.U.; et al. Pushing the detection limit of thin film magnetoelectric heterostructures. *J. Mater. Res.* **2017**, *32*, 1009–1019. [CrossRef]
5. Kneller, E. *Ferromagnetismus*; Springer: Berlin/Heidelberg, Germany, 1962; ISBN 9783642866951.
6. Livingston, J.D. Magnetomechanical properties of amorphous metals. *Phys. Status Solidi A* **1982**, *70*, 591–596. [CrossRef]
7. Lee, E.W. Magnetostriction and Magnetomechanical Effects. *Rep. Prog. Phys.* **1955**, *18*, 184–229. [CrossRef]
8. De Lacheisserie, T. *Theory and Application of Magnetoelasticity*; CRC Press: Boca Raton, FL, USA, 1993; ISBN 9780849369346.
9. Bou Matar, O.; Robillard, J.F.; Vasseur, J.O.; Hladky-Hennion, A.C.; Deymier, P.A.; Pernod, P.; Preobrazhensky, V. Band gap tunability of magneto-elastic phononic crystal. *J. Appl. Phys.* **2012**, *111*. [CrossRef]
10. Mazzamurro, A.; Dusch, Y.; Pernod, P.; Matar, O.B.; Addad, A.; Talbi, A.; Tiercelin, N. Giant magnetoelastic coupling in Love acoustic waveguide based on uniaxial multilayered TbCo2/FeCo nanostructured thin film on Quartz ST-cut. *Phys. Rev. Appl.* **2020**, *13*. [CrossRef]
11. Del Moral, A. Magnetostriction and magnetoelasticity theory: A modern view. In *Handbook of Magnetism and Advanced Magnetic Materials*; Kronmüller, H., Parkin, S., Eds.; Wiley-Interscience: Hoboken, NJ, USA, 2007; Volume 1, ISBN 9780470022177.
12. Atkinson, D.; Squire, P.T.; Gibbs, M.R.J.; Atalay, S.; Lord, D.G. The effect of annealing and crystallization on the magnetoelastic properties of Fe-Si-B amorphous wire. *J. Appl. Phys.* **1993**, *73*, 3411–3417. [CrossRef]
13. Barandiarán, J.M.; Gutiérrez, J.; García-Arribas, A.; Squire, P.T.; Hogsdon, S.N.; Atkinson, D. Comparison of magnetoelastic resonance and vibrating reed measurements of the large delta-E effect in amorphous alloys. *J. Magn. Magn. Mater.* **1995**, *140–144*, 273–274. [CrossRef]
14. Squire, P.T. Domain model for magnetoelastic behaviour of uniaxial ferromagnets. *J. Magn. Magn. Mater.* **1995**, *140–144*, 1829–1830. [CrossRef]
15. Gutiérrez, J.; García-Arribas, A.; Garitaonaindia, J.S.; Barandiarán, J.M.; Squire, P.T. ΔE effect and anisotropy distribution in metallic glasses with oblique easy axis induced by field annealing. *J. Magn. Magn. Mater.* **1996**, *157–158*, 543–544. [CrossRef]
16. Squire, P.T.; Atalay, S.; Chiriac, H. ΔE effect in amorphous glass covered wires. *IEEE Trans. Magn.* **2000**, *36*, 3433–3435. [CrossRef]
17. Ludwig, A.; Quandt, E. Optimization of the delta-E effect in thin films and multilayers by magnetic field annealing. *IEEE Trans. Magn.* **2002**, *38*, 2829–2831. [CrossRef]
18. Dong, C.; Li, M.; Liang, X.; Chen, H.; Zhou, H.; Wang, X.; Gao, Y.; McConney, M.E.; Jones, J.G.; Brown, G.J.; et al. Characterization of magnetomechanical properties in FeGaB thin films. *Appl. Phys. Lett.* **2018**, *113*, 262401. [CrossRef]
19. Yoshizawa, N.; Yamamoto, I.; Shimada, Y. Magnetic field sensing by an electrostrictive/magnetostrictive composite resonator. *IEEE Trans. Magn.* **2005**, *41*, 4359–4361. [CrossRef]
20. Nan, T.; Hui, Y.; Rinaldi, M.; Sun, N.X. Self-biased 215 MHz magnetoelectric NEMS resonator for ultra-sensitive DC magnetic field detection. *Sci. Rep.* **2013**, *3*, 1985. [CrossRef]
21. Hui, Y.; Nan, T.; Sun, N.X.; Rinaldi, M. High resolution magnetometer based on a high frequency magnetoelectric MEMS-CMOS oscillator. *J. Microelectromech. Syst.* **2015**, *24*, 134–143. [CrossRef]
22. Li, M.; Matyushov, A.; Dong, C.; Chen, H.; Lin, H.; Nan, T.; Qian, Z.; Rinaldi, M.; Lin, Y.; Sun, N.X. Ultra-sensitive NEMS magnetoelectric sensor for picotesla DC magnetic field detection. *Appl. Phys. Lett.* **2017**, *110*, 143510. [CrossRef]
23. Osiander, R.; Ecelberger, S.A.; Givens, R.B.; Wickenden, D.K.; Murphy, J.C.; Kistenmacher, T.J. A microelectromechanical-based magnetostrictive magnetometer. *Appl. Phys. Lett.* **1996**, *69*, 2930–2931. [CrossRef]
24. Gojdka, B.; Jahns, R.; Meurisch, K.; Greve, H.; Adelung, R.; Quandt, E.; Knöchel, R.; Faupel, F. Fully integrable magnetic field sensor based on delta-E effect. *Appl. Phys. Lett.* **2011**, *99*, 1–4. [CrossRef]
25. Kiser, J.; Finkel, P.; Gao, J.; Dolabdjian, C.; Li, J.; Viehland, D. Stress reconfigurable tunable magnetoelectric resonators as magnetic sensors. *Appl. Phys. Lett.* **2013**, *102*. [CrossRef]
26. Jahns, R.; Zabel, S.; Marauska, S.; Gojdka, B.; Wagner, B.; Knöchel, R.; Adelung, R.; Faupel, F. Microelectromechanical magnetic field sensor based on the delta-E effect. *Appl. Phys. Lett.* **2014**, *105*, 2012–2015. [CrossRef]
27. Zabel, S.; Kirchhof, C.; Yarar, E.; Meyners, D.; Quandt, E.; Faupel, F. Phase modulated magnetoelectric delta-E effect sensor for sub-nano tesla magnetic fields. *Appl. Phys. Lett.* **2015**, *107*. [CrossRef]
28. Zabel, S.; Reermann, J.; Fichtner, S.; Kirchhof, C.; Quandt, E.; Wagner, B.; Schmidt, G.; Faupel, F. Multimode delta-E effect magnetic field sensors with adapted electrodes. *Appl. Phys. Lett.* **2016**, *108*, 222401. [CrossRef]
29. Bian, L.; Wen, Y.; Li, P.; Wu, Y.; Zhang, X.; Li, M. Magnetostrictive stress induced frequency shift in resonator for magnetic field sensor. *Sens. Actuators A Phys.* **2016**, *247*, 453–458. [CrossRef]
30. Bennett, S.P.; Baldwin, J.W.; Staruch, M.; Matis, B.R.; Lacomb, J.; Van 'T Erve, O.M.J.; Bussmann, K.; Metzler, M.; Gottron, N.; Zappone, W.; et al. Magnetic field response of doubly clamped magnetoelectric microelectromechanical AlN-FeCo resonators. *Appl. Phys. Lett.* **2017**, *111*. [CrossRef]
31. Staruch, M.; Yang, M.-T.; Li, J.F.; Dolabdjian, C.; Viehland, D.; Finkel, P. Frequency reconfigurable phase modulated magnetoelectric sensors using ΔE effect. *Appl. Phys. Lett.* **2017**, *111*, 2–6. [CrossRef]
32. Bian, L.; Wen, Y.; Wu, Y.; Li, P.; Wu, Z.; Jia, Y.; Zhu, Z. A Resonant Magnetic Field Sensor with High Quality Factor Based on Quartz Crystal Resonator and Magnetostrictive Stress Coupling. *IEEE Trans. Electron Devices* **2018**, *65*, 2585–2591. [CrossRef]
33. Lukat, N.; Friedrich, R.M.; Spetzler, B.; Kirchhof, C.; Arndt, C.; Thormählen, L.; Faupel, F.; Selhuber-Unkel, C. Mapping of magnetic nanoparticles and cells using thin film magnetoelectric sensors based on the delta-E effect. *Sens. Actuators A Phys.* **2020**, *309*, 1–8. [CrossRef]

34. Spetzler, B.; Bald, C.; Durdaut, P.; Reermann, J.; Kirchhof, C.; Teplyuk, A.; Meyners, D. Exchange biased delta—E effect enables the detection of low frequency pT magnetic fields with simultaneous localization. *Sci. Rep.* **2021**, *11*, 1–14. [CrossRef]
35. Becker, R.; Döring, W. *Ferromagnetismus*; Springer: Berlin/Heidelberg, Germany, 1939; ISBN 9783642471124.
36. Atalay, S.; Squire, P.T. Magnetomechanical damping in FeSiB amorphous wires. *J. Appl. Phys.* **1993**, *73*, 871–875. [CrossRef]
37. Squire, P. Magnetostrictive materials for sensors and actuators. *Ferroelectrics* **1999**, *228*, 305–319. [CrossRef]
38. Charles, F.K. *Development of the Shear Wave*; Magnetometer; University of Bath: Bath, UK, 1992.
39. Kittmann, A.; Durdaut, P.; Zabel, S.; Reermann, J.; Schmalz, J.; Spetzler, B.; Meyners, D.; Sun, N.X.; McCord, J.; Gerken, M.; et al. Wide Band Low Noise Love Wave Magnetic Field Sensor System. *Sci. Rep.* **2018**, *8*, 1–10. [CrossRef]
40. Schmalz, J.; Spetzler, B.; Faupel, F.; Gerken, M. Love Wave Magnetic Field Sensor Modeling—From 1D to 3D Model. In Proceedings of the 2019 International Conference on Electromagnetics in Advanced Applications (ICEAA), Granada, Spain, 9–13 September 2019; pp. 765–769.
41. Schell, V.; Müller, C.; Durdaut, P.; Kittmann, A.; Thormählen, L.; Lofink, F.; Meyners, D.; Höft, M.; McCord, J.; Quandt, E. Magnetic anisotropy controlled FeCoSiB thin films for surface acoustic wave magnetic field sensors. *Appl. Phys. Lett.* **2020**, *116*. [CrossRef]
42. Schmalz, J.; Kittmann, A.; Durdaut, P.; Spetzler, B.; Faupel, F.; Höft, M.; Quandt, E.; Gerken, M. Multi-mode love-wave SAW magnetic-field sensors. *Sensors* **2020**, *20*, 3421. [CrossRef]
43. Sárközi, Z.; Mackay, K.; Peuzin, J.C. Elastic properties of magnetostrictive thin films using bending and torsion resonances of a bimorph. *J. Appl. Phys.* **2000**, *88*, 5827–5832. [CrossRef]
44. Staruch, M.; Matis, B.R.; Baldwin, J.W.; Bennett, S.P.; van 't Erve, O.; Lofland, S.; Bussmann, K.; Finkel, P. Large non-saturating shift of the torsional resonance in a doubly clamped magnetoelastic resonator. *Appl. Phys. Lett.* **2020**, *116*, 232407. [CrossRef]
45. Yarar, E.; Hrkac, V.; Zamponi, C.; Piorra, A.; Kienle, L.; Quandt, E. Low temperature aluminum nitride thin films for sensory applications. *AIP Adv.* **2016**, *6*, 075115. [CrossRef]
46. Durdaut, P.; Höft, M.; Friedt, J.-M.; Rubiola, E. Equivalence of Open-Loop and Closed-Loop Operation of SAW Resonators and Delay Lines. *Sensors* **2019**, *19*, 185. [CrossRef] [PubMed]
47. Spetzler, B.; Kirchhof, C.; Reermann, J.; Durdaut, P.; Höft, M.; Schmidt, G.; Quandt, E.; Faupel, F. Influence of the quality factor on the signal to noise ratio of magnetoelectric sensors based on the delta-E effect. *Appl. Phys. Lett.* **2019**, *114*, 183504. [CrossRef]
48. Durdaut, P.; Rubiola, E.; Friedt, J.; Muller, C.; Spetzler, B.; Kirchhof, C.; Meyners, D.; Quandt, E.; Faupel, F.; McCord, J.; et al. Fundamental Noise Limits and Sensitivity of Piezoelectrically Driven Magnetoelastic Cantilevers. *J. Microelectromech. Syst.* **2020**, 1–15. [CrossRef]
49. Spetzler, B.; Kirchhof, C.; Quandt, E.; McCord, J.; Faupel, F. Magnetic Sensitivity of Bending-Mode Delta-E-Effect Sensors. *Phys. Rev. Appl.* **2019**, *12*, 1. [CrossRef]
50. Kittel, C. On the theory of ferromagnetic resonance absorption. *Phys. Rev.* **1948**, *73*, 155–161. [CrossRef]
51. Spetzler, B.; Golubeva, E.V.; Müller, C.; McCord, J.; Faupel, F. Frequency Dependency of the Delta-E Effect and the Sensitivity of Delta-E Effect Magnetic Field Sensors. *Sensors* **2019**, *19*, 4769. [CrossRef]
52. Gebert, A.; McCord, J.; Schmutz, C.; Quandt, E. Permeability and Magnetic Properties of Ferromagnetic NiFe/FeCoBSi Bilayers for High-Frequency Applications. *IEEE Trans. Magn.* **2007**, *43*, 2624–2626.
53. Avdiaj, S.; Setina, J.; Syla, N. Modeling of the piezoelectric effect using the finite-element method (FEM). *Mater. Technol.* **2009**, *43*, 283–291.
54. Yang, J. An Introduction to the theory of piezoelectricity. In *Advances in Mechanics and Mathematics*, 2nd ed.; Springer International Publishing: Cham, Switzerland, 2018; Volume 9, ISBN 9783030031367.
55. IEEE Standards Board. IEEE Standard on Magnetostrictive Materials: Piezomagnetic Nomenclature. *IEEE Trans. Sonics Ultrason.* **1973**, *20*, 67–77. [CrossRef]
56. COMSOL AB. *Structural Mechanics Module User's Guide, COMSOL Multiphysics (TM) v. 5.4*; COMSOL AB: Stockholm, Sweden, 2018.
57. Mudivarthi, C.; Datta, S.; Atulasimha, J.; Evans, P.G.; Dapino, M.J.; Flatau, A.B. Anisotropy of constrained magnetostrictive materials. *J. Magn. Magn. Mater.* **2010**, *322*, 3028–3034. [CrossRef]
58. Gutiérrez, J.; Muto, V.; Squire, P.T. Induced anisotropy and magnetoelastic properties in Fe-rich metallic glasses. *J. Non. Cryst. Solids* **2001**, *287*, 417–420. [CrossRef]
59. McCord, J. Progress in magnetic domain observation by advanced magneto-optical microscopy. *J. Phys. D. Appl. Phys.* **2015**, *48*, 333001. [CrossRef]
60. Aharoni, A. Demagnetizing factors for rectangular ferromagnetic prisms. *J. Appl. Phys.* **1998**, *83*, 3432–3434. [CrossRef]
61. Varadan, V.K.; Vinoy, K.J.; Gopalakrishnan, S. *Smart Material Systems and MEMS: Design and Development Methodologies*; John Wiley & Sons: Hoboken, NJ, USA, 2007; ISBN 0470093617.
62. Krey, M.; Hähnlein, B.; Tonisch, K.; Krischok, S.; Töpfer, H. Automated parameter extraction of ScAlN MEMS devices using an extended euler–bernoulli beam theory. *Sensors* **2020**, *20*, 1001. [CrossRef]
63. Akiyama, M.; Umeda, K.; Honda, A.; Nagase, T. Influence of scandium concentration on power generation figure of merit of scandium aluminum nitride thin films. *Appl. Phys. Lett.* **2013**, *102*. [CrossRef]

64. Fichtner, S.; Wolff, N.; Krishnamurthy, G.; Petraru, A.; Bohse, S.; Lofink, F.; Chemnitz, S.; Kohlstedt, H.; Kienle, L.; Wagner, B. Identifying and overcoming the interface originating c-axis instability in highly Sc enhanced AlN for piezoelectric microelectromechanical systems. *J. Appl. Phys.* **2017**, *122*. [CrossRef]
65. Su, J.; Niekiel, F.; Fichtner, S.; Thormaehlen, L.; Kirchhof, C.; Meyners, D.; Quandt, E.; Wagner, B.; Lofink, F. AlScN-based MEMS magnetoelectric sensor. *Appl. Phys. Lett.* **2020**, *117*, 132903. [CrossRef]
66. Sharpe, W.N.; Turner, K.T.; Edwards, R.L. Tensile testing of polysilicon. *Exp. Mech.* **1999**, *39*, 162–170. [CrossRef]
67. Sharpe, W.N.; Yuan, B.; Vaidyanathan, R.; Edwards, R.L. Measurements of Young's modulus, Poisson's ratio, and tensile strength of polysilicon. *Proc. IEEE Micro Electro Mech. Syst.* **1997**, 424–429. [CrossRef]
68. Marc, J. *Madou Fundamentals of Microfabrication: The Science of Miniaturization*, 2nd ed.; CRC Press: Boca Raton, FL, USA, 2002; ISBN 0849308267.
69. Caro, M.A.; Zhang, S.; Riekkinen, T.; Ylilammi, M.; Moram, M.A.; Lopez-Acevedo, O.; Molarius, J.; Laurila, T. Piezoelectric coefficients and spontaneous polarization of ScAlN. *J. Phys. Condens. Matter* **2015**, *27*, 245901. [CrossRef] [PubMed]

Article

Teaching Magnetoelectric Sensing to Secondary School Students—Considerations for Educational STEM Outreach

Cara Broß [1,*], Carolin Enzingmüller [1,*], Ilka Parchmann [1] and Gerhard Schmidt [2]

1 IPN—Leibniz Institute for Science and Mathematics Education at Kiel University, 24118 Kiel, Germany; parchmann@leibniz-ipn.de
2 Institute for Electrical Engineering and Information Engineering, Kiel University, 24143 Kiel, Germany; gus@tf.uni-kiel.de
* Correspondence: bross@leibniz-ipn.de (C.B.); enzingmueller@leibniz-ipn.de (C.E.)

Abstract: A major challenge in modern society is the need to increase awareness and excitement with regard to science, technology, engineering and mathematics (STEM) and related careers directly or among peers and parents in order to attract future generations of scientists and engineers. The numbers of students aiming for an engineering degree are low compared to the options available and the workforce needed. This may, in part, be due to a traditional lack of instruction in this area in secondary school curricula. In this regard, STEM outreach programs can complement formal learning settings and help to promote engineering as well as science to school students. In a long-term outreach collaboration with scientists and engineers, we developed an outreach program in the field of magnetoelectric sensing that includes an out-of-school project day and various accompanying teaching materials. In this article, we motivate the relevance of the topic for educational outreach, share the rationales, objectives and aims, models and implementation strategies of our program and provide practical advice for those interested in outreach in the field of magnetoelectric sensing.

Keywords: public understanding/outreach; ME sensors; medical sensing; biomagnetic sensing; interdisciplinary/multidisciplinary

Citation: Broß, C.; Enzingmüller, C.; Parchmann, I.; Schmidt, G. Teaching Magnetoelectric Sensing to Secondary School Students—Considerations for Educational STEM Outreach. *Sensors* **2021**, *21*, 7354. https://doi.org/10.3390/s21217354

Academic Editor: Ahmed Toaha Mobashsher

Received: 21 September 2021
Accepted: 2 November 2021
Published: 5 November 2021

Publisher's Note: MDPI stays neutral with regard to jurisdictional claims in published maps and institutional affiliations.

Copyright: © 2021 by the authors. Licensee MDPI, Basel, Switzerland. This article is an open access article distributed under the terms and conditions of the Creative Commons Attribution (CC BY) license (https://creativecommons.org/licenses/by/4.0/).

1. Introduction

Attracting motivated and skilled workers in science, technology, engineering and mathematics (STEM) is difficult [1]. Adding to this problem, the ability to combine knowledge and skills from STEM disciplines will be increasing in demand as more and more institutions recognize innovations in integrated STEM areas as the key to the future economy and social progress [2–4]. Fostering these interdisciplinary competencies and skills takes time and careful planning. Often enough, schools do not have the capacity to fully prepare their students in this regard, as they traditionally focus on subject knowledge [5]. Furthermore, students' interest in certain STEM subjects tends to be low and continues to decline in developed countries [6], and misconceptions and stereotypes of science, scientists and engineers are rather common among students [7,8]. As a consequence, many school students do not want to pursue careers in these fields.

One way to address these issues could be by rethinking the way STEM is taught: Research indicates that STEM education can be a promising approach to strengthening students' abilities to combine knowledge and skills across disciplines as well as fostering their interest in STEM disciplines [2,5,9]. STEM education promotes the integration of the four disciplines, science, technology, engineering and mathematics, actively combining knowledge and competencies across disciplines within one context. Ideally, examples of authentic research and representations of researchers in the respective fields are integrated with STEM contents [10,11]. Combining the ideas of STEM education and authentic representations, university-led outreach initiatives can provide an excellent opportunity to deliver STEM education based on real-world interdisciplinary problems and authentic insights into modern science and engineering [9,12,13]. In doing so, they can complement formal school education and foster interdisciplinary skills and interest in STEM.

In this paper, we present a design-based STEM-integrated outreach activity centered around magnetoelectric sensing within the context of heart diagnostics. Besides STEM content, the program provides authentic insights into the interdisciplinary work of scientists and engineers. It was developed in collaboration with teachers, school students and researchers of the field of biomagnetic sensing and is based on our professional educational experience. The target group for the outreach activity consists of upper secondary school students in grades ten through thirteen (age fifteen and older) with prior knowledge in biology, chemistry and physics at the lower secondary level who have chosen the science profile as their A-level program.

2. Magnetoelectric Sensors for Medical Applications

2.1. Biomagnetic Sensing in the Collaborative Research Centre "Magnetoelectric Sensors"

Biomagnetic sensing is the measurement and analysis of magnetic fields of living organisms [14]. Such magnetic fields are induced by the same ionic currents that generate bioelectrical fields by flowing in and between cells. Biomagnetic sensing can therefore be used to monitor and analyze electrophysiological processes, such as cardiac electrical activity.

The most common method of measuring cardiac electrical activity is electrocardiography. Since its introduction well over a century ago, it has been optimized to derive information such as the rhythm of heartbeats or damage to the heart muscle from electrical potential differences on the body surface [15]. It was only in the last few decades that biomagnetic fields were discovered as a valuable source of medical information, introducing corresponding measurement approaches in the 1960s and 1970s [16]. These contactless measurements avoid inaccuracies due to poor electrical contact or the inhomogeneous conductivity of bodily tissue, which affect the electrical fields' propagation. They also have a higher spatial resolution and the possibility of providing sensors with better positioning, which leads to fewer exogenous signal artifacts [17].

However, biomagnetic signals are typically weak and can easily be polluted by environmental magnetic noise. The heart's magnetic field strength lies at amplitudes lower than 100 pT (which is about 500,000 times smaller than the earth's magnetic field's amplitude) and at frequencies between 0.01 and 100 Hz [14,17]. Currently, biomagnetic signals such as the heart's magnetic field can be measured with complex sensor technology such as superconducting quantum interference devices (SQUID) [18] or optically pumped magnetometers (OPMs) [19]. SQUID sensors require liquid cooling to operate and necessitate a magnetically shielded environment. OPMs are non-cryogenic but also require magnetic shielding for their use in medical contexts such as magnetocardiography. Both the necessary cooling and the magnetic shielding lead to high acquisition and operating costs and currently prevent mobile use.

The Collaborative Research Centre 1261 "Magnetoelectric Sensors" (short: CRC 1261) works on the development and evaluation of sensitive, low-cost and uncooled magnetoelectric (ME) sensors for medical contexts such as magnetocardiography (MCG) and magnetoencephalography (MEG). In contrast to SQUID sensors, which are made of single-phase magnetoelectric material, these new types of sensors are based on ME composites containing magnetostrictive and piezoelectric layers. By coupling the magnetostrictive strain to the piezoelectric phase (Figure 1), these sensors show a strong ME effect. One of the sensor principles that is the subject of research within the CRC 1261 is the bending beam principle.

The cantilever-type sensor is based on several measurement principles—the bending beam principle being one of them—combining knowledge of different disciplines (Figure 2). The sensor consists of a thin silicon layer as a substrate material on which at least two layers are applied: a piezoelectric and a magnetostrictive layer (Figure 1). Exposed to a magnetic field, the cantilever-type sensor bends similarly to a bimetal due to the magnetostrictive material. The bending of the sensor can be amplified using the sensor's (mechanical) self-resonance. Approaches from information technology (i.e., modulation and demodulation) can then compensate for the frequency shift involved in the effect amplification.

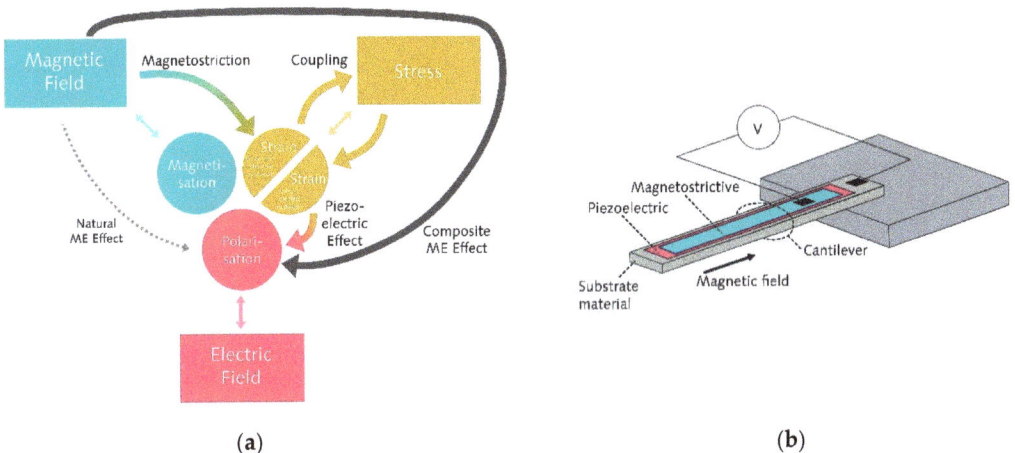

Figure 1. (a) Schematic representation of the composite ME effect and the natural ME effect. In this representation, only the direction that is used for magnetoelectric sensing is illustrated; (b) illustration of a ME cantilever beam sensor.

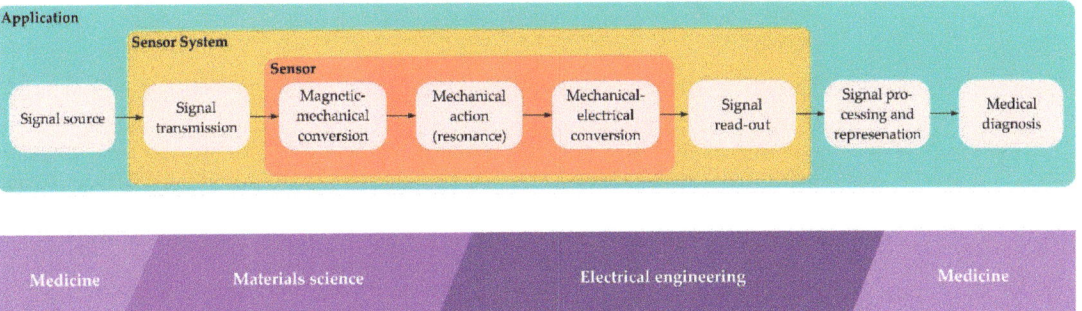

Figure 2. Biomagnetic sensing as a highly interdisciplinary application of magnetoelectric sensors (based on [20]).

2.2. Potential of Biomagnetic Sensing as a Topic for Outreach

STEM outreach is a form of science communication in which STEM institutions such as universities communicate findings and methods from current scientific research to the general public or schools. The goals of these outreach activities are often to increase STEM literacy, to foster interest in STEM and to promote positive attitudes towards STEM [12,21]. Biomagnetic sensing provides an excellent basis for STEM outreach activities, as it not only offers an ideal connection between students' personal frame of reference and interdisciplinary STEM content, but also is perfectly suited for the representation of authentic research.

2.2.1. Fostering Interdisciplinary Knowledge and Skills

Apart from fostering scientific knowledge within the separate STEM disciplines, biomagnetic sensing provides a context for fostering interdisciplinary knowledge and skills. In our modern society, there is an increasing demand for the ability to connect knowledge and skills from different disciplines, not only in the professional domain, but also in everyday life [2,3]. In order to prepare students for this demand, students have to be explicitly taught to build and use interdisciplinary skills and knowledge. A field of educational research that strives to achieve this is STEM education (e.g., see [9]). Its main objective is to actively link STEM subjects to one another and to combine knowledge across disciplines. Often, this interdisciplinary work is motivated by solving interdisciplinary problems or is embedded into contexts that require interdisciplinary actions such as health, biodiversity, or climate change. Research showed that this makes it possible to teach the skill of integrating knowledge across disciplines. Furthermore, successful integrated STEM education can foster knowledge within and interest in the separate disciplines [1,9].

Context-based learning provides a suitable framework for teaching and learning in this regard. Instead of teaching lists of information which seem to be irrelevant and disconnected for many students, contexts provide authentic problems to be explored for which knowledge is necessary and skills have to be further developed. This approach has been widely investigated and shows positive results, especially with regard to the interests of students [22].

In order to successfully integrate STEM subjects, a suitable context is crucial. On the one hand, the context has to be interesting and relevant to students, and on the other hand, its complexity has to correspond with the students' knowledge [2]. Biomagnetic sensing provides such a context. First, the medical context can illustrate the importance of basic physical-technical research for one's own life and society. Studies have repeatedly shown that medical and human biological contexts are seen as particularly attractive to a diverse group of students and can be used to generate interest and increase the perceived relevance of learning content [23–25]. Secondly, the aim of developing sensors for medical applications creates an authentic demand for interdisciplinary collaboration, e.g., in the form of knowledge integration processes that are necessary when designing an effective medical sensor. This further motivates the learning of interdisciplinary skills and knowledge.

2.2.2. Authentic Insights into the Nature of Science and Engineering

To understand the relevance and approaches of science for societal development, insights into what science is and how scientists and engineers work—concepts which are called the Nature of Science or Nature of Engineering, respectively [10,11]—are important in addition to content knowledge and skills. They prepare students to make informed decisions about the communication and application of science in their own lives and help to provide realistic experiences with STEM research that may result in them wanting to pursue a career within these fields [7]. Frameworks such as Programme for International Student Assessment (PISA) therefore incorporate the "knowledge about science" as well as the "knowledge of science" as elements of scientific literacy [26]. However, studies have shown that students often have a limited understanding of science and engineering as a way of knowing [10,27]. Learning about the nature of science requires students not only to simply engage in science and engineering activities, but also to explicitly address and discuss the nature of science and nature of engineering. That is to say, aspects of science or engineering are to be brought up in appropriate learning situations and illustrated by and discussed on the basis of those learning situations [11]. By doing so, one can convey insights into the inner workings of the respective disciplines, including questions such as "what is the nature of scientific knowledge?" or "how do engineers conduct their work?"

In the case of biomagnetic sensing research, interdisciplinary work is essential to developing a functional ME sensor suitable for medical applications. Portraying this interdisciplinarity not only allows one to discuss similarities and differences between disciplines such as materials science and electrical engineering, but also to highlight collaboration as an important feature of modern research. Studies show that social and cooperative components of modern research may have the potential to stimulate interest in science, especially among girls [28].

In summary, ME sensors and the context of biomagnetic sensing are very well suited to inspiring STEM learning. They can help students make connections across STEM disciplines and can be used to enrich knowledge about the nature of science and engineering through relevant contexts, interesting scientific principles that can be linked to the school curriculum, a variety of opportunities to integrate STEM disciplines, and the proximity to current research that allows portraying researchers and their work authentically.

3. Methods of Design

When designing educational interventions, tools and materials, a common and well-suited method is the so called design-based research approach [29]. Design-based research is led by two objectives [30]: first, the development of an effective practical intervention (e.g., an outreach event or learning materials), and second, the acquisition of theoretical insights concerning teaching and learning processes. It usually consists of an iterative process of design, implementation and evaluation [29].

In this outreach project, we used design-based research to first develop an out-of-school project day on biomagnetic sensing at a student laboratory, the Kiel Science Factory (https://www.forschungs-werkstatt.de/english/, accessed on 15 October 2021). An examination of the scientific content and the frame conditions such as students' prior knowledge and interests, an exploration of the student lab as well as ideas and feedback from CRC researchers formed the basis for the design of a prototype project day. Our goal was to develop and test a learning environment that provides insights into the interdisciplinary field of biomagnetic sensing involving scientists and engineers working in that area. From an educational point of view, the design framework applied the goals and principles of context-based learning and integrated STEM education by contextualizing the project day within a medical problem, by developing experiments that allow hands-on activities and by integrating media elements that directly showcase the work of the scientists and engineers. The focus for the accompanying research was to better understand students' interest and understanding in the science and scientists in the field of bio-magnetic sensing so that the design could be refined through several cycles. The major challenge in designing appropriate activities was to bridge the gap between traditional school science topics and the advanced content of the CRC. Explanations and hands-on experiments illustrating aspects of a real research process were required.

The whole process involves several iterations of design, implementation and evaluation to develop the project day (Figure 3). In a first step, the CRC scientists were involved in the process of developing prototype tasks and materials to ensure the relevance, correctness and authenticity of the content. We held a series of meetings with CRC scientists to discuss initial ideas and presented our progress regularly at retreat meetings in front of the entire CRC. The CRC scientists not only provided their expertise and feedback on teaching materials and experiments, but also actively contributed ideas that enriched the design. One example of this was the co-development of an experiment (see Box 1), a process that also led to a joint publication in a physics education journal [31].

After developing a prototype for the whole project day, we again consulted different experts: we invited pre-service teachers as experts for levels and structures of learning environments and doctoral students of the CRC as experts for research practices in the area of biomagnetic sensing to test and evaluate the project day. The pre-service teachers gave feedback on the didactic structure and the instructional tasks at the learning stations, while the doctoral students gave feedback on the authenticity of the materials and experiments. We refined the project day, taking into account feasibility, educational quality and authenticity based on the feedback received.

After these expert consultations, we tested the project day with our main target group, upper secondary school students. For this, we invited three classes ($N = 46$, grade 12–13) and their science teachers to our student laboratory. They had between four and seven years of physics experience and between two and five years of chemistry experience in school. A short questionnaire was designed to assess comprehensibility and attractiveness of the project day. Additionally, students were asked to write down what they liked, what they missed and what they learned during the project day. A final iteration with larger cohorts of students and statistical analyses of the effects on interest, insights into the nature of science and engineering, and an understanding of the content had been planned but could not yet be realized, due to the COVID-19 Pandemic.

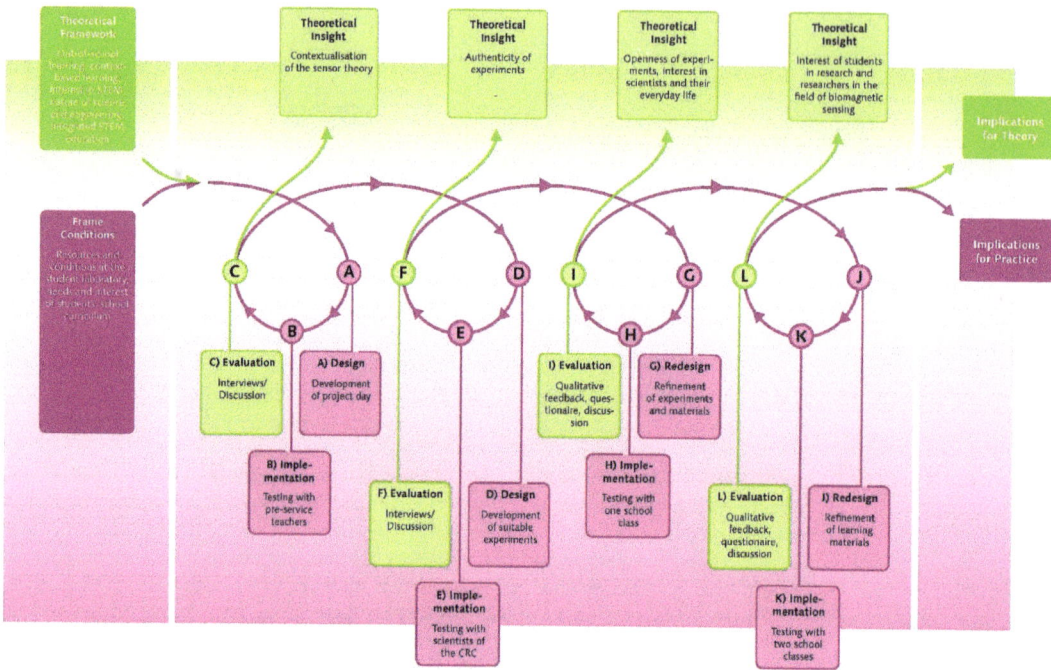

Figure 3. Illustration of the design-based research process that lead to the design of the project day.

4. Results of the First Design Process

The current design of the project day at the Kiel Science Factory is based on the research logic of the CRC 1261, so that the students can follow and understand the development of ME sensors for medical applications (Figure 4).

Introduction. At the beginning of the project day, the students get to know the medical context of heart diagnostics in order to capture their interest and increase their perception of relevance [23,25]. This is followed by a presentation of the objectives of the CRC 1261, which shows the benefits and added value of ME sensors for heart diagnostics. Both students and teachers confirmed that they enjoyed this introduction, which embedded the ME sensing approach in a medical context. After this introductory phase, the students begin to work in small groups with their own supervisor—university students that received content instruction beforehand—in order to encourage an active exchange regarding the respective sensor principles or concepts. In addition, this group setting ensures a high level of feedback, which further fosters the motivation and interest of the students and facilitates STEM learning [1,32]. This was also confirmed by the feedback of the students. They particularly appreciated the opportunity to ask questions at any time. In these small groups, the students go through four learning stations, at the end of which they build their own ME sensor based on information and introductions into the underlying principles.

Station 1: Magnetocardiography. At the first station, the supervisors explain the underlying principles of the electrophysiological activities of the heart as the basis for cardiac diagnostics. Students then apply this knowledge, record a classmate's electrocardiogram and learn to interpret it, drawing connections to their biology knowledge. Then, they are handed supplementary information on magnetocardiography and its advantages and disadvantages. Based on the knowledge acquired at this station, they compare the two diagnostic methods, electrocardiography and magnetocardiography.

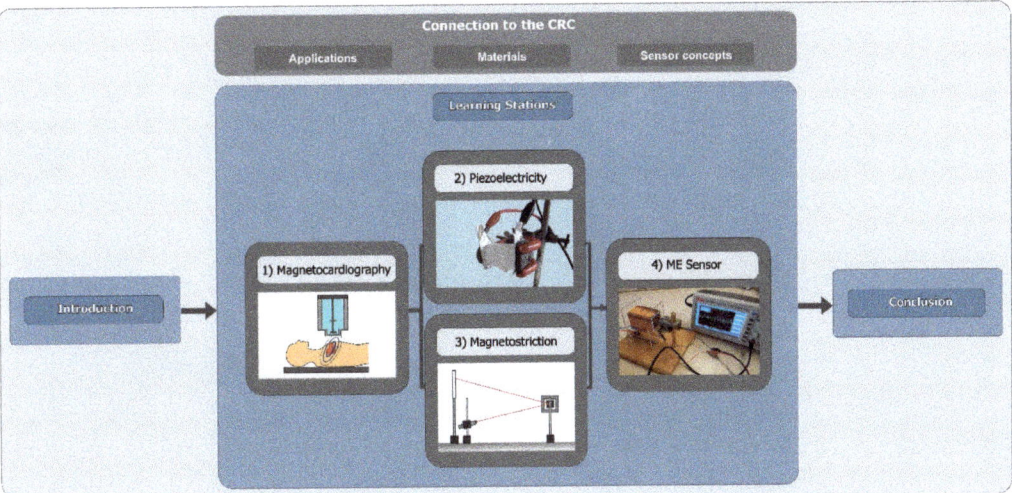

Figure 4. Structure of the project day at the Kiel Science Factory.

Stations 2 and 3: Piezoelectricity and Magnetostriction. At the following two stations, the students learn about piezoelectricity and magnetostriction as sensor material properties. Based on a model explanation of the piezoelectric phenomenon by the supervisor, students develop an experimental setup to measure the piezoelectric effect of piezoelectric crystals grown from Rochelle salt at station 2. For this purpose, they are provided with various materials, including piezoelectric crystals, aluminum foil as an electrode, a multimeter, cables, crocodile clips, different electrically isolating materials and stand equipment, which they can use for their experimental setup. They compare and discuss their results before moving on to station 3. Analogous to station 2, the supervisor of station 3 first explains the magnetostrictive phenomenon. Students then apply their knowledge by conducting a model experiment in which the magnetostrictive effect is made visible with the help of a magneto-optical sensor (for more information, refer to Box 1). Both stations with a focus on sensor materials were very well received, and the hands-on nature of these stations was highlighted particularly positively. Based on the feedback, we incorporated different levels of scaffolding—supporting students in acquiring knowledge—to allow students with different prior knowledge and skills to conduct their own inquiry.

Station 4: Magnetoelectric Sensor. At the last station, students combine the knowledge they have acquired about context, material properties and sensor principles in order to build a functioning ME sensor model from everyday materials. The ME sensor model is based on the basic principle of the ME cantilever beam sensor (Figure 1b), in which a piezoelectric, a magnetostrictive and a substrate material are combined. Strips cut from a CD serve as the substrate material, a piezoelectric element as the piezoelectric material and thin steel (or alternatively the magnetizable metal strip from the inside of security tags) as the magnetostrictive material (Figure 5). The coupling of these materials is achieved by using superglue. The ME sensor model can detect rather large magnetic fields generated by a coil with an AC voltage in the range of $U = 10\text{--}20$ V. This can be used as an opportunity to compare the sensitivity of the ME sensor models built by the students with the ME sensors professionally fabricated in the clean room by the CRC researchers. In the evaluation, building their own ME sensors was considered a highlight by the students. However, it was challenging to adapt the instructional support to the ability of the students. One class felt the sensor design process was too guided when they were given the materials needed for the sensor and were guided through the design process by their supervisor. However, after further opening up the design process by giving different sensor materials to choose from and scaling back support, a third of the student groups failed to develop a working ME sensor (since they chose a magnetostrictive material that was too thick) and were discouraged by the results. The supervisors were instructed to provide step-wise support when needed to resolve this issue.

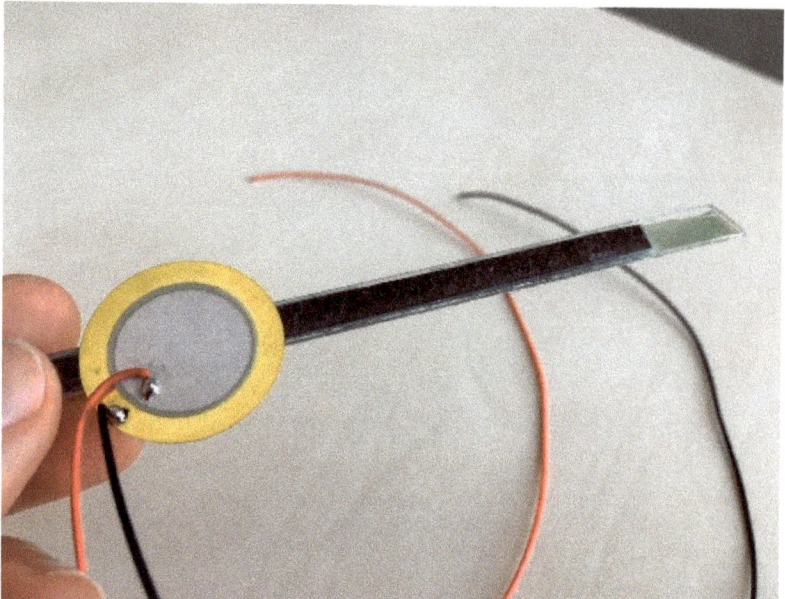

Figure 5. A student-built ME sensor model made out of a strip cut from a CD, a piezo-electric element, and thin steel.

Conclusion. At the end of the project day, the students compare their approach with the working process of the CRC researchers. This comparison was added to the project day after students noted that they would be interested in learning more about researchers and their daily lives at work. Another element included in the project day for the same reason proved ineffective and is therefore not part of the final design of the project day: We added a media station that provided insights into the work of CRC researchers through various media elements such as a 360-degree video and interviews with CRC researchers. This included, for instance, a description of a typical working day or insights into their personal motivation for their research. The 360-degree video depicted the sensor fabrication in the clean room.

Box 1. The Magnetostrictive Cantilever Beam Experiment.

> This experiment is part of the learning station 3 and illustrates the magnetostrictive effect using a cantilever-type sensor (for more detail see [31]). The magnetostrictive cantilever consists of a non-magnetostrictive substrate material and a magnetostrictive material attached to it. When this cantilever is exposed to a magnetic field parallel to its length, the length of the magnetostrictive material increases while the length of the non-magnetostrictive material remains constant. This causes the cantilever to bend and the tip of the cantilever to be displaced. The displacement is minimal and cannot be seen with the naked eye. We used an optical lever setup to make this effect visible for the students. A laser beam is directed onto the cantilever beam sensor and reflected onto a screen. The experimental setup includes the magnetostrictive cantilever, an electromagnet, a laser pointer and a screen (Figure 6, for more information on the materials see [31]).
> The cantilever beam is mounted so that it can be bent horizontally and is inserted inside a coil. A laser beam generated by the laser pointer is directed onto the tip of the reflective cantilever. When AC voltage is applied, the cantilever bends continuously, which in turn leads to the continuous displacement of the laser beam point on the screen. By applying alternating voltage with the self-resonance frequency of the cantilever beam, it is possible to amplify the movement of the cantilever and thus the laser beam point.
> This experiment can be integrated and modified in a number of ways for outreach purposes. For its use in the student lab, we focused on students' autonomous and exploratory learning. Therefore, the experimental setup and a demonstration of a magnetostrictive cantilever that oscillates in self-resonance frequency was presented to the students. Based on the definition of magnetostriction and—if necessary—with the help of prepared tips on cards students were to explain the oscillation of the cantilever. They were then able to use the experimental setup to determine the influence of different design aspects of the cantilever on the performance of the cantilever, while being exposed to different magnetic fields. Factors that could be compared included the thickness of the magnetostrictive material and the respective lengths of the magnetostrictive and non-magnetostrictive materials. These insights prepared the students for the design of their own ME sensor, which was to be manufactured from the materials they got to know in this experiment (with the addition of a piezoelectric material).

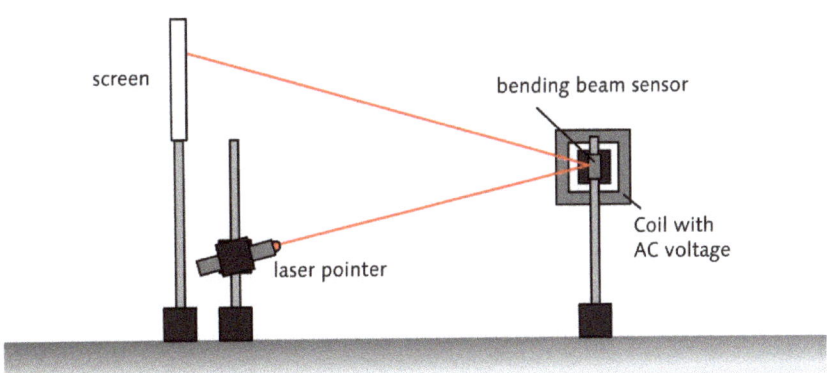

Figure 6. Experimental setup of the bending beam experiment.

Through the design-based research process, first insights about the design of a project day in the context of biomagnetic sensing and the interest of secondary school students regarding research and researchers in the field of biomagnetic sensing could be obtained. The results indicate that the design of the project day might be suitable to motivate the students and foster interest in aspects of biomagnetic sensing and the research of the CRC 1261. It seems that students are especially interested in the work of researchers and aspects of career orientation. Typical questions included: "What does a typical day at work look like? Is there a balance between theoretical and practical work? How do you become a researcher? Should I choose a certain university if I want to become a researcher? What is the best way to decide on a subject to study?" These findings can be used in order to obtain more detailed and robust results concerning the interest of students in researchers and research in the field of biomagnetic sensing in future iterations.

5. Discussion

In this paper, we have shared our ideas for designing an outreach activity, a project day, within the context of biomagnetic sensing and our experiences while developing and implementing it.

Biomagnetic sensing has proven a well-suited topic to inspire outreach activities, even though the content is unusual for classroom learning and is not part of the science curricula. Firstly, it is suitable for integrating STEM disciplines to foster students' interdisciplinary knowledge and skills. For instance, teaching about piezoelectric and magnetostrictive materials makes it possible to address the structure of matter in more detail, since both phenomena can be explained by the atomic structure of the respective materials. Understanding the structure of matter is an important part of both physics and chemistry education, and knowledge from both disciplines has to be combined to understand the phenomena. In countries with no curricular separation between the natural sciences, it can be stated that learning to connect knowledge within a discipline is as important for successful STEM education as the connection of knowledge across disciplines [1]. In addition, biomagnetic sensing provides a motivating context for school students by connecting the relevance of recent research with a medical problem. Secondly, biomagnetic sensing allows us to integrate aspects about both the nature of science and engineering. By working closely with researchers, it was possible to make authentic references to current research by including depictions of CRC researchers and their work in outreach materials. A facet of science that can be portrayed by the CRC is the facet of enterprising, since within the CRC there are projects that become independent enterprises and products that are introduced onto the market.

The chosen design approach allowed simultaneous research and design of the project day. The expert consultations we conducted during the design process were fruitful in that they provided new perspectives and led to several improvements before we invited groups of students to test the project day. In particular, the cooperation with CRC researchers proved valuable for our work as it helped to connect outreach activities to authentic research. Not only did the CRC researchers provide feedback on the final prototype of the project day but also contributed to the design of the prototype by, e.g., sharing ideas for model experiments. The feedback methods we used were well suited to resulting in valuable feedback on elements of the project day. We recommend using different evaluation methods such as short surveys, group discussions and interviews, as they provide a holistic insight into participants' perceptions of the project day.

There were a number of elements of the project day that were well-liked and received good feedback. For instance, the structure of the project day—embedding the sensor contents into the context of cardiac diagnostics and following the CRC research logic—was comprehensible for school students and seemed to foster their interest in ME sensors. The hands-on experiments were also very well-received. Students stated that they enjoyed exploring the properties of the sensor materials in more details and discovering the function of ME sensors while experimenting.

Improvements made to the project day included adapting the level of support during the experiments, the connections to CRC research, and the conclusion of the project day. Providing the right level of support during the experiments presented a challenge—with too little instruction, some students were overwhelmed and easily demotivated, with too much instruction, they quickly lost interest. Support staff, such as the learning station supervisors, can resolve this issue by providing adequate support when needed. In addition, the closing of the project day was adapted to the need of the students. During the design process, we learned that the students who tested the project day were interested in the life and work of CRC researchers. To provide insight into these areas, we included a phase in which we compared the development of the model ME sensor during the project day to the actual ME sensor development process in the CRC. To further link the CRC research to the project day, we set up a media station for students to visit during their breaks. Here, they could watch interviews with CRC researchers and a 360-degree video showing sensor development in the clean room during breaks. Since students did not fully utilize the media station, we plan to integrate the media elements in the regular learning stations.

In the future, the project day will be further evaluated with a larger sample of students, which should lead to further insights into which elements of the project day can particularly foster students' interest in STEM. Further research on students' specific interest in biomagnetic sensing will also be conducted to provide an even better foundation for future refinements of the project day as well as related outreach programs. Other plans for the future include establishing boundary activities that connect out-of-school measures such as the project day with in-school learning. Boundary activities increase the likelihood of creating lasting impacts on student interest and learning [33,34] and can take the form of preparation and/ or follow-up lessons at school that address concepts or ideas covered in the project day. Apart from boundary activities, a long-term goal of ours is to connect different outreach formats of the CRC to build a long-lasting, multi-faceted outreach program, making single outreach formats more sustainable.

6. Conclusions

In summary, the essential findings for the design of the project day were:

1. The didactic structure with a medical problem as the context and the development of an ME sensor as technological solution offers the students good orientation and has a motivating effect.
2. The experiments provide authentic insights, are easy to use and seem to motivate the students. When conducting the experiments, a balance has to be found between guidance and open inquiry. Therefore, the supervisors should be thoroughly prepared to provide support if necessary. In addition, the tasks should be designed to enable a stronger differentiation according to the performance of the students is possible.
3. The successful execution of the design of the ME sensor is crucial for the students' sense of competence. Supervisors can help at this stage by providing feedback if necessary. In addition, it is possible to give students more time to troubleshoot and develop a second version of their ME sensor if the first one did not work. In our experience, the problems of a poorly functioning sensor include either the use of the wrong material—an error that is easy to detect when comparing sensors between groups—or insufficient coupling. Insufficient coupling can be detected by closely inspecting the glue layer. In both cases, students are able to correct the errors by designing a second sensor that works if they have enough time and—if necessary—support from other students or their supervisor.
4. A separate media station has not proven effective in giving more insights into the life of scientists of the CRC. In the next design, we will consider integrating media elements that represent aspects of researchers and their work in the CRC directly into the learning stations.

Design-based research is widely described in the literature as a suitable method for designing evidence-based and effective outreach activities. Even though only the first cycles have been realized so far due to the pandemic, we could moderate a fruitful interaction between teachers, educational researchers and researchers in the field of biomagnetic sensing. Teachers were able to make connections to their science curricula, while student questions and feedback pointed out where step-wise guidance was necessary for different groups of students. CRC researchers helped to develop explanations that were appropriate for classroom learning and provided students with authentic scientific insights. We therefore encourage scientists and engineers in the field of magnetoelectric sensing to consider possible outreach activities in their own field and to actively collaborate with educational researchers in co-design processes.

Supplementary Materials: The following are available online at https://www.mdpi.com/article/10.3390/s21217354/s1.

Author Contributions: Conceptualization, C.B., C.E., I.P. and G.S.; methodology, C.B., C.E. and I.P.; validation, C.B., C.E. and I.P.; formal analysis, C.B.; investigation, C.B.; resources, C.E. and I.P.; data curation, C.B.; writing—original draft preparation, C.B.; writing—review and editing, C.B., C.E., I.P. and G.S.; visualization, C.B.; supervision, C.E., I.P. and G.S.; project administration, C.E. and I.P.; funding acquisition, C.E. and I.P. All authors have read and agreed to the published version of the manuscript.

Funding: This work was funded by the German Research Foundation (Deutsche Forschungsgemeinschaft, DFG), via the collaborative research center CRC 1261 *Magnetoelectric Sensors: From Composite Materials to Biomagnetic Diagnostics* (funding number: 286471992).

Institutional Review Board Statement: Ethical review and approval were waived for this study, due to the nature of the executed explorations: The participants of this study were informed of their role in this study which consisted of participating at the project day and giving feedback afterwards. No personal data were collected, and the participants remained anonymous.

Informed Consent Statement: Informed consent was obtained from all subjects involved in the study.

Acknowledgments: We would like to thank Daniel Laumann, Patrick Hayes and Klaus Boguschewski for their ideas and technical support with the experimental setups and Tobias Plöger for his help with the piezoelectric materials. We would like to thank the CRC researchers and the pre-service teachers for their feedback on the project day. Furthermore, we would like to thank the teachers and students who visited us and tested the project day.

Conflicts of Interest: The authors declare no conflict of interest. The funders had no role in the design of the study; in the collection, analyses, or interpretation of data; in the writing of the manuscript, or in the decision to publish the results.

References

1. National Research Council. *STEM Integration in K-12 Education: Status, Prospects, and an Agenda for Research*; National Academies Press: Washington, DC, USA, 2014.
2. Nadelson, L.; Seifert, A. Integrated STEM defined: Contexts, challenges, and the future. *J. Educ. Res.* **2017**, *110*, 221–223. [CrossRef]
3. European Commission. Council Recommendation on Key Competences for Lifelong Learning. Available online: https://ec.europa.eu/education/education-in-the-eu/council-recommendation-on-key-competences-for-lifelong-learning_en (accessed on 6 May 2021).
4. National Research Council (US). *Rising Above the Gathering Storm, Revisited: Rapidly Approaching Category 5*; National Academies Press (US): Washington, DC, USA, 2010; ISBN 9780309160971.
5. Maass, K.; Geiger, V.; Ariza, M.R.; Goos, M. The role of mathematics in interdisciplinary STEM education. *ZDM* **2019**, *51*, 869–884. [CrossRef]
6. Bybee, R.; McCrae, B. Scientific literacy and student attitudes: Perspectives from PISA 2006 science. *Int. J. Sci. Educ.* **2011**, *33*, 7–26. [CrossRef]
7. Fralick, B.; Kearn, J.; Thompson, S.; Lyons, J. How middle schoolers draw engineers and scientists. *J. Sci. Educ. Technol.* **2009**, *18*, 60–73. [CrossRef]
8. National Academy of Engineering. *Changing the Conversation: Messages for Improving Public Understanding of Engineering*; National Academies Press: Washington, DC, USA, 2008; ISBN 978-0-309-11934-4.
9. Martín-Páez, T.; Aguilera, D.; Perales-Palacios, F.J.; Vílchez-González, J.M. What are we talking about when we talk about STEM education? A review of literature. *Sci. Ed.* **2019**, *103*, 799–822. [CrossRef]
10. Pleasants, J.; Olson, J.K. What is engineering?: Elaborating the nature of engineering for K-12 education. *Sci. Ed.* **2019**, *103*, 145–166. [CrossRef]
11. Lederman, N.G.; Lederman, J.S. Teaching and learning nature of scientific knowledge: Is it Déjà vu all over again? *Discip. Interdscip. Sci. Educ. Res.* **2019**, *1*, 215. [CrossRef]
12. Millar, V.; Toscano, M.; van Driel, J.; Stevenson, E.; Nelson, C.; Kenyon, C. University run science outreach programs as a community of practice and site for identity development. *Int. J. Sci. Educ.* **2019**, *41*, 2579–2601. [CrossRef]
13. Vennix, J.; den Brok, P.; Taconis, R. Do outreach activities in secondary STEM education motivate students and improve their attitudes towards STEM? *Int. J. Sci. Educ.* **2018**, *40*, 1263–1283. [CrossRef]
14. Krause, H.-J.; Dong, H. Biomagnetic sensing. In *Label-Free Biosensing: Advanced Materials, Devices and Applications*; Schöning, M.J., Poghossian, A., Eds.; Springer International Publishing: Cham, Switzerland, 2018; pp. 449–474. ISBN 978-3-319-75220-4.

15. Macfarlane, P.W.; van Oosterom, A.; Pahlm, O.; Kligfield, P.; Janse, M.; Camm, A.J. (Eds.) *Comprehensive Electrocardiology*, 2nd ed.; Springer Ltd.: London, UK, 2010; ISBN 9781848820463.
16. Sternickel, K.; Braginski, A.I. Biomagnetism using SQUIDs: Status and perspectives. *Supercond. Sci. Technol.* **2006**, *19*, S160–S171. [CrossRef]
17. Reermann, J.; Elzenheimer, E.; Schmidt, G. Real-time biomagnetic signal processing for uncooled magnetometers in cardiology. *IEEE Sens. J.* **2019**, *19*, 4237–4249. [CrossRef]
18. Kwong, J.S.W.; Leithäuser, B.; Park, J.-W.; Yu, C.-M. Diagnostic value of magnetocardiography in coronary artery disease and cardiac arrhythmias: A review of clinical data. *Int. J. Cardiol.* **2013**, *167*, 1835–1842. [CrossRef] [PubMed]
19. Batie, M.; Bitant, S.; Strasburger, J.F.; Shah, V.; Alem, O.; Wakai, R.T. Detection of fetal arrhythmia using optically-pumped magnetometers. *JACC Clin. Electrophysiol.* **2018**, *4*, 284–287. [CrossRef] [PubMed]
20. Kampschulte, L.; Enzingmüller, C.; Wentorf, W.; Quandt, E.; Parchmann, I. Medizinische Sensoren entwickeln: Zusammenarbeit verschiedener Disziplinen. *Nat. Wiss. Unterr. Chem.* **2018**, *29*, 41–45.
21. Vennix, J.; den Brok, P.; Taconis, R. Perceptions of STEM-based outreach learning activities in secondary education. *Learn. Environ. Res.* **2017**, *20*, 21–46. [CrossRef]
22. Sevian, H.; Dori, Y.J.; Parchmann, I. How does STEM context-based learning work: What we know and what we still do not know. *Int. J. Sci. Educ.* **2018**, *40*, 1095–1107. [CrossRef]
23. Baram-Tsabari, A.; Sethi, R.; Bry, L.; Yarden, A. Identifiying students' interests in biology using a decade of self-generated questions. *Eurasia J. Math. Sci. Technol. Educ.* **2010**, *6*, 63–75. [CrossRef]
24. Swirski, H.; Baram-Tsabari, A.; Yarden, A. Does interest have an expiration date? An analysis of students' questions as resources for context-based learning. *Int. J. Sci. Educ.* **2018**, *40*, 1136–1153. [CrossRef]
25. Committee on Successful Out-of-School STEM Learning; Board on Science Education; Division of Behavioral and Social Sciences and Education; National Research Council. *Identifying and Supporting Productive STEM Programs in Out-of-School Settings*; National Academies Press: Washington, DC, USA, 2015; ISBN 978-0-309-37362-3.
26. OECD. *PISA 2018 Assessment and Analytical Framework*; OECD Publishing: Paris, France, 2019.
27. Lederman, N.G. Nature of science: Past, present, and future. In *Handbook of Research on Science Education: Environmental Education*; Lederman, N.G., Abell, S.K., Eds.; Routledge: Mahwah, NJ, USA, 2007; pp. 831–880. ISBN 9780203824696.
28. Dierks, P.O.; Höffler, T.N.; Blankenburg, J.; Peters, H.; Parchmann, I. Interest in science: A RIASEC-based analysis of students' interests. *Int. J. Sci. Educ.* **2016**, *38*, 238–258. [CrossRef]
29. Cobb, P.; Confrey, J.; diSessa, A.; Lehrer, R.; Schauble, L. Design experiments in educational research. *Educ. Res.* **2003**, *32*, 9–13. [CrossRef]
30. The Design-Based Research Collective. Design-based research: An emerging paradigm for educational inquiry. *Educ. Res.* **2003**, *32*, 5–8. [CrossRef]
31. Laumann, D.; Hayes, P.; Enzingmüller, C.; Parchmann, I.; Quandt, E. Magnetostriction measurements with a low-cost magnetostrictive cantilever beam. *Am. J. Phys.* **2020**, *88*, 448–455. [CrossRef]
32. Azevedo, F.S. Personal excursions: Investigating the dynamics of student engagement. *Int. J. Comput. Math. Learn.* **2006**, *11*, 57–98. [CrossRef]
33. Eshach, H. Bridging in-school and out-of-school learning: Formal, non-formal, and informal education. *J. Sci. Educ. Technol.* **2007**, *16*, 171–190. [CrossRef]
34. Fallik, O.; Rosenfeld, S.; Eylon, B.-S. School and out-of-school science: A model for bridging the gap. *Stud. Sci. Educ.* **2013**, *49*, 69–91. [CrossRef]

MDPI
St. Alban-Anlage 66
4052 Basel
Switzerland
Tel. +41 61 683 77 34
Fax +41 61 302 89 18
www.mdpi.com

Sensors Editorial Office
E-mail: sensors@mdpi.com
www.mdpi.com/journal/sensors

www.ingramcontent.com/pod-product-compliance
Lightning Source LLC
LaVergne TN
LVHW070723100526
838202LV00013B/1155